AF358832

Novel Insights in Horse Breeding and Genetics

Novel Insights in Horse Breeding and Genetics

Editors

Isabel Cervantes
Maria Dolores Gómez Ortiz

Basel • Beijing • Wuhan • Barcelona • Belgrade • Novi Sad • Cluj • Manchester

Editors
Isabel Cervantes
Universidad Complutense de
Madrid
Madrid
Spain

Maria Dolores Gómez Ortiz
Universidad de Sevilla
Sevilla
Spain

Editorial Office
MDPI AG
Grosspeteranlage 5
4052 Basel, Switzerland

This is a reprint of articles from the Special Issue published online in the open access journal *Animals* (ISSN 2076-2615) (available at: https://www.mdpi.com/journal/animals/special_issues/horse_breeding_genetics).

For citation purposes, cite each article independently as indicated on the article page online and as indicated below:

Lastname, A.A.; Lastname, B.B. Article Title. *Journal Name* **Year**, *Volume Number*, Page Range.

ISBN 978-3-7258-1453-4 (Hbk)
ISBN 978-3-7258-1454-1 (PDF)
doi.org/10.3390/books978-3-7258-1454-1

Cover image courtesy of Ekaterina Pustoshnaia

Contents

About the Editors

Isabel Cervantes

Isabel Cervantes is a Professor at the Department of Animal Production, Veterinary Faculty, Complutense University of Madrid. She defended her PhD in 2008. Her research focuses on the implementation of conservation and breeding programs, genetic variability management, and breeding value prediction in different species, primarily in horses. Isabel has participated in six competitive national projects and has been involved in more than 50 private research contracts with horse breeders' associations. She has published 69 papers included in *JCR* journals. She is currently the Vice-President of the EAAP Horse Commission. She won the EAAP Young Scientist Award in 2018.

Maria Dolores Gómez Ortiz

Maria Dolores Gómez Ortiz graduated in Veterinary Sciences and specialized in Genetics at the University of Córdoba (Spain), where she defended her PhD study on Spanish Trotter Horses in 2011. She is an honorary collaborator at the Department of Agronomic Engineering at the University of Seville and participates in doctoral courses at the University of Las Palmas de Gran Canaria. Her main role is as the Technical Director of the official Studbook of Purebred Menorca Horses, a position she has held since 2008. Her research focuses on horse breeding and selection within official conservation and breeding programs, breeding value prediction, and analysis of population genetic structure, among other areas. She is currently Treasurer of the Spanish Society of Equine Production and Secretary of the Spanish Society of Zoo-Ethnology.

Preface

In the complex and ever-evolving domain of equine genetics and breeding, the combination of scientific research and practical application reveals a rich tapestry of 'Novel Insights in Horse Breeding and Genetics', the focus of this Special Issue. Featuring 12 distinguished manuscripts of high scientific caliber and diverse perspectives, this collection explores various facets of equine genetics and genomics, providing valuable insights for scientific researchers, breeders, and equine enthusiasts alike.

Horses, individuals of unparalleled adaptability and morpho-functional performance in multiple domains (meat production, sportive, leisure…), are the focus of this collection of articles. From genomic selection strategies to innovations in traditional selection procedures, this Special Issue addresses the vast landscape of horse breeding, with papers of great interest to scientific readers, breeders and owners. These studies aim to elucidate the intricacies of genetic diversity, conservation, and the genetic basis of performance and welfare traits.

We can also divide the contributions into three different groups.

The first group, the most numerous and innovative, is related to the genes and genomics of horses; in this, we highlight works related to the SNP-based heritability of the pathology osteochondrosis dissecans in Hannoverian Horses (Zimmermann and Distl, 2023); a paper addressing the reconstruction of the major maternal and paternal lineages in feral New Zealand Kaimanawa Horses (Sharif et al., 2022); an analysis of the effects of selection on breed contributions in the crossbred Caballo de Deporte Español (Bartolomé et al., 2022); two publications related to copy number variations in different horse populations (Kim et al., 2022 and Laseca et al., 2022); a study that employs Y chromosome haplotypes to analyze the influence of origin and breeding on African Barb horses (Radovic et al., 2022); and an analysis of the genes related to white spotting (Rosa et al., 2022).

An analysis of the effect of mitochondrial DNA variations on the ability of horses to perform dressage and show jumping in the Holstein Horse Breed (Engel et al., 2022) acts as a link with the second group of contributions, which are related to performance and selection in horses using traditional procedures in different populations. This includes studies focusing on quarter horses, selected for their racing abilities (Faria et al., 2023); morphofunctional traits and their association with coat color in Pura Raza Menorquina horses (Perdomo-González et al., 2022); and shape space in Franches Montagnes horses (Gmel et al., 2022). The last group of studies includes an analysis of stress and welfare in Pura Raza Menorquina horses, which participate in traditional equestrian events (Olvera-Maneu et al., 2023).

The expertise shared by our contributors, coupled with the meticulous evaluations of our esteemed reviewers, forms the backbone of this compendium of publications with important and economic implications for equine management and selection; these insights can be applied to the specific population but also to the equine sector, contributing to adequate development and improvement.

We would like to extend our gratitude to the authors for their high-quality contributions and to the reviewers for their insightful assistance in reviewing and enhancing the publications. We also extend our gratitude to the assistants of the journal for their aid in the editorial work.

This collection of publications delves deep into the genetic landscape of equines, and advances our knowledge of horse breeding and genetics. We thus encourage all horse lovers to read and enjoy them.

Isabel Cervantes and Maria Dolores Gómez Ortiz

Editors

Article

SNP-Based Heritability of Osteochondrosis Dissecans in Hanoverian Warmblood Horses

Elisa Zimmermann and Ottmar Distl *

Institute for Animal Breeding and Genetics, University of Veterinary Medicine Hannover (Foundation), 30559 Hannover, Germany
* Correspondence: ottmar.distl@tiho-hannover.de

Simple Summary: The heritability of a trait is the proportion of phenotypic variance explained via genetic variance. Prior to the advent of genomics, heritability was estimated using extensive pedigree analyses. With the availability of genome-wide genotyping arrays, an alternative method became available to estimate heritability using genomic relationship matrices derived from genotype data. We used approaches that consider patterns of linkage disequilibrium and relatedness to estimate heritability of osteochondrosis dissecans in Hanoverian Warmblood horses based on genotype data from SNP arrays and imputed genotype data. Taking into account linkage disequilibrium patterns and relatedness in the data, heritability estimates on the linear scale for fetlock-, hock- and stifle-OCD were 0.41–0.43, 0.62–0.63, and 0.23–0.25, respectively, with standard errors of 0.11–0.14. In summary, SNP-based approaches are able to capture a greater proportion of additive genetic variance than previous estimates based on pedigree data.

Abstract: Before the genomics era, heritability estimates were performed using pedigree data. Data collection for pedigree analysis is time consuming and holds the risk of incorrect or incomplete data. With the availability of SNP-based arrays, heritability can now be estimated based on genotyping data. We used SNP array and 1.6 million imputed genotype data with different minor allele frequency restrictions to estimate heritabilities for osteochondrosis dissecans in the fetlock, hock and stifle joints of 446 Hanoverian warmblood horses. SNP-based heritabilities were estimated using a genomic restricted maximum likelihood (GREML) method and accounting for patterns of regional linkage disequilibrium in the equine genome. In addition, we employed GREML for family data to account for different degrees of relatedness in the study population. Our results indicate that we were able to capture a larger proportion of additive genetic variance compared to pedigree-based estimates in the same population of Hanoverian horses. Heritability estimates on the linear scale for fetlock-, hock- and stifle-osteochondrosis dissecans were 0.41–0.43, 0.62–0.63, and 0.23–0.25, respectively, with standard errors of 0.11–0.14. Accounting for linkage disequilibrium patterns had an upward effect on the imputed data and a downward impact on the SNP array genotype data. GREML for family data resulted in higher heritability estimates for fetlock-osteochondrosis dissecans and slightly higher estimates for hock-osteochondrosis dissecans, but had no effect on stifle-osteochondrosis dissecans. The largest and most consistent heritability estimates were obtained when we employed GREML for family data with genomic relationship matrices weighted through patterns of regional linkage disequilibrium. Estimation of SNP-based heritability should be recommended for traits that can only be phenotyped in smaller samples or are cost-effective.

Keywords: equid; osteochondrosis; genetic parameters; genomic relationship matrices; SNP-based heritability; linkage disequilibrium

Citation: Zimmermann, E.; Distl, O. SNP-Based Heritability of Osteochondrosis Dissecans in Hanoverian Warmblood Horses. *Animals* **2023**, *13*, 1462. https://doi.org/10.3390/ani13091462

Academic Editors: Isabel Cervantes and María Dolores Gómez Ortiz

Received: 28 February 2023
Revised: 24 March 2023
Accepted: 21 April 2023
Published: 25 April 2023

1. Introduction

Osteochondrosis (OC) is one of the most important orthopaedic diseases of the juvenile horse [1]. Due to disturbances in enchondral ossification, damage to the subchondral bone is the reason for the formation of intraarticular osteochondral fragments and subchondral bone cysts. If osteochondral fragments occur, the disease is referred to as osteochondrosis dissecans (OCD). Osteochondrosis occurs at certain predilection sites. Joints frequently affected are the metacarpophalangeal/metatarsophalangeal (fetlock), tarsocrural (hock) and femoropatellar joints (stifle) [2]. Therefore, it is of utmost interest to evaluate genetic parameters for OCD as precisely as possible in order to take breeding measures.

The aetiology of OCD appears to be multifactorial with a relevant genetic contribution [3,4]. There have been estimations regarding the heritability of OCD based on pedigree [5–21] and genotyping data [22,23]. Those estimates are shown in a previous review [1] and supplemented with results of more recent studies shown in Supplementary Table S1.

Before the genomics era, estimates of heritability were based on pedigree data. The introduction of SNP arrays enabled the estimation of heritability based on genotyping data. Genome-based heritability estimates offer many advantages through eliminating the need to collect extensive pedigree data. Apart from the time-consuming data collection, analysis of pedigree data poses the risk of biased results due to incorrect, incomplete, or varying depth of pedigrees. Heritability estimates between populations may vary because of population history, gene frequency, or environmental exposures [23].

There are various approaches to estimating heritability based on genotyping data. The fraction of phenotypic variance that can be explained using variants that have been identified as causal variants through genome-wide association studies (GWAS) is named h^2_{GWAS}. h^2_{GWAS} is limited because, for most complex diseases, only a small proportion of variants has already been identified [24]. h^2_{SNP} is the proportion of phenotypic variance explained using all SNPs on a genotyping platform. h^2_{SNP} is the upper limit for h^2_{GWAS} and can be a measurement of the proportion of already identified causal variants in the actual genetic variance of a trait [24]. The difference between h^2_{GWAS} and h^2_{SNP} is often referred to as missing heritability [25,26]. We want to estimate h^2_{SNP} and assess different methods using genomic REML algorithms (GREML). Heritability estimation methods based on genotyping data require certain assumptions regarding the population structure of the underlying population, indirect genetic effects, the presence of artificial or natural selection within the population, and linkage disequilibrium. These assumptions are specific for each population and trait and can severely bias the heritability estimates [27]. Different methods require certain assumptions [28]. The aim of heritability estimation using SNP array or Beadchip data is to approach h^2, which is the actual narrow sense heritability of a trait [29].

We estimated h^2_{SNP} using GCTA GREML [30], which is a single-component model to estimate heritability based on a genomic relationship matrix (GRM) and unrelated individuals [31]. As this approach is very sensitive to patterns of linkage disequilibrium (LD) [32], we used a similar single-component approach that is implemented in the software LDAK and weights SNP effect sizes according to regional LD patterns to construct the GRM [32]. LD describes the non-random association of alleles at two or more loci. LD varies because of factors such as population history, natural or artificial selection, mutation, and other forces that cause changes in allele frequency [33]. It can cause upwards biased estimates of heritability due to repeated tagging of SNPs [34].

As Beadchip arrays are usually based on common SNPs, we want to use imputed SNPs for heritability estimation to capture the effects of more causal variants [27,35–37]. However, it is recommendable to prune for minor allele frequency (MAF) because rare variants are imputed less accurately [36,38].

In a previous study on osteochondrosis in horses, SNP-based heritability for osteochondrosis in the hock was estimated in a population of 479 North American Standardbred Horses using REML analysis in GCTA and LDAK with the weighted GRM in an imputed data set with ~1.25 million SNPs. The OCD frequency in this study population was 0.27. The analyses were repeated using a smaller study population, with individuals pruned for relatedness at 0.25. SNP-based estimates seemed to be biased upwards via LD, which implies the need to account for LD in heritability estimations in horse populations [20].

Zaitlen et al. [24] proposed a method to estimate heritability based on a population with different degrees of kinship that avoids the need to remove closely related individuals from the study population. This method is implemented in GCTA and is known as GREML analysis for family data. It provides estimations of SNP-based heritability in family data as well as narrow sense heritability h_f^2, which enables the quantification of genetic effects resulting from kinship and, thus, enables the detection of higher amounts of heritability [24]. This method has not yet been used in horse populations with very specific relatedness structures.

The aim of this work is to estimate the heritability of the trait OCD in the fetlock, hock, and stifle joints based on SNP data. We conducted a GREML analysis, an LDAK analysis using an LD-weighted GRM based on unrelated individuals, and a GREML analysis for family data using two simultaneously constructed GRMs for individuals with different relatedness structures. Beside the effects of the different GRMs, we will observe the effects of imputation and the different MAF restrictions, comparing the heritability estimates with previous pedigree-based analyses using a similar study sample [5].

2. Materials and Methods

2.1. Animals

The horses included in this study were a subset of the study population previously analyzed by Hilla et al. [5]. For the present study, 446 four-year-old Hanoverian warmblood horses were included. The inclusion criteria were as follows: only one horse per sire and maternal sire was allowed, either as a control or a case. The controls and cases were randomly distributed among the sires and maternal sires. The control horses had to be free of all diseases found during the veterinary health examination for pre-selection at auctions, at licensing, or during the purchase examination. The cases were horses with OCD only and free of any other disease recorded in the veterinary health check. The veterinary health check included clinical and radiographic examination of all four limbs. Only osteochondral fragments at the specific predilection sites of the fetlock, hock, and stifle joints were classified as OCD [5,39]. Osteochondral fragments plantar in the fetlock joints and at the insertion sites of the short sesamoid ligaments at the proximal phalanx of the hindlimbs were classified as plantar and dorsodistal fragments of the fetlock joints; thus, there were not considered OCD. Distal and proximal interphalangeal joints, fetlock joints, hock joints, and stifle joints were evaluated for contour changes stemmed from periarticular osteophytes or exostoses and for a narrow or absent joint space. These changes were classified as osteoarthroses. Radiographic changes in the shape, symmetry, contour, and structure of the navicular bone and the shape, size, number, and location of the canales sesamoidales were scored on a scale of 1–4 [40]. Only horses with a score of 1 were considered free of radiographic changes to the navicular bone. Horses with the presence of a sidebone were also scored as not being free of radiographic changes. After removing all horses affected by diseases other than OCD, the data set was filtered for cases and controls. The strict inclusion criteria resulted in the final data set comprising 446 horses. Traits were encoded as 0/1 variates for each joint. We did not consider an overall score for OCD because genetic correlations of OCD between the different joints were moderately negative (fetlock-OCD with hock- and stifle-OCD: -0.12 and -0.18) or moderately positive (hock-OCD with stifle-OCD: 0.17) [5].

The phenotypic and genotype data were provided by the Association of Hanoverian Warmblood breeders (Hannoveraner Verband e.V., Verden, Germany). The frequencies of OCD were 0.2489 ($n = 111$), 0.3139 ($n = 140$), and 0.0291 ($n = 13$) in the fetlock, hock, and stifle joints, respectively. Relationships expressed through the contingency coefficient between the frequencies of fetlock, hock, and stifle OCD were close to zero because 96, 125, and 11 horses represented the sole cases of fetlock, hock, and stifle joint OCD, respectively.

2.2. Methods

Genome-wide genotyping data was obtained using the GGP Equine (71 589 SNPs) genotyping array. Descriptive statistics of the population were calculated with SAS, version 9.4 (Statistical Analysis System, Cary, NC, USA, 2023). The SNP data have been imputed to 1,617,270 SNPs with an information score of 0.95 using BEAGLE 5.4 [41] and publicly available whole genome sequencing data for horses (Supplementary Table S2). Subsequently, the imputed and non-imputed data sets were pruned at MAFs of 0.01, 0.025, or 0.05 using PLINK 1.9 [42,43], resulting in six different data sets. Using all six data sets, heritabilities for OCD of the fetlock, hock, and stifle joints were estimated using the GREML analysis implemented in GCTA (genome-wide complex trait analysis) 1.94.1 [30] with one GRM [44] resulting in SNP-based heritability (h^2_{SNP}). Subsequent estimations were performed using the LD-weighted genomic relatedness matrix as implemented in LDAK 5.2 [45] and the integrated REML analysis [32], resulting in estimations of h^2_{SNPw}. Using the GREML analysis for family data with two GRMs simultaneously, based on all pairs of individuals and related individuals [24] implemented in GCTA 1.94.1 [30], we estimated h^2_f. The GRM based on all pairs of individuals captured information on the sharing of causal variants tagged using SNPs. The second GRM considered only individuals who were identical-by-state above a certain threshold (0.05) and, consequently, only related individuals. Hence, it captured information on shared causal variants that could not be tagged using SNPs [24,29]. Both GRMs fitted into a mixed linear model and were supposed to provide estimates of $h^2_{SNP-all-pairs}$ from the first GRM and the missing heritability $h^2_{SNP-related}$ from the second GRM. Those values were summed up to h^2_f [24].

We obtained heritability estimates (h^2_{fw} and $h^2_{SNP-all-pairs-w}$) by implementing the LD-weighted genomic relationship matrix estimated with LDAK [32] in the GREML analysis for family data [24]. Sex was included in all analyses as a covariate. As we used 0/1-data, all heritability estimates were transformed onto the liability scale using the prevalence option of GCTA. The study design for heritability estimations is illustrated in Supplementary Figure S1.

3. Results

The results of our heritability estimates for osteochondrosis in the fetlock joint are given in Table 1. Additionally, estimates for $h^2_{SNP-all-pairs}$ and $h^2_{SNP-all-pairs-w}$ are given in Supplementary Table S3. The SNP-based heritabilities estimated with GREML revealed that the heritability estimates decreased in the imputed data set compared with the original data set. The SNP-based heritabilities estimated with GREML and LDAK for fetlock-OCD differed in several aspects. Accounting for regional LD patterns increased heritability estimates for the imputed genotype data but slightly decreased heritability estimates for the original data sets. Heritability estimates using GREML analysis for family data resulted in higher estimates for both data sets, the original and imputed genotype data, as well as when regional LD patterns were considered. The effects of using different MAFs had only small effects when we used LD-weighted genomic relatedness matrices with LDAK. Standard errors for heritability estimates using GREML analysis for family data slightly increased from 0.11–0.12 to 0.13–0.14 on the linear scale.

After transformation onto the liability scale, we obtained fairly high estimates for heritability and their standard errors.

When comparing the GREML analysis for the original and imputed data sets, the same trends were observed for the heritability estimates for hock- and fetlock-OCD (Table 2, Supplementary Table S4). However, the increase in heritability estimates was much smaller when GREML analysis was applied to family data than to fetlock-OCD. The consistency and magnitude of the heritability estimates were highest when we used GREML analysis for family data with LDAK.

The most consistent heritability estimates were obtained for stifle-OCD for the analysis accounting for family data and LD patterns for both data sets (the original and imputed genotype data) (Table 3, Supplementary Table S5). The effect of family structure on heritability estimates was small, while LD patterns had slightly larger effects. In general, differences between the different approaches were relatively small. Transformation onto the liability scale gave meaningless estimates >1 due to the low frequency of cases.

Table 1. Heritability estimates with their standard errors ($h^2 \pm SE$) for OCD in fetlock joint estimated with GCTA GREML, LDAK, and GCTA GREML for family data and GCTA GREML for family data and a LD-weighted genomic relationship matrix; the original and imputed SNP data and, at minor allele frequencies (MAF) of 0.01, 0.025 and 0.05, is included with transformation to the liability scale (Obs = observed scale, Liab = liability scale).

Approach	Data Set	MAF 0.01		MAF 0.025		MAF 0.05	
		Obs	Liab	Obs	Liab	Obs	Liab
GREML ($h^2_{SNP} \pm SE$)	Original	0.34 ± 0.12	0.64 ± 0.22	0.33 ± 0.12	0.61 ± 0.22	0.32 ± 0.12	0.60 ± 0.22
	Imputed	0.31 ± 0.11	0.58 ± 0.21	0.30 ± 0.11	0.56 ± 0.20	0.28 ± 0.11	0.53 ± 0.20
LDAK ($h^2_{SNPw} \pm SE$)	Original	0.33 ± 0.12	0.61 ± 0.22	0.33 ± 0.12	0.61 ± 0.22	0.32 ± 0.12	0.60 ± 0.22
	Imputed	0.34 ± 0.12	0.64 ± 0.23	0.34 ± 0.12	0.63 ± 0.22	0.33 ± 0.12	0.62 ± 0.22
GREML fam ($h^2_f \pm SE$)	Original	0.43 ± 0.14	0.80 ± 0.26	0.42 ± 0.14	0.78 ± 0.26	0.44 ± 0.14	0.81 ± 0.26
	Imputed	0.41 ± 0.13	0.76 ± 0.24	0.40 ± 0.13	0.74 ± 0.24	0.38 ± 0.13	0.71 ± 0.23
GREML fam LD-weighted ($h^2_{fw} \pm SE$)	Original	0.41 ± 0.14	0.76 ± 0.26	0.41 ± 0.14	0.77 ± 0.26	0.41 ± 0.14	0.76 ± 0.26
	Imputed	0.43 ± 0.14	0.79 ± 0.26	0.42 ± 0.14	0.79 ± 0.26	0.43 ± 0.14	0.81 ± 0.26

Table 2. Heritability estimates with their standard errors ($h^2 \pm SE$) for osteochondrosis dissecans in hock joint estimated with GCTA GREML, LDAK, and GCTA GREML for family data and GCTA GREML for family data; a LD-weighted genomic relationship matrix, with the original and imputed SNP data and at minor allele frequencies (MAF) of 0.01, 0.025 and 0.05, is also included with transformation to the liability scale (Obs = observed scale, Liab = liability scale).

Approach	Data Set	MAF 0.01		MAF 0.025		MAF 0.05	
		Obs	Liab	Obs	Liab	Obs	Liab
GREML ($h^2_{SNP} \pm SE$)	Original	0.60 ± 0.11	1.02 ± 0.19	0.60 ± 0.11	1.01 ± 0.18	0.57 ± 0.11	0.98 ± 0.18
	Imputed	0.54 ± 0.10	0.93 ± 0.18	0.52 ± 0.10	0.90 ± 0.18	0.50 ± 0.10	0.85 ± 0.17
LDAK ($h^2_{SNPw} \pm SE$)	Original	0.59 ± 0.11	1.01 ± 0.19	0.59 ± 0.11	1.00 ± 0.19	0.58 ± 0.11	1.00 ± 0.19
	Imputed	0.62 ± 0.11	1.06 ± 0.19	0.61 ± 0.11	1.05 ± 0.19	0.60 ± 0.11	1.03 ± 0.19
GREML fam ($h^2_f \pm SE$)	Original	0.62 ± 0.12	1.07 ± 0.21	0.63 ± 0.12	1.09 ± 0.21	0.63 ± 0.12	1.09 ± 0.21
	Imputed	0.57 ± 0.12	0.97 ± 0.21	0.56 ± 0.12	0.95 ± 0.21	0.53 ± 0.12	0.91 ± 0.20
GREML fam LD-weighted ($h^2_{fw} \pm SE$)	Original	0.63 ± 0.12	1.08 ± 0.21	0.63 ± 0.12	1.08 ± 0.21	0.62 ± 0.12	1.07 ± 0.21
	Imputed	0.63 ± 0.12	1.09 ± 0.21	0.62 ± 0.12	1.07 ± 0.21	0.63 ± 0.12	1.07 ± 0.21

Table 3. Heritability estimates with their standard errors ($h^2 \pm SE$) for osteochondrosis dissecans in stifle joint estimated with GCTA GREML, LDAK, AND GCTA GREML for family data and GCTA GREML for family data; a LD-weighted genomic relationship matrix, with the original and imputed SNP data and at minor allele frequencies (MAF) of 0.01, 0.025 and 0.05, is also included with transformation to the liability scale (Obs = observed scale, Liab = liability scale).

Approach	Data Set	MAF 0.01		MAF 0.025		MAF 0.05	
		Obs	Liab	Obs	Liab	Obs	Liab
GREML ($h^2_{SNP} \pm SE$)	Original	0.25 ± 0.11	1.60 ± 0.69	0.24 ± 0.11	1.55 ± 0.68	0.23 ± 0.11	1.47 ± 0.68
	Imputed	0.19 ± 0.10	1.25 ± 0.64	0.17 ± 0.10	1.11 ± 0.62	0.16 ± 0.10	1.04 ± 0.61
LDAK ($h^2_{SNPw} \pm SE$)	Original	0.23 ± 0.11	1.50 ± 0.69	0.23 ± 0.11	1.49 ± 0.68	0.23 ± 0.11	1.47 ± 0.68
	Imputed	0.21 ± 0.11	1.37 ± 0.68	0.20 ± 0.11	1.29 ± 0.67	0.20 ± 0.10	1.27 ± 0.67
GREML fam ($h^2_f \pm SE$)	Original	0.26 ± 0.12	1.66 ± 0.75	0.24 ± 0.12	1.55 ± 0.75	0.23 ± 0.12	1.49 ± 0.74
	Imputed	0.24 ± 0.11	1.53 ± 0.70	0.22 ± 0.11	1.40 ± 0.70	0.23 ± 0.11	1.45 ± 0.70
GREML fam LD-weighted ($h^2_{fw} \pm SE$)	Original	0.25 ± 0.12	1.63 ± 0.75	0.25 ± 0.12	1.63 ± 0.75	0.25 ± 0.12	1.60 ± 0.75
	Imputed	0.23 ± 0.12	1.48 ± 0.74	0.23 ± 0.11	1.48 ± 0.73	0.23 ± 0.11	1.45 ± 0.73

4. Discussion

According to the findings of previous studies, it seems to be recommendable to account for linkage disequilibrium when estimating heritability based on SNP data [20,29,31,32,34,46]. Horses have long-range linkage disequilibria, which is why SNPs can show effects of a risk variant as far away as 1 Mb [47]. Additionally, LD is higher within breeds than across breeds [48], which is important since population data are usually used for heritability estimations. In general, REML-based estimates, such as those obtained from GREML analysis in GCTA, are sensitive to patterns of LD [32]. The linkage disequilibrium between SNPs is used to create the GRM and the LD between SNPs; causal variants can cause bias in heritability estimation [32,36]. As the intensity of linkage disequilibrium varies regionally along the genome, LDAK weights the SNPs according to local patterns of LD [32]. While we cannot observe a large impact of LD using our original data set, we see slight differences between GREML analysis and LDAK analysis in the imputed data set. The difference between heritability estimates increases with increasing MAF restrictions, which is attributable to the fact that less genetic variation is captured with SNPs when lower frequencies are recorded. Additionally, allele frequency and linkage disequilibrium are dependent on each other [49], which explains why the estimations conducted with LDAK are able to compensate changes in MAF. We assume that linkage disequilibrium does not play a major role in our study population. One possible explanation could be that we included many individuals with diverse LD structures, meaning that they outweighed each other in our analysis.

One cause of undetected heritability could be that rare variants, and eventually even variants with large effects, may not be mapped on the available genotyping arrays that mainly include common SNPs [36]. Therefore, it is recommended to perform heritability estimations on imputed data sets [29]. To capture as much variation as possible, we imputed our Beadchip data to 1,617,270 SNPs, which corresponds to the recommendations given by Evans et al. [29] for heritability estimations. When comparing the results of the original and imputed data sets, we observe for all traits analyzed the most consistent estimates when family data and LD patterns are accounted for. Even the differences between the original and imputed data shrink or are no longer present. In the present data set, imputation had no or very little effect on SNP-based heritabilities; thus, we were unable to detect variation due to rare alleles.

The single-component analyses in GCTA and LDAK calculated GRM based on the available SNP data to subsequently estimate heritability. For those analyses, it was recommended to prune for relatedness to eradicate bias caused by common environmental or other non-additive genetic effects [29]. The resulting unrelated individuals are by definition distantly related individuals because they share distant ancestors [50]; however, they are assumed to provide random genetic variance [28]. The need for pruning for relatedness arises from the model assumption in the GREML analysis that all measured genetic effects are direct effects. If related individuals were included, the indirect genetic effects between those individuals would be counted as direct effects and, thus, inflate heritability estimates [28]. Indirect genetic effects may result from genetic maternal effects [28]. The idea of the GREML analysis for family data was to find a way to circumvent pruning for relatedness in a study population and, thus, ensure a larger study population, which in turn should lead to lower standard errors. Additionally, the GREML analysis for family data estimates h_f^2 and, thus, is able to capture higher heritability [24]. While h_f^2 provides an unbiased estimate of the heritability of the trait, the proportions of the single components do not always seem to be assessed correctly [24,29]. We only observed this phenomenon in the imputed data set for stifle-OCD when we employed GREML for family data without an LD-weighted genomic relationship matrix. In all other analyses, we could not observe imbalanced contributions to the heritability estimates resulting from the two different GRMs. The most likely reason for this issue is the very low frequency of cases for stifle-OCD.

In our analyses for fetlock-OCD, we can confirm that we detected higher estimates of heritability with GREML for family data than with the single-component REML algorithms.

whereas for hock- and stifle-OCD the increase in heritability estimates was rather small. We assume that in our population a significant amount of heritability for fetlock-OCD may be due to indirect genetic effects that are captured examining the genetic effects between individuals with varying degrees of relatedness. This is the first analysis with GREML for family data that has been performed in a horse population. The most consistent heritability estimates were obtained using GREML for family data and an LD-weighted genomic relationship matrix for the original and imputed genotype data. With frequencies of cases closer to 0.5 in the population under study, differences between the original and imputed data sets diminished. However, we have to note that GREML for family data is designed for human populations with their specific relatedness structure and significantly larger available data sets [24,31]. While in a human population, full siblings are common, in horse population, full-siblings are uncommon.

Since OCD is defined as a binary trait, all results of the REML analyses have been transformed onto the liability scale as recommended [30,51–53]. In agreement with previous studies, upward bias may occur, particularly when estimates on the linear scale are high and more frequencies deviate from 0.5 [11,12].

The present study used data from Hilla et al. [5]. We selected horses as representatively as possible and avoided including closely related animals, such as paternal half-siblings and maternal sire half-siblings. The results obtained from the present study allow us to assume that analyses using GREML for family data and an LD-weighted genomic relationship matrix result in higher heritability estimates compared to estimates based on pedigree data (h^2_{ped}) in a Hanoverian Warmblood horse population. However, larger genotype data sets should be available to reach lower standard errors. Nevertheless, heritability estimations based on SNP-based methods may give reliable results even in much smaller data sets compared to pedigree-based estimates.

Similar results were reported for hock-OCD in a population of North American Standardbred horses [20]. Compared to our results, this previous study showed larger increases in heritability estimates when taking into account LD patterns compared to the standard GRM. Thus, we assume stratification based on families and breed history may have contributed to this result. In addition, the LDAK version used by McCoy et al. [20,32] was a less improved version of the software, which could have had an effect on the results. Heritability estimates are specific for populations because of different familial structures and selective signatures in the genome [1,33], which may also contribute to the difference in our results. In the present study, standard errors on the linear scale were at 0.11–0.14, resulting in 95% confidence intervals from ±0.22 to ±0.27, while standard errors on the linear scale were at 0.12 and 0.16 in the previous study on US Standardbreds.

In summary, we recommend the use of GREML for family data with an LD-weighted genomic relatedness matrix to estimate heritabilities, particularly for traits which are difficult or very costly to record. Due to restrictions in sampling and varyingly strong LD patterns in populations, the approach as provided by LDAK should be implemented in the estimation procedure. The pursuit of more precise heritability estimates is worthwhile means of achieving estimated breeding values with higher reliabilities and a higher selection response in health traits.

5. Conclusions

Estimation of heritabilities based on SNP arrays is recommended because reasonably high accuracy of estimates can be achieved in smaller samples compared to pedigree-based studies with similar sample sizes. The use of genomic REML analysis for family data with LD-weighted genomic relationship matrices allows the capture of most of the additive genetic variance and provides the most consistent estimates at different MAFs. The present study yielded higher heritability estimates with reasonable standard errors than a previous study for the same population. Further studies with larger data sets should be performed to validate these results.

Supplementary Materials: The following supporting information can be downloaded at: https://www.mdpi.com/article/10.3390/ani13091462/s1, Figure S1. Survey on the study design to estimate heritabilities using different approaches and genotype data. Table S1. Estimates for heritability for equine osteochondrosis in previous studies. Table S2. Publicly available whole genome sequencing data of horses. Table S3. Estimates for $h^2_{SNP-all-pairs}$ and $h^2_{SNP-all-pairs-w}$ with their standard errors for osteochondrosis dissecans in fetlock joint estimated with GCTA GREML for family data and GCTA GREML for family data with a LD-weighted genomic relationship matrix with original and imputed SNP data and at minor allele frequencies of 0.01, 0.025, and 0.05, including transformation to liability scale (obs = observed scale, liab = liability scale). Table S4. Estimates for $h^2_{SNP-all-pairs}$ and $h^2_{SNP-all-pairs-w}$ with their standard errors for osteochondrosis dissecans in hock joint estimated with GCTA GREML for family data and GCTA GREML for family data with a LD-weighted genomic relationship matrix with the original and imputed SNP data and at minor allele frequencies of 0.01, 0.025 and 0.05, including transformation to liability scale (obs = observed scale, liab = liability scale). Table S5. Estimates for $h^2_{SNP-all-pairs}$ and $h^2_{SNP-all-pairs-w}$ with their standard errors for osteochondrosis dissecans in stifle joint estimated with GCTA GREML for family data and GCTA GREML for family data with a LD-weighted genomic relationship matrix with original and imputed SNP data and at minor allele frequencies of 0.01, 0.025, and 0.05, including transformation to liability scale (obs = observed scale, liab = liability scale).

Author Contributions: Conceptualization, O.D. and E.Z.; methodology, O.D. and E.Z.; software, O.D.; validation, O.D. and E.Z.; formal analysis, O.D.; investigation, E.Z. and O.D.; resources, O.D.; data curation, O.D.; writing—original draft preparation, E.Z.; writing—review and editing, O.D.; visualization, O.D. and E.Z.; supervision, O.D.; project administration, O.D. All authors have read and agreed to the published version of the manuscript.

Funding: This Open Access publication was funded by the Deutsche Forschungsgemeinschaft (DFG German Research Foundation)—491094227 "Open Access Publication Funding" and the University of Veterinary Medicine Hannover, Foundation.

Institutional Review Board Statement: Not applicable.

Informed Consent Statement: Not applicable.

Data Availability Statement: Restrictions apply to the availability of these data. Data and genotypes were provided by the Association of Hanoverian Warmblood breeders (Hannoveraner Verband e.V. Verden, Germany) and are available on reasonable request from the authors with the permission of the Association of Hanoverian Warmblood breeders.

Acknowledgments: We would like to thank the Association of Hanoverian Warmblood breeders (Hannoveraner Verband e.V., Verden, Germany) for support in data collection and the involved veterinarians for providing radiographic examinations.

Conflicts of Interest: The authors declare no conflict of interest.

References

1. Distl, O. The genetics of equine osteochondrosis. *Vet. J.* **2013**, *197*, 13–18. [CrossRef]
2. van Weeren, R. Fifty years of osteochondrosis. *Equine Vet. J.* **2018**, *50*, 554–555. [CrossRef] [PubMed]
3. Naccache, F.; Metzger, J.; Distl, O. Genetic risk factors for osteochondrosis in various horse breeds. *Equine Vet. J.* **2018**, *50*, 556–563. [CrossRef] [PubMed]
4. Ahmadi, F.; Mirshahi, A.; Mohri, M.; Sardari, K.; Sharifi, K. Osteochondrosis dissecans (OCD) in horses: Hormonal and biochemical study (19 cases). *Vet. Res. Forum.* **2021**, *12*, 325–331. [CrossRef]
5. Hilla, D.; Distl, O. Heritabilities and genetic correlations between fetlock, hock and stifle osteochondrosis and fetlock osteochondral fragments in Hanoverian Warmblood horses. *J. Anim. Breed. Genet.* **2013**, *131*, 71–81. [CrossRef] [PubMed]
6. Schougaard, H.; Ronne, J.F.; Phillipson, J. A radiographic survey of tibiotarsal osteochondrosis in a selected population of trotting horses in Denmark and its possible genetic significance. *Equine Vet. J.* **1990**, *22*, 288–289. [CrossRef]
7. Grøndahl, A.M.; Engeland, A. Influence of radiographically detectable orthopedic changes on racing performance in standardbred trotters. *J. Am. Vet. Med. Assoc.* **1995**, *206*, 1013–1017.
8. Philipsson, J.; Andréasson, E.; Sandgren, B.; Dalin, G.; Carlsten, J. Osteochondrosis in the tarsocrural joint and osteochondral fragments in the fetlock joints in Standardbred trotters. II. Heritability. *Equine Vet. J.* **1993**, *25*, 38–41. [CrossRef]
9. Teyssèdre, S.; Dupuis, M.C.; Guérin, G.; Schibler, L.; Denoix, J.M.; Elsen, J.M.; Ricard, A. Genome-wide association studies for osteochondrosis in French Trotter horses1. *J. Anim. Sci.* **2012**, *90*, 45–53. [CrossRef]

0. Pieramati, C.; Pepe, M.; Silvestrelli, M.; Bolla, A. Heritability estimation of osteochondrosis dissecans in Maremmano horses. *Livest. Prod. Sci.* **2003**, *79*, 249–255. [CrossRef]
1. Stock, K.F.; Hamann, H.; Distl, O. Estimation of genetic parameters for the prevalence of osseous fragments in limb joints of Hanoverian Warmblood horses. *J. Anim. Breed. Genet.* **2005**, *122*, 271–280. [CrossRef]
2. Stock, K.F.; Distl, O. Genetic correlations between osseous fragments in fetlock and hock joints, deforming arthropathy in hock joints and pathologic changes in the navicular bones of Warmblood riding horses. *Livest. Sci.* **2006**, *105*, 35–43. [CrossRef]
3. Winter, D.B.E.; Glodek, P.; Hertsch, B. Genetic disposition of bone diseases in sport horses. *Züchtungskunde* **1996**, *68*, 92–108.
4. Willms, F.; Röhe, R.; Kalm, E. Genetic analysis of different traits in horse breeding by considering radiographic findings - 1st communication: Importance of radiographic findings in breeding sport horses. *Züchtungskunde* **1999**, *71*, 330–345.
5. Schober, M. *Schätzung von Genetischen Effekten Beim Auftreten von Osteochondrosis Dissecans Beim Warmblutpferd*; Georg-August-Universität Göttingen: Göttingen, Germany, 2003.
6. van Grevenhof, E.M.; Schurink, A.; Ducro, B.J.; van Weeren, P.R.; van Tartwijk, J.M.; Bijma, P.; van Arendonk, J.A. Genetic variables of various manifestations of osteochondrosis and their correlations between and within joints in Dutch warmblood horses. *J. Anim. Sci.* **2009**, *87*, 1906–1912. [CrossRef]
7. Jönsson, L.; Dalin, G.; Egenvall, A.; Näsholm, A.; Roepstorff, L.; Philipsson, J. Equine hospital data as a source for study of prevalence and heritability of osteochondrosis and palmar/plantar osseous fragments of Swedish Warmblood horses. *Equine Vet. J.* **2011**, *43*, 695–700. [CrossRef]
8. Wittwer, C.; Hamann, H.; Rosenberger, E.; Distl, O. Genetic parameters for the prevalence of osteochondrosis in the limb joints of South German Coldblood horses. *J. Anim. Breed. Genet.* **2007**, *124*, 302–307. [CrossRef]
9. Russell, J.; Matika, O.; Russell, T.; Reardon, R.J. Heritability and prevalence of selected osteochondrosis lesions in yearling Thoroughbred horses. *Equine Vet. J.* **2017**, *49*, 282–287. [CrossRef] [PubMed]
0. McCoy, A.M.; Norton, E.M.; Kemper, A.M.; Beeson, S.K.; Mickelson, J.R.; McCue, M.E. SNP-based heritability and genetic architecture of tarsal osteochondrosis in North American Standardbred horses. *Anim. Genet.* **2019**, *50*, 78–81. [CrossRef]
1. Wypchło, M.; Korwin-Kossakowska, A.; Bereznowski, A.; Hecold, M.; Lewczuk, D. Polymorphisms in selected genes and analysis of their relationship with osteochondrosis in Polish sport horse breeds. *Anim. Genet.* **2018**, *49*, 623–627. [CrossRef]
2. Lewczuk, D.; Hecold, M.; Ruść, A.; Frąszczak, M.; Bereznowski, A.; Korwin-Kossakowska, A.; Kamiński, S.; Szyda, J. Single nucleotide polymorphisms associated with osteochondrosis dissecans in Warmblood horses at different stages of training. *Anim. Prod. Sci.* **2017**, *57*, 608–613. [CrossRef]
3. Falconer, D.S. *Introduction to Quantitative Genetics*, 4th ed.; Pearson Education Limited: Essex, England, 1996.
4. Zaitlen, N.; Kraft, P.; Patterson, N.; Pasaniuc, B.; Bhatia, G.; Pollack, S.; Price, A.L. Using Extended Genealogy to Estimate Components of Heritability for 23 Quantitative and Dichotomous Traits. *PLoS Genet.* **2013**, *9*, e1003520. [CrossRef]
5. Manolio, T.A.; Collins, F.S.; Cox, N.J.; Goldstein, D.B.; Hindorff, L.A.; Hunter, D.J.; McCarthy, M.I.; Ramos, E.M.; Cardon, L.R.; Chakravarti, A.; et al. Finding the missing heritability of complex diseases. *Nature* **2009**, *461*, 747–753. [CrossRef] [PubMed]
6. Maher, B. Personal genomes: The case of the missing heritability. *Nature* **2008**, *456*, 18–21. [CrossRef]
7. Speed, D.; Cai, N.; Johnson, M.R.; Nejentsev, S.; Balding, D.J. Reevaluation of SNP heritability in complex human traits. *Nat. Genet.* **2017**, *49*, 986–992. [CrossRef]
8. Barry, C.S.; Walker, V.M.; Cheesman, R.; Davey Smith, G.; Morris, T.T.; Davies, N.M. How to estimate heritability: A guide for genetic epidemiologists. *Int. J. Epidemiol.* **2023**, *52*, 624–632. [CrossRef] [PubMed]
9. Evans, L.M.; Tahmasbi, R.; Vrieze, S.I.; Abecasis, G.R.; Das, S.; Gazal, S.; Bjelland, D.W.; de Candia, T.R.; Goddard, M.E.; Neale, B.M.; et al. Comparison of methods that use whole genome data to estimate the heritability and genetic architecture of complex traits. *Nat. Genet.* **2018**, *50*, 737–745. [CrossRef]
0. Yang, J.; Lee, S.H.; Goddard, M.E.; Visscher, P.M. GCTA: A tool for genome-wide complex trait analysis. *Am. J. Hum. Genet.* **2011**, *88*, 76–82. [CrossRef] [PubMed]
1. Yang, J.; Zeng, J.; Goddard, M.E.; Wray, N.R.; Visscher, P.M. Concepts, estimation and interpretation of SNP-based heritability. *Nat. Genet.* **2017**, *49*, 1304–1310. [CrossRef]
2. Speed, D.; Hemani, G.; Johnson, M.R.; Balding, D.J. Improved heritability estimation from genome-wide SNPs. *Am. J. Hum. Genet.* **2012**, *91*, 1011–1021. [CrossRef]
3. Slatkin, M. Linkage disequilibrium—Understanding the evolutionary past and mapping the medical future. *Nat. Rev. Genet.* **2008**, *9*, 477–485. [CrossRef]
4. Ren, D.; Cai, X.; Lin, Q.; Ye, H.; Teng, J.; Li, J.; Ding, X.; Zhang, Z. Impact of linkage disequilibrium heterogeneity along the genome on genomic prediction and heritability estimation. *Genet. Sel. Evol.* **2022**, *54*, 47. [CrossRef]
5. Lee, S.H.; DeCandia, T.R.; Ripke, S.; Yang, J.; Sullivan, P.F.; Goddard, M.E.; Keller, M.C.; Visscher, P.M.; Wray, N.R. Estimating the proportion of variation in susceptibility to schizophrenia captured by common SNPs. *Nat. Genet.* **2012**, *44*, 247–250. [CrossRef]
6. Yang, J.; Bakshi, A.; Zhu, Z.; Hemani, G.; Vinkhuyzen, A.A.; Lee, S.H.; Robinson, M.R.; Perry, J.R.; Nolte, I.M.; van Vliet-Ostaptchouk, J.V.; et al. Genetic variance estimation with imputed variants finds negligible missing heritability for human height and body mass index. *Nat. Genet.* **2015**, *47*, 1114–1120. [CrossRef]
7. Mancuso, N.; Rohland, N.; Rand, K.A.; Tandon, A.; Allen, A.; Quinque, D.; Mallick, S.; Li, H.; Stram, A.; Sheng, X.; et al. The contribution of rare variation to prostate cancer heritability. *Nat. Genet.* **2016**, *48*, 30–35. [CrossRef] [PubMed]

38. Lee, S.H.; Yang, J.; Chen, G.-B.; Ripke, S.; Stahl, E.A.; Hultman, C.M.; Sklar, P.; Visscher, P.M.; Sullivan, P.F.; Goddard, M.E.; et al. Estimation of SNP Heritability from Dense Genotype Data. *Am. J. Hum. Genet.* **2013**, *93*, 1151–1155. [CrossRef]
39. Hilla, D.; Distl, O. Prevalence of osteochondral fragments, osteochondrosis dissecans and palmar/plantar osteochondral fragments in Hanoverian Warmblood horses. *Berl. Munch. Tierarztl. Wochenschr.* **2013**, *126*, 236–244. [PubMed]
40. Hilla, D.; Distl, O. Genetic parameters for osteoarthrosis, radiographic changes of the navicular bone and sidebone, and their correlation with osteochondrosis and osteochondral fragments in Hanoverian warmblood horses. *Livest. Sci.* **2014**, *169*, 19–26. [CrossRef]
41. Browning, B.L.; Zhou, Y.; Browning, S.R. A One-Penny Imputed Genome from Next-Generation Reference Panels. *Am. J. Hum. Genet.* **2018**, *103*, 338–348. [CrossRef] [PubMed]
42. Chang, C.C.; Chow, C.C.; Tellier, L.C.A.M.; Vattikuti, S.; Purcell, S.M.; Lee, J.J. Second-generation PLINK: Rising to the challenge of larger and richer datasets. *GigaScience* **2015**, *4*, 7. [CrossRef]
43. Purcell, S.; Chang, C. *PLINK 1.9*; Purcell Lab: Melbourne, Australia, 2023.
44. Yang, J.; Benyamin, B.; McEvoy, B.P.; Gordon, S.; Henders, A.K.; Nyholt, D.R.; Madden, P.A.; Heath, A.C.; Martin, N.G.; Montgomery, G.W.; et al. Common SNPs explain a large proportion of the heritability for human height. *Nat. Genet.* **2010**, *42*, 565–569. [CrossRef] [PubMed]
45. Zhang, Q.; Privé, F.; Vilhjálmsson, B.; Speed, D. Improved genetic prediction of complex traits from individual-level data or summary statistics. *Nat. Commun.* **2021**, *12*, 4192. [CrossRef] [PubMed]
46. Min, A.; Thompson, E.; Basu, S. Comparing heritability estimators under alternative structures of linkage disequilibrium. *G3* **2022**, *12*, jkac134. [CrossRef]
47. McCoy, A.M.; Beeson, S.K.; Splan, R.K.; Lykkjen, S.; Ralston, S.L.; Mickelson, J.R.; McCue, M.E. Identification and validation of risk loci for osteochondrosis in standardbreds. *BMC Genom.* **2016**, *17*, 41. [CrossRef]
48. McCue, M.E.; Bannasch, D.L.; Petersen, J.L.; Gurr, J.; Bailey, E.; Binns, M.M.; Distl, O.; Guérin, G.; Hasegawa, T.; Hill, E.W.; et al. A High Density SNP Array for the Domestic Horse and Extant Perissodactyla: Utility for Association Mapping, Genetic Diversity, and Phylogeny Studies. *PLoS Genet.* **2012**, *8*, e1002451. [CrossRef]
49. Wray, N.R. Allele Frequencies and the r2 Measure of Linkage Disequilibrium: Impact on Design and Interpretation of Association Studies. *Twin. Res. Hum. Genet.* **2005**, *8*, 87–94. [CrossRef] [PubMed]
50. Tang, M.; Wang, T.; Zhang, X. A review of SNP heritability estimation methods. *Brief. Bioinform.* **2022**, *23*, bbac067. [CrossRef] [PubMed]
51. Visscher, P.M.; Hill, W.G.; Wray, N.R. Heritability in the genomics era—Concepts and misconceptions. *Nat. Rev. Genet.* **2008**, *9*, 255–266. [CrossRef]
52. Zhu, H.; Zhou, X. Statistical methods for SNP heritability estimation and partition: A review. *Comput. Struct. Biotechnol. J.* **2020**, *18*, 1557–1568. [CrossRef]
53. Falconer, D.S. The inheritance of liability to certain diseases, estimated from the incidence among relatives. *Ann. Hum. Genet.* **1965**, *29*, 51–76. [CrossRef]

 *animals* 

Article

Reconstruction of the Major Maternal and Paternal Lineages in the Feral New Zealand Kaimanawa Horses

Muhammad Bilal Sharif [1,2,3], Robert Rodgers Fitak [4], Barbara Wallner [5], Pablo Orozco-terWengel [6], Simone Frewin [7], Michelle Fremaux [8] and Elmira Mohandesan [1,2,*]

1 Department of Evolutionary Anthropology, University of Vienna, Djerassiplatz 1, A-1030 Vienna, Austria
2 Human Evolution and Archaeological Sciences (HEAS), University of Vienna, Djerassiplatz 1, A-1030 Vienna, Austria
3 Vienna Doctoral School of Ecology and Evolution (VDSEE), University of Vienna, Djerassiplatz 1, A-1030 Vienna, Austria
4 Department of Biology, Genomics and Bioinformatics Cluster, University of Central Florida, 4110 Libra Dr, Orlando, FL 32816, USA
5 Institute of Animal Breeding and Genetics, Veterinary University of Vienna, Veterinärplatz 1, A-1210 Vienna, Austria
6 Cardiff School of Biosciences, Cardiff University, The Sir Martin Evans Building, Museum Avenue, Cardiff CF10 3AX, Wales, UK
7 Feed2U Ltd., 19 Wairere Valley Road, Paparoa 0571, New Zealand
8 InfogeneNZ (EPAGSC), School of Agriculture and Environment, Massey University, 1 Drysdale Drive, Palmerston North 4410, New Zealand
* Correspondence: elmira.mohandesan@univie.ac.at

Citation: Sharif, M.B.; Fitak, R.R.; Wallner, B.; Orozco-terWengel, P.; Frewin, S.; Fremaux, M.; Mohandesan, E. Reconstruction of the Major Maternal and Paternal Lineages in the Feral New Zealand Kaimanawa Horses. *Animals* **2022**, *12*, 3508. https://doi.org/10.3390/ani12243508

Academic Editors: Isabel Cervantes and María Dolores Gómez Ortiz

Received: 6 October 2022
Accepted: 8 December 2022
Published: 12 December 2022

Publisher's Note: MDPI stays neutral with regard to jurisdictional claims in published maps and institutional affiliations.

Simple Summary: New Zealand has the fourth largest feral horse population (Kaimanawa and Far North horses) in the world. The Kaimanawas (KHs) are feral horses descended from various domestic horse breeds released into the Kaimanawa ranges in the 19th and 20th centuries. Over time, the population size has fluctuated dramatically due to hunting, large-scale farming and forestry. Currently, the herd is managed by an annual round-up, limiting the number to 300 horses to protect the rare and unique native flora in this region. Here, we examined 96 KHs to investigate their genetic similarity with respect to other domestic horse breeds, using uniparental markers (mitochondrial DNA, Y-chromosome). Our results indicate that although six maternal and six paternal lineages contributed to the KH gene pool, the current population is dominated by few ancestral lineages, and possibly represents two KH sub-populations. We show that three horse breeds, namely Welsh ponies, Thoroughbred and Arabian horses had a major influence in the genetic-makeup of the extant KH population. Moreover, our results suggest that mitochondrial genetic diversity in KHs is closer to the Sable Island horses, and less than other feral horse populations around the world. Our current findings, combined with ongoing research will provide insight into the KH population-specific genetic variations and level of inbreeding. This will advance equine genomic research and improve the management strategies to conserve these treasured New Zealand horses.

Abstract: New Zealand has the fourth largest feral horse population in the world. The Kaimanawas (KHs) are feral horses descended from various domestic horse breeds released into the Kaimanawa ranges in the 19th and 20th centuries. Over time, the population size has fluctuated dramatically due to hunting, large-scale farming and forestry. Currently, the herd is managed by an annual round-up, limiting the number to 300 individuals to protect the native ecosystem. Here, we genotyped 96 KHs for uniparental markers (mitochondrial DNA, Y-chromosome) and assessed their genetic similarity with respect to other domestic horses. We show that at least six maternal and six paternal lineages contributed unequally to the KH gene pool, and today's KH population possibly represents two sub-populations. Our results indicate that three horse breeds, namely Welsh ponies, Thoroughbreds and Arabian horses had a major influence in the genetic-makeup of the extant KH population. We show that mitochondrial genetic diversity in KHs ($\pi = 0.00687 \pm 0.00355$) is closer to that of the Sable Island horses ($\pi = 0.0034 \pm 0.00301$), and less than other feral horse populations around the world. Our current findings, combined with ongoing genomic research, will provide insight into the population-specific genetic variation and inbreeding among KHs. This will largely advance equine research and improve the management of future breeding programs of these treasured New Zealand horse.

Keywords: feral horses; SNP genotyping; mitochondrial DNA; Y-chromosome; Kaimanawa; New Zealand; genomic resources; conservation value

1. Introduction

Feral horse populations are found throughout the world [1]. They exist in a natural state and differ from their domesticated counterparts as they reproduce independently of human interventions. Feral horses represent valuable genomic resources that may be important for the future of horse breeding. Maintaining feral genetic diversity is especially important at this time when the genetic diversity of domestic horses (*Equus caballus*) is dramatically eroding and no truly wild horses exist in the world [2,3]. Although feral horse populations have been studied in their social organization, behavior and demography (reviewed in [4]), there is a lack of comprehensive and comparative genomic analysis on most feral horses. There are only three exceptional cases, in which comparative genome-wide SNP genotyping analyses were performed on feral Andean horses, American Mustangs [5] and the Canadian Sable Island and Alberta Foothills horses [6,7]. In these studies, the results were contrasted with various domestic breeds to understand genetic adaptation to high altitude [5], genetic diversity, phylogenetic relationships [6], and genomic consequences of inbreeding [7]. Except the studies mentioned above, most genetic studies on feral horses have been restricted to only a few populations from North America and often employ a limited number of specific genetic markers such as microsatellites and mitochondrial DNA (mtDNA) to investigate their genetic diversity and phylogenetic relationships to domestic horse breeds (e.g., [8–12]).

Currently, with an estimated number of 2,500 horses across the country, New Zealand's feral horse population (Kaimanawa horses (KHs) and Far North horses) is the fourth largest in the world, after Australia (Brumbies; ~1 million), United States of America (Mustangs ~120,000) and Canada (e.g., Sable Island and the Alberta Foothills; ~3500) [13]. Historically domestic horses were first introduced to New Zealand in 1814 from New South Wales (AU) by Reverend Samuel Marsden [14]. Subsequently, more horses arrived with European travelers, settlers, explorers, and cavalry as well as those which were owned and traded by the Māori people (indigenous Polynesian people of mainland New Zealand) [14]. KHs were first seen and reported in the Kaimanawa mountain ranges located in central North Island of New Zealand in 1876 [14,15]. These feral horses descended from various domestic horse breeds released into this region in the 19th and 20th centuries (Figure 1A). Over the last 150 years, KHs have lived freely at altitudes up to 1,500 m above sea level in a unique ecological region devoid of predators and competitors. Over time, the population size fluctuated dramatically due to hunting, large-scale farming and forestry. By 1979, only 174 horses remained [16]. After this dramatic decline, the New Zealand Wildlife Act (1953) was amended to give KHs legal protection within a limited geographic region [14]. This protection resulted in a rapid population growth (10-fold in ~14 years) of horses in poor health conditions. The high stocking density consequently led to threatening the rare and unique native flora by trampling and overgrazing [17]. Since 1993, KH numbers have been controlled through systematic mustering (i.e., round-up) programs that keep the population size congruent with its botanical environment. The mustered horses were sent to slaughter or sold at auction. Today's management strategies have been humanely improved and have been decided by the Kaimanawa Wild Horse Advisory Group. The herd continues to be managed by the New Zealand Department of Conservation, restricting the population census size to 300 individuals via annual helicopter mustering (Figure 1B,C).

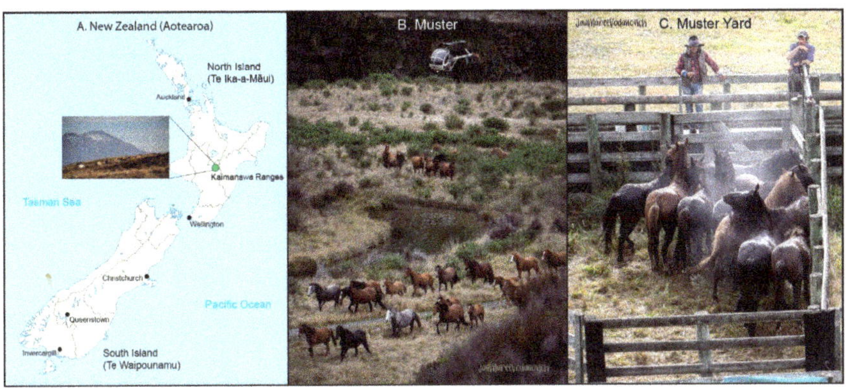

Figure 1. Geographical location and mustering process in Kaimanawa horses (KHs). (**A**) The location of the Kaimanawa mountain ranges, the habitat of feral KHs. (**B**) Annual mustering with helicopter, by which the horses are taken away from their natural habitat. (**C**) Mustering yard where animals are collected and examined by a veterinarian for their health and suitability for re-homing. *Image credit: Jan Maree Vodanovich.*

The majority of mustered horses are offered to the public for rehoming, depending on their good health, suitability of the new home, and a taming plan having been put in place. To avoid further breeding in the domestic setting, and to facilitate the taming process, all young stallions and colts get castrated. Recently (May 2022), and after decades of research, immune-contraception was applied to 150 female mustered horses as an alternative and complementary option to rehoming for controlling the population growth [18]. This change has been marked as the start of a new era in the management of KHs, allowing the vaccinated mares, foals, and stallions to return to their home territories and re-form their family bands, rather than being broken apart and rehomed.

The KH's uncertain origin beyond the historical reports raises questions regarding their phylogenetic relationship with other horse breeds, genetic diversity, long-term genetic health and viability. The first and only genetic study on KHs was conducted in the 1990s [19] and focused on comparing KH's allele frequencies across 16 red blood cell and plasma protein markers with those in more populous equine breeds in New Zealand, i.e., pure-bred hot blood (Thoroughbreds, Arabians) and warm blood (Standardbreds, Station Hacks "work horses with no particular breed"), as well as a representative of a heavy cold blooded horse line (Shire Horse) as a more distant breed. The main conclusion was that KHs are not genetically homogenous. They more closely resembled to Thoroughbred and Station Hack horses that made up most of the military and farm horses of the early 20th century in New Zealand, than Shire Horses. Although this study was the first one to shed light on the genetic relationship of KHs to other domestic horse breeds, it lacks resolution in the phylogenetic tree due to the few numbers of markers. In addition, at the time of this study, KHs have been present in New Zealand for only ~10 generations, which is not an adequate amount of time for evolutionary forces to accumulate markedly different genetic variation between KHs and the breeds from which they developed.

Since the late 1990's, no significant genetic studies have been conducted in KHs. Here, we infer for the first time the current maternal and paternal genetic diversity of KHs, by means of uniparental markers (mtDNA and Y-chromosome (Y-chr)). Previous studies on domestic horses have shown high levels of genetic diversity and lack of phylogeographic structure in mtDNA [3,20,21]. Although mtDNA is a powerful marker to study the contribution of various ancestral maternal lineages into a horse population, it is limited in terms of making a correspondence between mtDNA haplotypes (HTs), breeds and geography. In contrast, the paternally inherited Y-chr is an informative genetic marker for investigating the origin and influence of very recent paternal lineages. Here, for the first time, we use a combination of mtDNA and Y-chr markers to, (*i*) examine the number of mtDNA haplogroups/types (HGs/Ts) in KHs as an indirect indication of the population's current genetic diversity, and (*ii*) determine Y-chr HG/T and compare them to the other domestic breeds to understand the origin and influence of different patrilines in the KH gene pool.

In addition, pictorial records obtained from the largest archival photos of the Kaimanawa horses, suggest the existence of two KH herds within the muster area [13]. The grey horses are exclusively documented in the southern zones, whereas liver chestnut horses are found more in northern zones [13]. Although, we have no geographic information from the mustered KHs (southern vs. northern zones), the coat color information was provided by KH owners. Here, we attempt to genetically investigate whether there is population structure within the documented area, using a combination of uniparental genetic markers and coat color information.

2. Materials and Methods

2.1. Sample Collection

We collected hair follicles from a total of 96 KHs (53 female, 43 males), mustered during the period of 1993–2019 from different zones in the Kaimanawa ranges and rehomed by private owners or sampled for research purposes (Table S1). Among which, there is one sample from a male animal mounted on display at the Auckland War Memorial Museum. To avoid any stress and discomfort to animals during sampling, the owners followed two main strategies: firstly, sampling was performed over the summer period when hair follicles are most relaxed, and secondly, only a few hairs were pulled out at any attempt with several repeats to complete the sampling.

2.2. DNA Extraction and SNP Genotyping

Genomic DNA was extracted from hair follicles of 96 KHs, following a magnetic bead-based DNA extraction protocol by the technical personnel at Neogen. The individuals were genotyped with the Neogen GeneSeek Genomic Profiler (GGP) Equine v4. array (75K), following the manufacturer's instruction. This array contains ~71K SNP markers evenly distributed throughout the equine genome (autosomal and X-chr: ~70,231, mtDNA 1187 and Y-chr: 170). The raw data files from Illumina's GenomeStudio Final Report were converted to plink input files (lgen, fam and map), using a custom python script. Following [22], genotypes with bad quality calls (GenCall score < 0.15) were marked as missing data.

2.3. Relatedness Calculation

Prior to relatedness estimation, the following SNPs and individuals were excluded using PLINK v1.90 [23]: (*i*) SNPs that were missing in >5% of individuals; (*ii*) SNPs with minor allele frequency < 5%; (*iii*) SNPs with Hardy–Weinberg equilibrium deviation (–hwe) with *p-value* $< 1 \times 10^{-5}$ (as recommend in [24,25]); (*iv*) SNPs that were in linkage disequilibrium (–indep-pairwise 50 5 0.5; randomly one SNP from each pair was removed) and (*v*) individuals with >5% missing data.

As pedigree information (first-degree relatives) was only available for a few samples, we performed the identity-by-descent (IBD) analysis to evaluate the relatedness of the individuals. We calculated the genome sharing value (pi-hat) based on autosomal SNPs using PLINK v1.90 (–genome; [23]). The filtered dataset included 91 individuals and a total of 24,943 autosomal SNPs. Individuals with pi-hat $\geq$ 0.25 (second-degree relatives and closer) were further marked as related (Table S2).

2.4. Mitochondrial DNA Analysis

To evaluate mtDNA genetic diversity, we extracted 1187 mt-variants from the KH genotype data. The distribution of these SNP markers along the entire length of equine mtDNA is shown in Figure S1. To filter the mtDNA-variants dataset, we used PLINK v1.90 [23] to exclude: (*i*) individuals with missing data > 5%; and (*ii*) SNPs with missing data in more than one individual. We further excluded: (*iii*) maternally related individuals (pi-hat $\geq$ 0.25 & same mtDNA HT) (Table S2). Our final mtDNA dataset consisted of 1081 SNPs genotyped in 71 KHs (Table S3). Moreover, we constructed a comparative dataset from published complete mtDNA sequences (79 domestic and two Przewalski's horses, NCBI accessions: JN398377-457) reported in [20] as well as from Donkey mtDNA (*Equus asinus*: NC_001788.1) (Table S3), by extracting the overlapping SNPs positions as explained in Supplementary Text.

The genetic diversity indices including the total number of haplotypes (h), total number of polymorphic (segregating) sites (S), average number of nucleotide differences (k), haplotype diversity (H_d), nucleotide diversity (π) and theta estimator based on the segregating sites (θ_W) were calculated by DnaSP v6.12.03 [26] (Table 1). In addition, to understand whether the nucleotide diversity estimation is biased in the presence of the missing data (e.g., SNP genotyping), we calculated this value based on the complete mtDNA sequences, as well as 1,081 extracted positions in the global modern horse dataset [20], and compared the values (Table 1, Figure S2).

Table 1. Mitochondrial DNA (mtDNA) genetic diversity indices calculated for the KHs and the global modern horse dataset [20] based on the 1081 SNPs and complete mtDNA sequences (16,600 bp). The standard deviation values are shown in italics.

Genetic Diversity Indices	Kaimanawa Horse	Global Modern Horse [20]	
	1081 SNPs	1081 SNPs	Complete mtDNA
Number of samples (N)	71	81	81
Number of polymorphic (segregating) sites (S)	126	334	659
Number of haplotypes (h)	6	66	79
Average number of nucleotide differences (k)	27.3	45	83
Haplotype diversity (H_d)	0.518 *(0.043)*	0.994 *(0.003)*	0.999 *(0.002)*
Nucleotide diversity (π)	0.025 *(0.002)*	0.042 *(0.001)*	0.0049 *(0.0002)*
Theta estimator based on the segregating sites (θ_W)	0.024 *(0.006)*	0.062 *(0.016)*	0.008 *(0.002)*

To determine the maternal phylogenetic relationship between KHs and other domestic horse breeds, we constructed a maximum likelihood (ML) tree using MEGA v.10.2.5 [27] with the TN93 (Tamura-Nei) as the best fit substitution model based on the Bayesian Information Criterion (BIC) (Figure S3). In addition, to investigate the genetic similarity between KHs and other domestic horses, a median-joining (MJ) network [28] was constructed, using NETWORK v10.2.0.0 software (https://www.fluxus-engineering.com, accessed on 6 October 2022), by applying uniform substitution weights and without activating the external rooting option (Figure 2). The use of MJ network in inferring evolutionary relationships has been criticized by [29], due to two major shortcomings. Firstly, it is a distance-based approach based on the overall similarity between two sequences without taking the evolutionary model of DNA sequences into consideration, and secondly, it lacks the evolutionary direction (no rooting). Therefore, in this study we only used this approach to show the uniparental haplotype distributional in KHs, when compared to already established phylogenetic tree in the global modern horse dataset. The mtDNA HG nomenclature was adopted from [20], and the frequencies of HTs were calculated by direct counting (Figure 2).

To compare the genetic diversity in KHs with the other feral horse populations, we calculated the diversity indices (Table S5) and constructed a MJ network (Figure S4), using available sequences from partial mtDNA D-loop (255-bp, containing 28 SNPs). The comparative dataset included seven feral horse populations namely Assateague (NCBI accessions: GU014400-02), Fluoride Cracker (NCBI accessions: AY997150-51, AY997192), Grand Turk (NCBI accessions: HQ593024-34), Mustang (NCBI accessions: AJ413746-822), Sable Island (NCBI accessions: HQ592901-13, HQ592951-56, HQ593035, HQ593063), Saint-Pierre et Miqueleon (NCBI accessions: HQ592920-29, HQ593044) and Theodore National Park (NCBI accessions: MG761995-97) [10,12,30–32] (Figure S4).

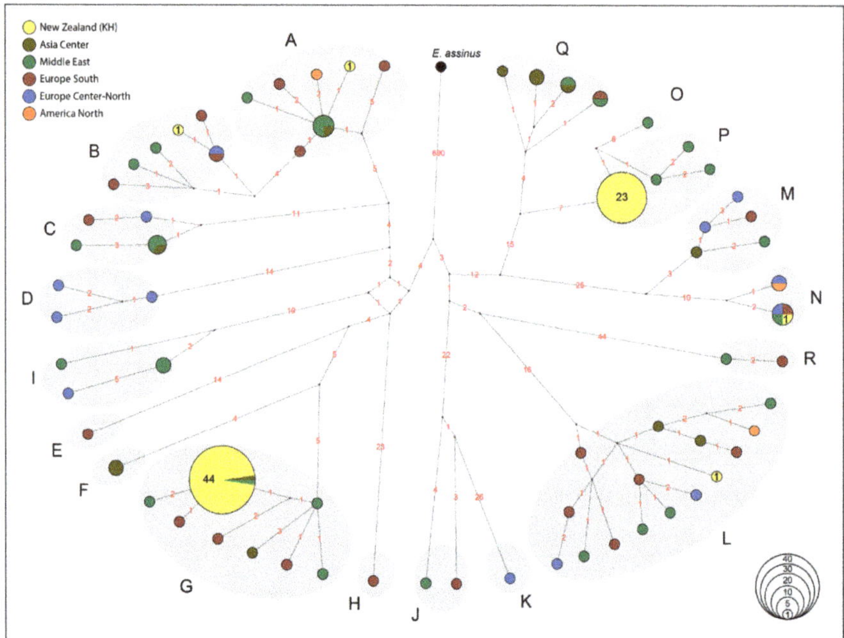

Figure 2. Median Joining network depicting mitochondrial DNA (mtDNA) haplogroup (HG) distribution (A–R) in KHs in relation to the global modern horses [20], based on 1081 SNPs. Haplotypes (HT) are indicated as circles, with size proportional to their frequency. The exact number of KHs are shown inside the circles. The colors indicate the geographical origin of the samples. The number of nucleotide differences between different haplotypes are shown on each branch. Donkey (*E. assinus* NC_001788.1) mitogenome was used as an outgroup. The MJ network [28] was constructed using NETWORK v10.2.0.0 software (https://www.fluxus-engineering.com, accessed on 6 October 2022) by applying uniform substitution weights and without activating the external rooting option.

2.5. Y-Chromosome Analysis

To investigate the paternal ancestry and evaluate the Y-chr HT diversity in the KH population, 170 genetic variants, which were previously described as "male-specific Y-chr (MSY) variants" by [33], were genotyped on the GGP Equine array, and analyzed in male KHs (n = 43). Out of 170 variants, 121 determine the HTs in the so-called "crown" group (detected in the majority of modern domestic breeds), while 49 define non-crown HTs (detected in Asian and some northern European breeds) [33,34]. Five additional variants (rAF, rDT, qCU, qCR and qW), which are not included in the current version (v4) of the Illumina GGP Equine array and provide resolution within the clade A of the crown group, were inferred by determining the allelic states with LGC KASP assays [35] (Tables S8 and S9), as described in [33]. The positions of all Y-chr variants in the GGP Equine array (v4) are according to the LipY764 (GCA_002166905.2) Y-chr assembly [33]. Information about MSY markers, variant type, indicative haplogroups, and ancestral and derived alleles is presented in Table S6.

To filter the KH GGP array Y-chr SNPs dataset, we used PLINK v1.90 [23] to exclude (*i*) individuals with missing data > 10%; (*ii*) variants with heterozygous calls in males and (*iii*) variants with missing data > 5%. The remaining missing calls were imputed by introducing the allelic states observed in the samples belonging to the same HG, following the method described in [36] (Table S6). We further excluded: (*iv*) paternally related individuals (pi-hat ≥ 0.25 & same Y-chr HT) (Table S2). Our final Y-chr dataset consists of 157 SNPs genotyped in 37 male KHs.

To investigate the phylogenetic relationship between KHs and other domestic horse breeds, a similar dataset of overlapping 157 SNPs was constructed from published Y-chr sequences (156 domestic, and one Przewalski's horse, NCBI BioProjects accessions PRJNA430351 and PRJNA787432) reported in [34] (Table S6). A ML phylogenetic tree was generated using MEGA v.10.2.5 [27] with the K2 (Kimura-2) as the best fit substitution model based on the Bayesian Information Criterion (BIC) (Figure S5). Following the maximum parsimony trees constructed in [33,34], the topology and the number of horses exhibiting each Y-chr HT were visualized as a MJ network [28] generated with NETWORK v10.2.0.0 software (https://www.fluxus-engineering.com, accessed on 6 October 2022), using uniform substitution weights and without activating the external rooting option (Figure 3). As mentioned in Section 2.4, MJ network is only used to investigate the paternal haplotype distribution, and similarity between KHs and modern horse breeds without inferring phylogenetic relationships. The HG nomenclature was adopted from [34] and the frequencies of each HT were calculated by direct counting.

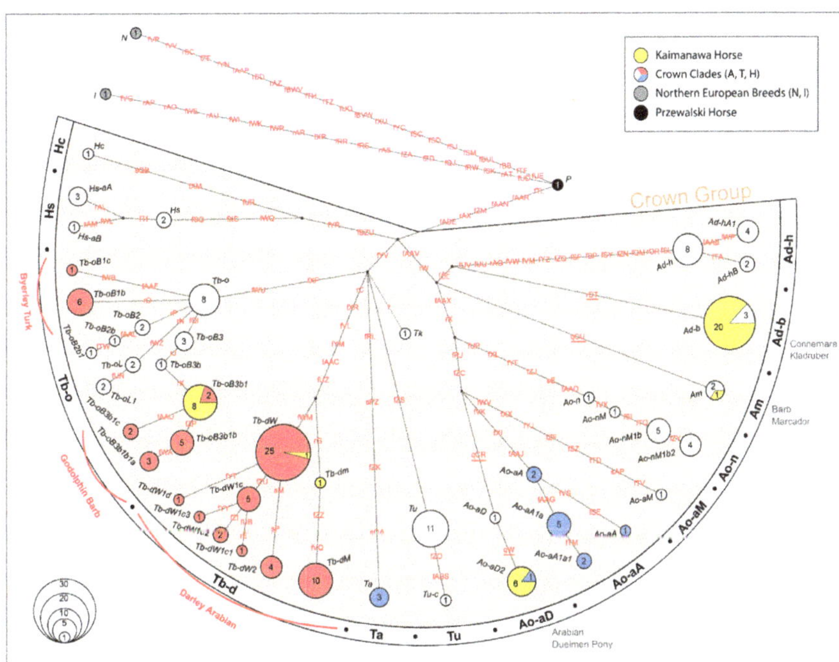

Figure 3. Median Joining network depicting Y-chromosome (Y-chr) haplogroup distribution in KHs in relation to the global modern horse breeds [34] based on 157 Y-chr SNPs. Haplotypes are indicated as circles and the numbers inside represent the number of individuals. The name of the variants observed in each haplotype are shown in red. The variants genotyped using KASP are distinguished by underline. The MJ network [28] was constructed using NETWORK v10.2.0.0 software (https://www.fluxus-engineering.com, accessed on 6 October 2022), by applying uniform substitution weights and without activating the external rooting option.

2.6. Population Structure and Genetic Distance within the KH Population

To evaluate the possibility of population structure within the sampled KHs, we performed nuclear (157 Y variants) F_{ST} (Fixation Index) [37] pairwise comparison, using ARLEQUIN v3.5.2.2 [38] with 1000 permutations. Since there is a lack of geographic information for the sampled KHs, we defined two sub-populations based on the observed dominant maternal HGs (G and P) (see results Section 3.2). In group-1, we included 27 male samples with maternal HG G, exhibiting paternal HGs Ad-b, Tb-oB3b1, Am, Ao-aD2 and

Td-dm. In group-2, we included 13 male samples with maternal HG P, exhibiting paterna HGs Ao-aD2, Ad-b and Tb-oB3b1 (Table S10). We further cross checked our genetic data with phenotypic information (coat color) provided by KH owners (Table S10).

3. Results

3.1. Relatedness in KH population

Based on the IBD evaluation using autosomal SNPs, we show that ~53% of the individuals in the KH population have one or more second degree or closer relatives (pi-hat $\geq$ 0.25 in the sampled population. We excluded 22 maternally and 6 paternally related individuals from the mtDNA and Y-chr analysis, respectively (Table S2).

3.2. Mitochondrial DNA Genetic Diversity and Haplogroup Distribution

In this study, we successfully genotyped 1,081 mtDNA SNPs in 71 unrelated KHs (Table S3). We constructed a haplotype network to determine the genetic similarity between the KHs and the global mtDNA dataset representing 27 horse breeds across Asia, Europe Middle East and the Americas [20]. KHs are clustered within six previously defined mtDNA HGs (A, B, G, L, N and P) (Figure 2, Table S4), with nucleotide diversity (π) = 0.025 $\pm$ 0.002 and haplotype diversity (H_d) = 0.518 $\pm$ 0.043 (Table 1). The HGs G and P are the most frequent ones (G: 62%; P: 32%), while A, B, L and N were each represented by only single individuals (Figure 2, Table S4).

Furthermore, using the global modern horse dataset, our comparison between nucleotide diversity estimates using genotypic data (π = 0.042 $\pm$ 0.001) versus complete mtDNA sequences (π = 0.0049 $\pm$ 0.0002) indicate that in the face of missing data, the nucleotide diversity can be an overestimation of the true value as invariant sites are missing in SNPs genotype data and hence omitted from the calculations (Table 1, Figure S2).

In addition, we calculated the genetic diversity in KHs based on 28 SNPs retrieved from the partial mtDNA D-loop region (225-bp) and compared the value with the overlapping data from seven feral horse populations (Table S5). The mtDNA D-loop haplotype diversity in KHs (H_d = 0.11 $\pm$ 0.051) was comparable with the estimates obtained for the Sable Island (H_d = 0.095 $\pm$ 0.084) and Grand Turk (H_d = 0.182 $\pm$ 0.144) feral horse populations, while lower than other feral horse populations such as Assateague, Fluoride Cracker, Mustangs, Saint-Pierre et Miquleon and Theodore National Park. Moreover, our data show that even though the value of nucleotide diversity in KHs (π = 0.00687 $\pm$ 0.00355) is lower than any other feral horse populations, it is still closer to the Sable Island (π = 0.0034 $\pm$ 0.00301) population (Table S5) Among the feral horse populations, Saint-Pierre et Miqueon show the highest haplotype diversity (H_d = 0.782 $\pm$ 0.093) and nucleotide diversity (π = 0.12597 $\pm$ 0.01716) based on D-loop analysis. In addition, we observed slightly lower haplotype diversity and nucleotide diversity in a combined feral horse dataset (H_d = 0.714 $\pm$ 0.017; π = 0.0636 $\pm$ 0.002), compared to the domestic ones (H_d = 0.89 $\pm$ 0.02; π = 0.08166 $\pm$ 0.00586) (Table S5).

3.3. Y-Chromosome Genetic Diversity and Haplogroup Distribution

Our analysis on a combined dataset (157 SNPs) of male KHs (n = 37), and published Y-chr dataset (156 domestic horses and a single przewalski horse) retrieved from [34] (Table S6), indicate that KHs haplotypes are clustered within six previously defined MSY crown HTs (Ad-b, Am, Ao-aD2, Tb-dM, Tb-dW and Tb-oB3b1) (Figure 3, Table S7). All HTs detected in KHs have been previously reported in British ponies, Arabian and Thorough bred horses.

The HT Ad-b is represented by more than half of our samples (Ad-b: 20/37), and it has been reported in breeds other than Arabians and Thoroughbreds, such as Connemara pony (Ireland) and Kladruber (Baroque type breed from Czech Republic). The HT Am is represented by only one KH sample (Am: 1/37), and it has been previously detected in Barb (North Africa) and Marchador (Brazil) horses. Among the three typical Arabian HTs (i.e., Ao-aA, Ta and Ao-aD2) commonly observed in occidental and local (Iran, Syria Arabian lines, we only observed the Ao-aD2 in our KH samples (Ao-aD2: 6/37) (Figure 3)

Within the clade T, we have detected genetic signatures of two of the three influential Thoroughbred foundation sire lines in KHs, namely the Darley Arabian (HG Tb-d), and the Godolphin Barb (sub-HG Tb-oB3b1) [33]. Within the Darley Arabian line in KHs (2/37), the HT Tb-dW (Thoroughbreds) is represented by one KH, and another KH does not cluster with any HT defined by the samples reported in [34]. This sample contains the basal variant rG and have the reference allele for variants fZZ and fVQ (which are diagnostic to the HT Tb-dM), hence shown as HT *Tb-dm* here (Figure 3).

While the Darley Arabian line was represented by only two KHs, the Godolphin Barb line is more common (HT Tb-oB3b1: 8/37). We have not detected any signature of the sire line lineage related to Byerley Turk (sub-HG Tb-oB1), which has previously been reported in Thoroughbreds, Akhal Teke/Turkomans, and in a small set of Arabian horses with unknown ancestry [34].

3.4. Population Structure and Genetic Distance within the KH Population

We observed that among male horses with maternal HG G, the majority have British and Thoroughbred paternal ancestry (Ad-b: 18/27; Tb-oB3b1: 6/27), while the majority of KHs with maternal HG P have Arabian paternal ancestry (Ao-aD2: 7/13) (Table S10). F_{ST} pairwise genetic distance based on the Y-variants in KHs, indicate a significant (0.128, *p-value* < 0.05) and moderate structure within the male KH population harboring ancestral maternal G and P lineages. In addition, our results indicate that all grey KHs sampled in the Kaimanawa range (7/7) harbor maternal HG P, while chestnut horses (22/30) are predominantly found in horses with maternal HG G (Table S11).

4. Discussion

4.1. Genetic Diversity of Maternal Lineages in KHs

Although we observed the contribution of at least six ancestral maternal lineages in the KH gene pool, the majority of samples contributed to only two major HGs (G and P). These HGs are frequently observed in the Middle East and European horse breeds (Figure 2). The HG P has been previously found exclusively in samples from Middle East such as an unspecified Iranian breed, Caspian pony and Arabian horses, while HG G was found in central Asia (Akhal-Teke, Naqu), the Middle East (unspecified Iranian and Syrian horse, Arabian) and southern Europe (Giara horse, unspecific Italian breed) [20]. The unequal contribution of maternal lineages in today's KHs can be explained by the expected effect of genetic drift in such a small, isolated population.

Mitochondrial haplotype and nucleotide diversity in KHs (H_d = 0.518 ± 0.043; π = 0.025 ± 0.002) is lower than the corresponding values obtained for the global modern horse dataset (H_d = 0.994 ± 0.003; π = 0.042 ± 0.001). This can, in principle, be explained by a smaller proportion of ancestral mitochondrial diversity in KH population, or by an effect of genetic drift in a small, isolated population. In addition, SNPs genotype data have ascertainment bias towards known variants in certain horse breeds. Therefore, any rare variants private to the KHs remain undetected. Although these unrepresented SNPs are more likely to be rare, they can be common in such a small, isolated population. In the absence of the complete mtDNA data, that bias leads to underestimating the diversity indices within the small, isolated KH population.

Furthermore, this genotyping bias limits the power of traditional statistical framework for neutrality test (e.g., Tajima's *D* [39], and Fu and Li's *F* [40]) to accurately evaluate the genomic signature of past evolutionary processes such as selection, demographic expansion or contraction. Therefore, we highlight the importance of performing whole genome sequencing in KHs and other feral horse populations to accurately infer the effect of past demographic events on the current genetic diversity of feral horses around the world, with greater application in management strategies.

Moreover, we attempted to compare the mitochondrial genetic diversity in KHs with other feral horses based on the overlapping partial D-loop region. However, it is worth noting that even though high levels of genetic diversity in horse mtDNA D-loop region (~350–650 bp) have been previously detected [21,32,41–43], recurrent mutations tend to blur the structure of the phylogenetic tree as examined in [20] (Figure S4). As shown in [20] and confirmed by our mt-variants dataset, these few polymorphic positions (28 SNPs) retrieved from D-loop region are not informative enough to reconstruct the mtDNA haplotypes network (Figure S4), nevertheless they indicate the incorporation of a wide range of matrilineal lineages into the feral horses around the world.

4.2. Genetic Diversity of Paternal Lineages in KHs

We have observed an unequal contribution of at least six ancestral paternal lineages in the KH gene pool, dominated by three HTs (Ad-b, Ao-aD and Tb-oB3b1) (Figure 3). Regardless of this unequal paternal ancestry, all detected HTs belong to the crown group which is a recently expanded horse MSY HG, predominant in modern domestic breeds. This confirms the historical reports indicating that the KH population was recently established from various modern domestic breeds imported to New Zealand, such as British ponies, Arabian horses and Thoroughbreds. Thus, beside the Thoroughbred ancestry shown in [19], our data add and confirm the historical contribution of imported horses with British and Arabian genetic influences in today's extant KHs.

Based on written reports by James Boyd in the mid 1990's [15], one of the main founders of the KH population were the Exmoor ponies which were imported to the Hawkes Bay area in New Zealand by Major George Gwavas Carlyon in the middle 19th century. These ponies were crossed with local horses resulting in the Carlyon Pony breed. The Carlyon ponies were later crossed with two Welsh pony stallions, which were imported by Sir Donald McLean, and a breed known as the Comet was produced. Apparently in the 1870s, McLean released a Comet stallion and several mares into the Kaingaroa Plains where KHs were roaming [15,19]. As Ad-b has been reported in Welsh ponies [44], our data support the major historical influence of these two Welsh ponies into KH male gene pool.

Considering all British and Irish pony breeds share a very similar breeding and lineage history—originated from Celtic ponies and were refined with oriental, Iberian and English Thoroughbred stallions [45]—we suggest that Exmoor ponies might exhibit the Ad-b, alike other British pony breeds. However, with no available information on the MSY variations in Exmoor ponies, their influence in KHs remains unresolved.

Among the three typical Arabians HGs/Ts (Ao-aA, Ta and Ao-aD2) reported in [34], we observed the HT Ao-aD2 in our KHs (Figure 3). When it comes to the origin of the Arabian line in KHs, it is documented that it was not until the 1920's that Arabian horses were first introduced into New Zealand, from India [46]. In addition, it has also been reported [15] that Nicholas Koreneff released an Arabian stallion into the Argo Valley region during the 1960's, when he was forced to sell his farm. Since there is no information on whether this stallion was the only Arabian horse released in the Argo area, we cannot conclude that Ao-aD2 comes from this stallion. Nevertheless, our data for the first time demonstrate the influence of Arabian horses into the KH gene pool.

In addition to the Arabian and British pony breeds stated above, KHs show footprints of two of the three influential Thoroughbred foundation sires, namely the Darley Arabian (HG Tb-d) and the Godolphin Barb (sub-HG Tb-oB3b1) [33]. This major influx of Thoroughbreds or Thoroughbred descent horses into the KH population could be explained by a historical strangles epidemic which threatened the horses at the Waiouru Mounted Cavalry Stables in 1941 [14]. Some of these military horses died of the infection, whilst others survived and joined the feral KH herd.

These diverse and unequal contributions of different paternal lines in KHs can be the result of founder events, selection and genetic drift in a small, isolated population. Moreover, it can be explained by the natural social structure in horses, which is remarkably uniform in feral horses around the world [4,47,48]. There are stable breeding groups called "bands", each containing a single stallion with one or more mares and their offspring. Mares and stallions within bands are unlikely to be closely related as both male and female offspring disperse from their natal band [4,47,48]. Other stallions live as bachelors alone or in groups with no breeding mares or juvenile females in mixed sex peer groups [4,47,48]. Therefore, the mating success of the ancestral foundation stallions would determine the proportion of the HGs observed in today's KH population. Lastly, a consequence of this behavioral limitation is the further reduction of genetic variation that exacerbates the small effective population size that uniparental markers have. Consequently, if the effective population size is strongly reduced there is also a limit to the time in the past about which these markers have phylogeographic information (e.g., [49,50]). However, while these uniparental markers are relevant to establish sex biased evolutionary processes, exploiting the wealth of genealogical history in the nuclear DNA (e.g., via SNP arrays, ddRAD or whole genome sequencing) should be the next step to clearly understand the evolutionary origin of the KH.

4.3. Loss of Maternal and Paternal Lineages in the KH Population

We predict that the rare mtDNA HGs in KHs, such as A and B, will be lost in a few generations as they are observed only in one 14-year- and 20-year-old female, respectively (Table S10). In addition, we have only detected HGs L and N in two advanced age male individuals (18- and 16-year-old) (Table S10). Therefore, we conclude that either the maternal HG L and N has been already lost in the population, or they are in a very low frequency, and hence more susceptible to being lost in future as a result of genetic drift.

It is worth noting that maternal HG L, and paternal Tb-dw—which are common worldwide—were both observed only once and as a combination in one 18-year-old male individual with tan coat color. Tan coat color is very rare in the KH population and has been observed only in this sample in our collection. To reduce the sampling bias, we have almost sampled 32% of the KH population mustered during the period of 1993–2019 from different zones in the Kaimanawa ranges. Considering the small census size (~300 individuals) of the population and high level of relatedness (~50%), we can assume that our data exhibit a realistic picture of the maternal and paternal haplotype contributions in the KH population. Nevertheless, we cannot fully rule out the possibility of existing very rare and hence undetected haplotypes in the KH population.

4.4. Population Structure within KH Population: Parallel Contribution of European and Middle Eastern Parental Ancestry

We observed that among male horses with maternal HG G, the majority have British and Thoroughbred paternal ancestry, while the majority of KHs with maternal HG P have Arabian paternal ancestry (Table S10). This co-occurrence suggests a parallel contribution of these combinations of parental lineages, which can be explained by the effect of sampling from particular regions in the Kaimanawa ranges where initially different escaped and/or imported horses with European (G; Ad-b and Tb-oB3b1) and Middle Eastern (P; Ao-aD2) origins were settled. Although we have no geographical information from the sampled KHs (northern vs. southern), we further evaluated these results by adding phenotypic information (coat color) provided by KH owners. In agreement with pictorial evidence, our combined genetic and phenotypic data suggest the existence of two genetically isolated sub-populations in the Kaimanawa mountain ranges. There is one in the southern zones, where grey horses are exclusively found, and one in the northern area, where the population is dominated by bay and chestnut horses.

Whether this ancestral parental combination is only a reflection of the past founder groups, or today's KH population splits into two genetically isolated sub-populations, would have a major impact on effective population size and genetic viability of the population. In the case of the later, it would not be a one herd of 300 horses, but rather an equal split of each approximately 150, which has to be carefully considered in management strategies to control the population size. Obtaining whole genomic data, and geographic information of sampled KHs would be the crucial future step to accurately evaluate the population structure of today's KHs and to conduct informed management decisions.

5. Conclusions

In summary, our data indicate that although a diverse genetic ancestry contributed to the KH gene pool, today's population is dominated by few maternal and paternal lineages. Therefore, to ensure the long-term viability of a genetically healthy population, minimizing the loss of the current genetic diversity in KHs must be a priority. This has already been initiated by returning the KHs with more rare and interesting colors (e.g., palomino, grey, tan) into the range, and excluding such mares from the immune contraception intervention [9]. In addition, our genetic data in combination with phenotypic information provide support for the possibility of the existence of two genetically isolated sub-populations (northern vs. southern) of feral horses in the Kaimanawa mountain range.

Whether KHs harbor rare and population-specific (private) functional variations as a result of adaptation to their natural environment, can only be confirmed or rejected with more comprehensive genome-wide data. As previously shown [51], whole genome sequencing holds great potential to detect a large number of novel variants, many of which might be specific to the population under study. Ongoing genomic research on KHs would shed light on their genomic variation enabling them to adapt to their unique and harsh environment, and investigate the genomic footprints of past demographic events and future conservation strategies in this population.

Supplementary Materials: The following supporting information can be downloaded at: https://www.mdpi.com/article/10.3390/ani12243508/s1, Figure S1: The coverage of mitochondrial SNP markers available on the GGP Equine v4. array (75K); Figure S2: Nucleotide diversity variation along the length of horse mtDNA shown for global modern horses and KHs; Figure S3: mtDNA ML phylogenetic tree constructed for KHs and global modern horses; Figure S4: Partial mtDNA D-loop (225 bp) HG network in KHs in relation to modern domestic and feral horses; Figure S5: Y-chr ML phylogenetic tree constructed for KHs and global modern horses; Table S1: Sample information on New Zealand feral KHs collected for this study; Table S2: Information on the relatedness among KHs; Table S3: Information on 1081 mtDNA SNPs genotyped in KHs and retrieved from global modern horses; Table S4: mtDNA HG affiliation in KHs and global modern horses; Table S5: Partial mtDNA D-loop (255-bp) genetic diversity in KHs compared with other feral horse populations; Table S6: Information on 157 MSY variants genotyped in male KHs using Equine GGP array and KASP genotyping techniques, and retrieved from global modern horses; Table S7: MSY HG affiliation in male KHs and global modern horses; Table S8: KASP Experimental setup used for genotyping five MSY variants in KHs; Table S9: Cluster plots for the five MSY variants genotyped using Bio-Rad CFX Manager 3.1; Table S10: Summary of mtDNA and Y-chr HG affiliation in male KHs, including their age at the time of sampling (~2020); Table S11: The number of KHs with different coat colors in relation to mtDNA and Y-chr HGs. References: [52–54].

Author Contributions: Conceptualization, E.M.; Methodology, E.M., M.B.S. and B.W.; Software, E.M. and M.B.S.; Validation, E.M., M.B.S. and B.W.; Formal Analysis, M.B.S. and E.M.; Investigation, E.M.; Resources, E.M., S.F., M.F. and B.W.; Data Curation, M.B.S.; Writing—Original Draft Preparation, E.M. and M.B.S.; Writing—Review and Editing, E.M., M.B.S., R.R.F., P.O.-t. and B.W.; Visualization, M.B.S.; Supervision, E.M.; Project Administration, E.M., S.F. and M.F.; Funding Acquisition, E.M. All authors have read and agreed to the published version of the manuscript.

Funding: This project has received funding from the international Austrian Science Fund (FWF: I3838-B29)—German Research Foundation (DFG: 401205971) joint grant. E.M. and M.B.S. are supported by the Austrian Science Fund (FWF), granted to E.M.

Institutional Review Board Statement: The study was discussed and approved by the institutional ethics and welfare committee of the University of Vienna and Kaimanawa Heritage Horse welfare society.

Informed Consent Statement: The horse owners provided informed consent for contributing samples into this genetic study.

Data Availability Statement: mtDNA and MSY SNPs for KHs, and comparative datasets are provided in Supplementary Tables S3 and S6, respectively.

Acknowledgments: The authors would like to thank Sue Rivers and the Kaimanawa Heritage Horse society (https://kaimanawaheritagehorses.org/, accessed on 6 October 2022) for their valuable collaboration on this project, providing us with permission to study New Zealand feral horses. We would like to thank Simone Frewin, and Kelly Wilson as representatives of Kaimanawa horse owners, facilitating the sampling procedure. We are grateful to all the Kaimanawa horse owners (Armstrong, R.; Armstrong, S.; Baish, L.; Belcher, D.; Bradley, J.; Brooks, T.; Buckby, A.; Cameron, F.; Clarck, S.; Conway, K.; Cottam, T.; Croucher, T.; Dalley, J.; Duncan, K.; Fenwick, J.; Flintoff-Baker, F.; Flintoff-Baker, W.; Fremaux, M.; Fryer, A.; Fryer, D.; Fryer, E.; Fryer, H.; Garrod, L.; Haigh, C.; Haines, C.; Hall, G.; Horan, K.; Houghton, K.; Johansson, K.; Jones, K.; King, E.; Moleta, L.; Moore, R.; O'leary, B.; Owen, S.; Pedersen, J.; Phillips-Harris, C.; Rayner, M.; Rolleston, N.; Rowe, R.; Secrest, S.; Seddon, A.; Sinclair-Morgan, C.; Stradling, J.; Stroud, V.; Sutherland, D.; Sutherland, L.; Thompson, T.; van-de-Molen, S.; Vandervegte, S.; Walker, K.; Walker, S.; Wildbore, L.; Wilson, K.; Wreaks, D. and Wright, J.) for providing samples, photos and information on subject horses. We thank Matt Rayner for sample donation from the Auckland War Memorial Museum. We thank the photographer Jan Maree Vodanovich for providing us high-quality and beautiful photos from the muster process. We acknowledge general support from G. Weber, Department of Evolutionary Anthropology, University of Vienna, Austria. We thank the CUBE platform, hosted by the Division of Computational System Biology at the University of Vienna (https://cube.univie.ac.at/, accessed on 6 October 2022) for generous computational resources. We thank Lara Radović for her technical support in KASP genotyping assay, and Maeve Nic Samhradain for her assistance in English edition and proof reading. Open Access Funding by the Austrian Science Fund (FWF).

Conflicts of Interest: The authors declare no conflict of interest.

References

1. Lever, C. *Naturalized Mammals of the World*; Longman: London, UK, 1985.
2. Fages, A.; Hanghøj, K.; Khan, N.; Gaunitz, C.; Seguin-Orlando, A.; Leonardi, M.; Constantz, C.M.; Gamba, C.; Al-Rasheid, K.A.; Albizuri, S. Tracking Five Millennia of Horse Management with Extensive Ancient Genome Time Series. *Cell* **2019**, *177*, 1419–1435. [CrossRef] [PubMed]
3. Librado, P.; Gamba, C.; Gaunitz, C.; Der Sarkissian, C.; Pruvost, M.; Albrechtsen, A.; Fages, A.; Khan, N.; Schubert, M.; Jagannathan, V.; et al. Ancient Genomic Changes Associated with Domestication of the Horse. *Science* **2017**, *356*, 442–445. [CrossRef] [PubMed]
4. Cameron, E.; Linklater, W.; Minot, E.; Stafford, K. Population Dynamics 1994–1998, and Management of Kaimanawa Wild Horses. *Sci. Conserv.* **2001**, *171*, 1173–2946.
5. Hendrickson, S.L. A Genome Wide Study of Genetic Adaptation to High Altitude in Feral Andean Horses of the Páramo. *BMC Evol. Biol.* **2013**, *13*, 1–13. [CrossRef]
6. Colpitts, J.; McLoughlin, P.D.; Poissant, J. Runs of Homozygosity in Sable Island Feral Horses Reveal the Genomic Consequences of Inbreeding and Divergence from Domestic Breeds. *BMC Genom.* **2022**, *23*, 501. [CrossRef] [PubMed]
7. Tollett, C.M. Genomic Diversity and Origins of the Feral Horses (*Equus ferus caballus*) of Sable Island and the Alberta Foothills. Master's Thesis, University of Saskatchewan, Saskatoon, SK, Canada, 2018.
8. King, S.R.; Schoenecker, K.A.; Fike, J.A.; Oyler-McCance, S.J. Feral Horse Space Use and Genetic Characteristics from Fecal DNA. *J. Wildl. Manag.* **2021**, *85*, 1074–1083. [CrossRef]
9. Kobayashi, I.; Akita, M.; Takasu, M.; Tozaki, T.; Kakoi, H.; Nakamura, K.; Senju, N.; Matsuyama, R.; Horii, Y. Genetic Characteristics of Feral Misaki Horses Based on Polymorphisms of Microsatellites and Mitochondrial DNA. *J. Vet. Med. Sci.* **2019**, *81*, 707–711. [CrossRef] [PubMed]
10. Ovchinnikov, I.V.; Dahms, T.; Herauf, B.; McCann, B.; Juras, R.; Castaneda, C.; Cothran, E.G. Genetic Diversity and Origin of the Feral Horses in Theodore Roosevelt National Park. *PLoS ONE* **2018**, *13*, e0200795. [CrossRef]
11. Conant, E.; Juras, R.; Cothran, E. A Microsatellite Analysis of Five Colonial Spanish Horse Populations of the Southeastern United States. *Anim. Genet.* **2012**, *43*, 53–62. [CrossRef]
12. Prystupa, J.M.; Hind, P.; Cothran, E.G.; Plante, Y. Maternal Lineages in Native Canadian Equine Populations and Their Relationship to the Nordic and Mountain and Moorland Pony Breeds. *J. Hered.* **2012**, *103*, 380–390. [CrossRef]
13. Wilson, K. *Wild Horses of the World*; Penguin Random House: Auckland, New Zealand, 2020; ISBN 978-0-14-377291-0.
14. Fleury, W. *Kaimanawa Wild Horse Herd—Draft Management Strategy*; Department of Conservation: Wanganui, New Zealand, 1991.
15. Boyd, J. Kaimanawa Heritage Horses Welfare SocietyHorse History. Available online: https://kaimanawaheritagehorses.org/horsehistory/ (accessed on 5 November 2022).
16. Aitken, V.; Kroef, P.; Pearson, A.; Ricketts, W. *Observations of Feral Horses in the Southern Kaimanawas*; Report to the Kaimanawa Wild Horse Committee; Massey University: Palmerston North, New Zealand, 1979.
17. Rogers, G.M. Kaimanawa Feral Horses and Their Environmental Impacts. *N. Z. J. Ecol.* **1991**, *15*, 49–64.

18. Kaimanawa Heritage Horses Welfare Society Immunocontraception. Available online: https://kaimanawaheritagehorses.org/immunocontraception/ (accessed on 12 September 2022).

19. Halkett, R.J. A Genetic Analysis of the Kaimanawa Horses and Comparisons with Other Equine Types. Master's Thesis, Massey University, Palmerston North, New Zealand, 1996.

20. Achilli, A.; Olivieri, A.; Soares, P.; Lancioni, H.; Kashani, B.H.; Perego, U.A.; Nergadze, S.G.; Carossa, V.; Santagostino, M.; Capomaccio, S. Mitochondrial Genomes from Modern Horses Reveal the Major Haplogroups That Underwent Domestication. *Proc. Natl. Acad. Sci. USA* **2012**, *109*, 2449–2454. [CrossRef] [PubMed]

21. Vilà, C.; Leonard, J.A.; Gotherstrom, A.; Marklund, S.; Sandberg, K.; Lidén, K.; Wayne, R.K.; Ellegren, H. Widespread Origins of Domestic Horse Lineages. *Science* **2001**, *291*, 474–477. [CrossRef] [PubMed]

22. Petersen, J.L.; Mickelson, J.R.; Cothran, E.G.; Andersson, L.S.; Axelsson, J.; Bailey, E.; Bannasch, D.; Binns, M.M.; Borges, A.S.; Brama, P.; et al. Genetic Diversity in the Modern Horse Illustrated from Genome-Wide SNP Data. *PLoS ONE* **2013**, *8*, e54997. [CrossRef] [PubMed]

23. Chang, C.C.; Chow, C.C.; Tellier, L.C.; Vattikuti, S.; Purcell, S.M.; Lee, J.J. Second-Generation PLINK: Rising to the Challenge of Larger and Richer Datasets. *GigaScience* **2015**, *4*, s13742-015-0047-8. [CrossRef]

24. Zhao, S.; Jing, W.; Samuels, D.C.; Sheng, Q.; Shyr, Y.; Guo, Y. Strategies for Processing and Quality Control of Illumina Genotyping Arrays. *Brief. Bioinform.* **2018**, *19*, 765–775. [CrossRef]

25. Guo, Y.; He, J.; Zhao, S.; Wu, H.; Zhong, X.; Sheng, Q.; Samuels, D.C.; Shyr, Y.; Long, J. Illumina Human Exome Genotyping Array Clustering and Quality Control. *Nat. Protoc.* **2014**, *9*, 2643–2662. [CrossRef]

26. Rozas, J.; Ferrer-Mata, A.; Sánchez-DelBarrio, J.C.; Guirao-Rico, S.; Librado, P.; Ramos-Onsins, S.E.; Sánchez-Gracia, A. DnaSP 6: DNA Sequence Polymorphism Analysis of Large Data Sets. *Mol. Biol. Evol.* **2017**, *34*, 3299–3302. [CrossRef]

27. Kumar, S.; Stecher, G.; Li, M.; Knyaz, C.; Tamura, K. MEGA X: Molecular Evolutionary Genetics Analysis across Computing Platforms. *Mol. Biol. Evol.* **2018**, *35*, 1547–1549. [CrossRef]

28. Bandelt, H.-J.; Forster, P.; Röhl, A. Median-Joining Networks for Inferring Intraspecific Phylogenies. *Mol. Biol. Evol.* **1999**, *16*, 37–48. [CrossRef]

29. Kong, S.; Sánchez-Pacheco, S.J.; Murphy, R.W. On the Use of Median-Joining Networks in Evolutionary Biology. *Cladistics* **2016**, *32*, 691–699. [CrossRef] [PubMed]

30. Eggert, L.S.; Powell, D.M.; Ballou, J.D.; Malo, A.F.; Turner, A.; Kumer, J.; Zimmerman, C.; Fleischer, R.C.; Maldonado, J.E. Pedigrees and the Study of the Wild Horse Population of Assateague Island National Seashore. *J. Wildl. Manag.* **2010**, *74*, 963–973. [CrossRef]

31. Luís, C.; Bastos-Silveira, C.; Cothran, E.G.; Oom, M. do M. Iberian Origins of New World Horse Breeds. *J. Hered.* **2006**, *97*, 107–113. [CrossRef] [PubMed]

32. Jansen, T.; Forster, P.; Levine, M.A.; Oelke, H.; Hurles, M.; Renfrew, C.; Weber, J.; Olek, K. Mitochondrial DNA and the Origins of the Domestic Horse. *Proc. Natl. Acad. Sci. USA* **2002**, *99*, 10905–10910. [CrossRef] [PubMed]

33. Felkel, S.; Vogl, C.; Rigler, D.; Dobretsberger, V.; Chowdhary, B.P.; Distl, O.; Fries, R.; Jagannathan, V.; Janečka, J.E.; Leeb, T.; et al. The Horse Y Chromosome as an Informative Marker for Tracing Sire Lines. *Sci. Rep.* **2019**, *9*, 1–12. [CrossRef] [PubMed]

34. Remer, V.; Bozlak, E.; Felkel, S.; Radovic, L.; Rigler, D.; Grilz-Seger, G.; Stefaniuk-Szmukier, M.; Bugno-Poniewierska, M.; Brooks, S.; Miller, D.C.; et al. Y-Chromosomal Insights into Breeding History and Sire Line Genealogies of Arabian Horses. *Genes* **2022**, *13*, 229. [CrossRef]

35. He, C.; Holme, J.; Anthony, J. SNP Genotyping: The KASP Assay. In *Crop Breeding*; Springer: Berlin/Heidelberg, Germany, 2014; pp. 75–86.

36. Lippold, S.; Xu, H.; Ko, A.; Li, M.; Renaud, G.; Butthof, A.; Schröder, R.; Stoneking, M. Human Paternal and Maternal Demographic Histories: Insights from High-Resolution Y Chromosome and MtDNA Sequences. *Investig. Genet.* **2014**, *5*, 1–17. [CrossRef]

37. Wright, S. The Genetical Structure of Populations. *Ann. Eugen.* **1949**, *15*, 323–354. [CrossRef]

38. Excoffier, L.; Lischer, H.E. Arlequin Suite Ver 3.5: A New Series of Programs to Perform Population Genetics Analyses under Linux and Windows. *Mol. Ecol. Resour.* **2010**, *10*, 564–567. [CrossRef]

39. Tajima, F. Statistical Method for Testing the Neutral Mutation Hypothesis by DNA Polymorphism. *Genetics* **1989**, *123*, 585–595. [CrossRef]

40. Fu, Y.-X.; Li, W.-H. Statistical Tests of Neutrality of Mutations. *Genetics* **1993**, *133*, 693–709. [CrossRef] [PubMed]

41. Kavar, T.; Dovč, P. Domestication of the Horse: Genetic Relationships between Domestic and Wild Horses. *Livest. Sci.* **2008**, *116*, 1–14. [CrossRef]

42. Cieslak, M.; Pruvost, M.; Benecke, N.; Hofreiter, M.; Morales, A.; Reissmann, M.; Ludwig, A. Origin and History of Mitochondrial DNA Lineages in Domestic Horses. *PLoS ONE* **2010**, *5*, e15311. [CrossRef] [PubMed]

43. Gurney, S.; Schneider, S.; Pflugradt, R.; Barrett, E.; Forster, A.C.; Brinkmann, B.; Jansen, T.; Forster, P. Developing Equine MtDNA Profiling for Forensic Application. *Int. J. Leg. Med.* **2010**, *124*, 617–622. [CrossRef]

44. Mühlberger, M. Genetische Untersuchung Zu Paternalen Linien in Kaltblutpferden Und Kleinpferden Europas. Master's Thesis, Veterinärmedizinischen Universität Wien, Wien, Austria, 2019.

45. Willmann, R. Das Exmoor-Pferd: Eines Der Ursprünglichsten Halbwilden Pferde Der Welt. *Nat. Und Mus.* **1999**, *129*, 389–407.

46. Allfrey, S. *Horses & Foals*; WHSmith: London, UK, 1980; ISBN 9780706342130.

7. Duncan, P.; Boy, V.; Monard, A.-M. The Proximate Mechanisms of Natal Dispersal in Female Horses. *Behaviour* **1996**, *133*, 1095–1124. [CrossRef]
8. Linklater, W.L. Adaptive Explanation in Socio-Ecology: Lessons from the Equidae. *Biol. Rev.* **2000**, *75*, 1–20. [CrossRef]
9. OROZCO-terWENGEL, P.; Corander, J.; Schlötterer, C. Genealogical Lineage Sorting Leads to Significant, but Incorrect Bayesian Multilocus Inference of Population Structure. *Mol. Ecol.* **2011**, *20*, 1108–1121. [CrossRef]
10. Hein, J.; Schierup, M.; Wiuf, C. *Gene Genealogies, Variation and Evolution: A Primer in Coalescent Theory*; Oxford University Press: New York, NY, USA, 2004.
11. Jagannathan, V.; Gerber, V.; Rieder, S.; Tetens, J.; Thaller, G.; Drögemüller, C.; Leeb, T. Comprehensive Characterization of Horse Genome Variation by Whole-Genome Sequencing of 88 Horses. *Anim. Genet.* **2019**, *50*, 74–77. [CrossRef]
12. Katoh, K.; Standley, D.M. MAFFT Multiple Sequence Alignment Software Version 7: Improvements in Performance and Usability. *Mol. Biol. Evol.* **2013**, *30*, 772–780. [CrossRef]
13. Hall, T.; Biosciences, I.; Carlsbad, C. BioEdit: An Important Software for Molecular Biology. *GERF Bull. Biosci.* **2011**, *2*, 60–61.
14. Letunic, I.; Bork, P. Interactive Tree Of Life (ITOL) v5: An Online Tool for Phylogenetic Tree Display and Annotation. *Nucleic Acids Res.* **2021**, *49*, W293–W296. [CrossRef] [PubMed]

Article

Effects of Selection on Breed Contribution in the Caballo de Deporte Español

Ester Bartolomé [1,*], Mercedes Valera [1], Jesús Fernández [2] and Silvia Teresa Rodríguez-Ramilo [3]

[1] Departamento de Agronomía, ETSIA, Universidad de Sevilla, 41013 Sevilla, Spain; mvalera@us.es
[2] Departamento de Mejora Genética Animal, Instituto Nacional de Investigación y Tecnología Agraria y Alimentaria-Centro Superior de Investigaciones Científicas, 28040 Madrid, Spain; jmj@inia.csic.es
[3] GenPhySE, Université de Toulouse, Institut National de Recherche pour l'Agriculture, l'Alimentation et l'Environnement, Ecole Nationale Vétérinaire de Toulouse, 31326 Castanet-Tolosan, France; silvia.rodriguez-ramilo@inrae.fr
* Correspondence: ebartolome@us.es

Simple Summary: The Caballo de Deporte Español (CDE) is a sport horse breed, which originated from crosses between different sport horse breeds in the search for a good sport aptitude for Dressage, Eventing and Show Jumping disciplines. The main aim of this study was to determine the effects of 15 years of selection on this breed and find out whether it has been effective and adequate regarding the CDE main breeding objectives. The whole known pedigree was used, comprising 47,884 animals (18,799 males and 29,085 females). Pedigree analyses were performed in order to check the population structure, origins and evolution over the years. For the analyses, animals were divided into fourteen breed groups. Performance data used in the analyses were the estimated breeding values (EBV) of the Show Jumping, Dressage and Eventing sport disciplines from the routine genetic evaluations of the CDE breeding programme. Results showed some degree of subdivision within the population and, therefore, inbred matings. Regarding the evolution of breeding values, we found that EBVs of offspring were higher than the EBVs of parents showing that genetic gain has actually occurred. Moreover, selection decisions seem to be taken by breeders based, to certain extent, on the genetic evaluations.

Abstract: The equine breeding industry for sport's performance has evolved into a fairly profitable economic activity. In particular, the Caballo de Deporte Español (CDE) is bred for different disciplines with a special focus on Show Jumping. The main aim of this study was to determine the effects of 15 years of selection and to find out whether it has been effective and adequate regarding the CDE main breeding objectives. The whole pedigree of 19,045 horses registered as CDE was used comprising 47,884 animals (18,799 males and 29,085 females). An analysis performed to check for the pedigree completeness level yielded a number of equivalent complete generations (t) equal to 1.95, an average generation interval (GI) of 10.87 years, mean inbreeding coefficient (F) of 0.32%, an average relatedness coefficient (AR) of 0.09% and an effective population size (Ne) of 204. For the analyses, animals were divided into fourteen breed groups. Additionally, in order to study the evolution of these breeds over time and their influence on CDE pedigree, five different periods were considered according to the year of birth of the animals. Performance data used in the analyses were the estimated breeding values (EBV) of the Show Jumping sport discipline of 12,197 horses in the CDE pedigree, available from the 2020 routine genetic evaluations of the CDE breeding program (starting in 2004). Dressage and Eventing EBV values were also assessed. Results showed values of F higher than expected under random mating; this pointed to some degree of inbred matings. With regard to the evolution of breeding values, we found that, in general, EBVs of offspring were higher than the EBVs of parents. Notwithstanding, there is still a need for improvement in population management and the coordination of the breeders to get higher responses but controlling the loss of genetic diversity in the CDE breed.

Citation: Bartolomé, E.; Valera, M.; Fernández, J.; Rodríguez-Ramilo, S.T. Effects of Selection on Breed Contribution in the Caballo de Deporte Español. *Animals* **2022**, *12*, 1635. https://doi.org/10.3390/ani12131635

Academic Editor: Cristina Sartori

Received: 16 April 2022
Accepted: 21 June 2022
Published: 25 June 2022

Publisher's Note: MDPI stays neutral with regard to jurisdictional claims in published maps and institutional affiliations.

Keywords: pedigree; open breed; breeding program; founders; show jumping; equine

1. Introduction

The equine breeding industry for sports performance has evolved in recent years into a fairly profitable economic activity. Thus, horse breeds have been selected to find the animal that is best suited for the equestrian discipline they participate in. This is true for pure breeds as well as for mixed breeds. In fact, different horse breeds, such as the Irish Sport Horse, Brazilian Sport Horse, Polish Sport Horse or, more recently, the Caballo de Deporte Español, were created as composite breeds, with the horses' capacity for a certain sports performance as the objective or selection criterion [1]. In particular, the Caballo de Deporte Español (Spanish Sport Horse; CDE) is bred for Dressage, Eventing and Endurance disciplines with a special focus on Show Jumping discipline, taking advantage of the probable heterosis effect that results from this multiple breed combination [2,3]. The breed is managed by the Caballo de Deporte Español Breeders Association (ANCADES, www.ancades.com (accessed on 25 February 2022)), and its breeding program was officially approved in 2004 and updated in 2020 [4]. Its main goal is to select a horse with a suitable functional conformation, temperament and health, able to attain a high performance at either national or international sports events in which it participates. When the CDE Official Stud Book was created, an animal could be registered in the Foundational Registry (and thus be identified as a CDE animal) up until 2004 if it came from any of the "permitted" cross-breeds (established by Spanish laws APA/3318/2002 and APA/1646/2004) and was born between 1992 and 1998. Since the Foundational Registry was closed in 2004, an animal is considered to be a CDE either when both parents are registered as CDE, when only one of them is a CDE with the other parent from any of the "permitted" breeds included in these laws, or when both parents belong to any of these "permitted" breeds (even when both parents are from foreign breeds, their offspring could be inscribed as CDE at the owner's request if they are not inscribed in any other official Stud Book). Although CDE is an open breed that allows animals of different breeds on its Stud Book [1], since the establishment of its breeding program, breeders are being encouraged to use animals already registered as CDE as reproducers. Notwithstanding, although heterosis falls in the F2 generation if purebred animals are no longer incorporated, an advantage in the phenotype of animals with a more "mixed" genetic background can still be manifested, with crosses being directed to obtain CDE horses showing a lot of heterosis [5]. However, no studies have been developed before about CDE breeders' real preferences when selecting horses to create the next generation, i.e., whether they choose animals with the best estimated breeding values (EBV) for sport performance or just animals from the geographically nearest studs. Different criteria may have different repercussions on the outcome of the CDE breeding program. The selection process and mating are a consequence of a conscious decision, usually intended only for short-term achievements in intensively managed domesticated species [6]. In a crossbreed like the CDE horse, decisions are also influenced by breed preferences when choosing the individuals selected for breeding and, consequently, the evolution of performance and of the genetic parameters across time could also show this effect. The main aim of this study was to determine/characterize the effects of 15 years of selection on this breed and find out whether it has been effective and adequate regarding the main CDE breeding objectives.

2. Materials and Methods

2.1. Population Description

Since the creation of the CDE Official Stud Book (in 2002) until 2004, an animal could have been inscribed in the Foundational Registry (and thus be identified as a CDE animal) if it fulfilled two requirements: (i) coming from any of the admitted horse breeds (established by Spanish laws APA/3318/2002 and APA/1646/2004); and (ii) being born between 1992 and 1998. Since the closing of the Foundational Registry in 2004, not only animals with both parents registered as CDE can be included in the CDE Official Stud Book, but horses with only one or neither parent registered as CDE can also be included and considered CDE. In the latter cases, the "external" parents (i.e., non-CDE horses) should belong to any of the admitted breeds established in the CDE Stud Book. Obviously, their offspring can be inscribed in the CDE Stud Book if they are not already registered in any other official Stud Book. Furthermore, the non-CDE parents can be inscribed in the Auxiliary Registry of the CDE Stud Book as Foreign CDE (CDEx), and from that point, their descendants would be registered as CDE directly in the Principal Registry of the CDE Stud Book.

The pedigree information used in this study was gathered from the Caballo de Deporte Español official Stud Book, provided by ANCADES. For analysis purposes, no differences between CDE registrations were considered, thus counting 21,163 CDE horses for this study (19,045 registered as CDE in the Principal Registry and 2118 registered as Foreign CDE in the Auxiliary Registry of the Stud Book), representing 44.7% males and 55.3% females. The whole pedigree included 47,884 animals (18,799 males and 29,085 females).

According to the composite nature of this breed, different horse breeds were included in the CDE Stud Book via the relatives of the CDE animals (55.8% of the whole pedigree). These breeds were considered independently according to their representation in the CDE pedigree. Breeds with fewer than 500 animals (less than 1.5%) in the pedigree were grouped together in an "Other Horse Breeds" group (OHB). A total of 14 breed groups were assessed, 12 of them corresponding to majority horse breeds (>500 animals) present in the CDE Stud Book, together with the OHB and the CDE horses groups: Belgian Warmblood breed (BWP; 624 animals), Oldenburger breed (OLDBG; 642 animals), Westphalian breed (WESTF; 653 animals), Lusitano breed (LUS; 875 animals), Holsteiner breed (HOLST; 1269 animals), Anglo-Arab Horse breed (AAH; 1561 animals), Hanoverian breed (HANN; 1742 animals), Arab Horse breed (AH; 1835 animals), Dutch Warmblood (KWPN; 2120 animals), Selle Français (SF; 2311 animals), Thoroughbred (TH; 2924 animals), Other Horse Breeds group (OHB; 4731 animals), Pura Raza Española (PRE; 5434 animals) and the Caballo de Deporte Español breed group (CDE; 21163).

In order to study the evolution of these breeds over time (in terms of the proportion of individuals belonging to each breed) and their influence on CDE pedigree, five different periods were considered according to the year of birth of the animals. The first period (P0), including animals born before 2000, was settled before the Foundational Registry of the CDE Stud Book was created in 2002. Consequently, P0 only has animals from the Foundational Registry, and thus, no CDE animals were found in this period. Then, the second period (P1) included animals born between 2000 and 2004, the third period (P2) between 2005 and 2009, the fourth period (P3) involved individuals born between 2010 and 2014, and finally, the fifth and last period (P4), between 2015 and 2021. Figure 1 shows the distribution across these periods of the breed groups considered, showing different size and color of the circles according to the number of animals represented in each period. Furthermore, in order to confirm that all breeds considered for this study were genetically different between them, a principal component analysis was developed, using a matrix with the proportion of the individual genome coming from each original founder's breed, calculated from the pedigree-derived relationship matrix (Figure A1).

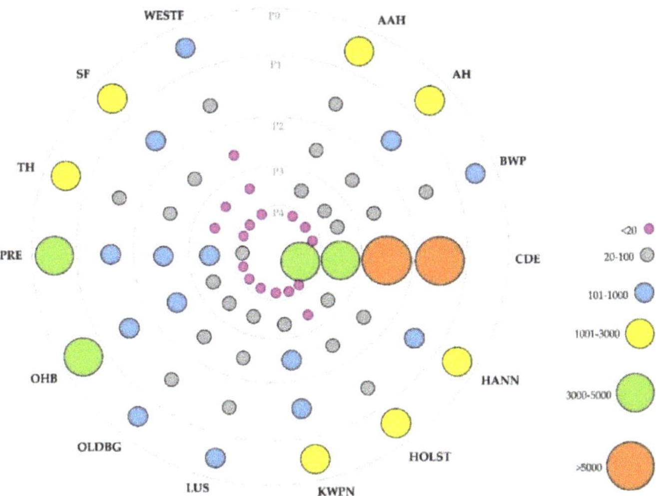

Figure 1. Distribution of horse breeds from the Caballo de Deporte Español Official Stud Book according to the period considered. Different sizes and colors of the circles indicate the number of animals of a certain breed within periods. Where AAH is Anglo-Arab Horse, AH is Arab Horse, BWP is Belgian Warmblood, CDE is Caballo de Deporte Español, HANN is Hanoverian Horse, HOLST is Holsteiner Horse, KWPN is KWPN Horse, LUS is Lusitano Purebred, OLDBG is Oldenburg Horse, OHB is Other Horse Breeds, PRE is Pura Raza Española, TH is Thoroughbred, SF is Selle Français, WESTF is Westphalian Horse; P0 is the period with animals born before 2000, P1 between 2000 and 2004, P2 between 2005 and 2009, P3 between 2010 and 2014, and P4 after 2015.

2.2. Pedigree Analyses

Pedigree analyses were computed using the program ENDOG (v.4.8, [7]). In order to ascertain the circumstances affecting the genetic history of the CDE horse breed population, some parameters were calculated to determine the amount of available information, the pedigree completeness level, the number of complete generations (t) and the generation interval (GI) as a measure of the speed of genetic transmission due to the physiology of the species and the particular management performed. The first one was assessed as the proportion of ancestors known per generation for each offspring [8]. The t value was computed as the sum of $(1/2)^n$, where n is the number of generations separating the individual from each known ancestor [9]. The GI was calculated as the average age of parents at the birth of their offspring kept for reproduction [10]. The three parameters were computed for the whole population and for a reference population, including only animals born in the last 10 years (between 2011 and 2021), which mostly corresponds to animals from the last generation. These will also be referred to as "active animals".

In order to monitor the animal genetic resources for this breed, the effective population size (Ne) was also computed. PopRep 1.0. software's (PopRep.tzv.fal.de (accessed on 15 June 2022)) [11] decision tree was used for picking the best method to calculate Ne for our data. After the analyses, the Ne calculated by individual increase in inbreeding, following Ref [12], appeared as the method that best fitted our data. Ne was calculated as $Ne = \frac{1}{2*\Delta F}$, *where* $\Delta F = \frac{F_t - F_{t-1}}{1 - F_{t-1}}$, where F_t is the inbreeding coefficient of the offspring, and F_{t-1} is the inbreeding coefficient of the parents.

2.3. Performance Data

The EBV of 12,197 horses for Show Jumping, 5253 horses for Eventing and 7790 horses for Dressage disciplines, from the CDE pedigree, were available for this study, including the performance data of animals from 2004 to 2020 [13].

Genetic parameters for the Show Jumping, Eventing and Dressage sport disciplines used in this study were obtained from the 2020 routine genetic evaluation information of the CDE breeding program. Genetic evaluation implies a multivariate BLUP analysis using the following genetic model:

$$y = Xb + Za + Wp + Qr + e,$$

where **y** was the vector of observations for the analyzed traits of each discipline; **b** was the vector of fixed effects; **a** was the vector of additive genetic effects; **p** the vector of rider horse interaction effect (for Show Jumping discipline) or the rider effect (for Eventing and Dressage disciplines); **r** the vector of permanent environmental effects (for Show Jumping and Dressage disciplines); and **e** the vector of random residual terms, **X, Z, W** and **Q** were the incidence matrices assigning observations to the fixed, animal, rider-horse interaction or rider and permanent environmental effects, respectively.

In respect of the Show Jumping discipline, the analyzed traits were (i) penalty score transformed to a positive scale (ranging from 50 to 100, with 100 representing the highest punctuation in the competition "0 penalties" and 50 the lowest "maximum penalties") and (ii) weighted total ranking (calculated on a positive points scale by assigning a value of 100 to the first classified animal (within the same event, penalty scale and competition level) and a value of 0 to the last) for Show Jumping. The fixed effects considered were: age, breed, gender, event, height of fences and category type of test interaction. The general index for this discipline combined both variables considered with a weight of 50% each.

For the Endurance discipline, the analyzed traits were (i) penalty score transformed to a positive scale of the Show Jumping exercise, (ii) final points obtained in the Dressage exercise and (iii) penalty score transformed to a positive scale of the Cross exercise. The fixed effects considered in this genetic model were: age, breed, gender, event, level of the event and owner, whereas the general index combined with a weight of 25% for (i), 35% for (ii) and 40% for (iii).

Finally, in the Dressage discipline, the analyzed traits were (i) points obtained at the dressage reprise, (ii) points for walk, (iii) points for trot, (iv) points for gallop, (v) points for submission and vi) points for general impression. This model considered the following fixed effects: age, breed, gender, event, level of the event and stud. Additionally, all traits were combined in the following general index: 70% for (i), 10% for (ii) and 10% for (iii), 5% for (iv), 2.5% for (v) and 2.5% for (vi).

It must be highlighted that, in all genetic evaluations, EBVs were standardized for an interval of 80–120 with a population average of 100. All BLUP genetic evaluations were carried out with the VCE software—version 6.0.2 (Nashville Video Production Company/Podcast Studio, Nashville, TN, USA) [14]. In addition, for the official genetic evaluations of the three disciplines considered, the accuracy of EBVs (r), was calculated as $r = \sqrt{1 - \frac{PEV}{\delta_a^2}}$, where PEV is the prediction error variance and δ_a^2 was the additive genetic variance.

2.4. Effect of Genetic Selection on Inbreeding and Average Relatedness Coefficient of the Caballo de Deporte Español Population

The level of genetic variability in the whole CDE population and within each breed group defined was assessed considering the following parameters:

Mean inbreeding coefficient (F), defined as the probability that an individual has two identical alleles by descent in any locus [15].

Average relatedness coefficient (AR) of each individual, interpreted as the representation of the animal in the whole pedigree, or the amount of the animal genetic information it shares. Thus, it is equivalent to the average relationship of an individual with the rest of the population, i.e., the marginal of the relationship matrix divided by the number of individuals. Furthermore, the average of all AR values is also equivalent to twice 1—the expected heterozygosity, which is a common measure of genetic diversity [16,17]. When

AR is calculated for a founder, it expresses the proportion of the genetic information arising from it, and thus, its relevance in the development of the pedigree. In order to check the effect that breeders' decisions have had when selecting the breeding stock on the genetic merit of the CDE population, the genetic variability was assessed for the whole population and also according to a hypothetical breeding scenario, where only CDE horses with CDE breed parents were allowed to register as CDE in the CDE official Stud Book.

Furthermore, using results from the Pedigree analyses developed before with the program ENDOG (v.4.8, [7]), the evolution across time of F and AR values for animals with best EBVs for the performance ability for Show Jumping, was also evaluated. The EBVs values used were described previously in Section 2.3.

2.5. Probability of Gene Origin of the CDE Horse Breed

Ancestors with both unknown parents in the available database were considered founders, as it is a common practice (unless molecular information exists for those animals).

In order to check the genetic representation of the founder individuals in the CDE population and in all the breed groups considered, the genetic contribution of founders to the descendant gene pool of the population [18] was assessed, classified by breed of origin and per period. This way the evolution of the relative influence of each breed in the composition of the CDE population could be studied. Additionally, for the 10 most contributing founders on each period (P0 to P4) their mean EBV for Show Jumping performance and that of their descendants on each period was computed to determine the genetic trend with time.

3. Results

3.1. Population Description

Breed groups' distribution appearing in Figure 1 showed PRE and OHB as majority horse breeds in P0, with 4387 (19.2%) and 4076 (17.8%) animals born in this period (percentage within period), respectively. Afterwards, from P1 to P4, CDE was the majority horse breed, showing an increasing tendency from 79.4% (8804) animals in P1 to 98.2% (3986) in P4. On the other hand, PRE, OHB, KWPN, HANN, AH and SF breeds showed higher representation after CDE in P1, with 5.4% (596), 4.0% (442), 2.5% (276), 1.5% (169), 1.3% (144) and 1.1% (123) animals, respectively. All of them showed a decreasing tendency over the periods, with PRE still showing the highest representation in every period (after CDE), with 310 animals in P2 (4.9%), 121 in P3 (3.4%) and 20 in P4 (0.5%). The rest of the breeds showed fewer than 100 animals in P3 and P4 periods.

3.2. Pedigree Analyses

In order to study the quality of the pedigree information, for both the individuals included in the entire CDE Stud Book and those in the defined reference population (active animals), the pedigree completeness level was assessed. When the whole CDE population was considered, tracing back just two generations yielded a completeness level of 50%. The same level (50%) was obtained when analyzing the reference population, even when four generations were accounted for. Therefore, the quality of the pedigree information is higher for the reduced set of animals.

The number of animals, number of equivalent complete generations, average generation interval and effective population size, computed via individual increase in inbreeding for both the entire and the reference populations considered, were included in Table 1. The reference population represented 14.1% of the entire CDE pedigree and showed 1.91 more equivalent complete generations (3.86) and a generation interval 1.02 years longer (11.89) than the whole pedigree (with values of 1.95 and 10.87, respectively). On the other hand, Ne was 4% smaller in the reference population considered (Ne = 196) than in the whole pedigree (Ne = 204).

Table 1. Number of animals, number of equivalent complete generations (t), average generation interval (GI) and effective population size computed via individual increase in inbreeding (Ne) for the whole Caballo de Deporte Español Studbook and for the animals born between 2011 and 2021 (reference population).

Population	Number of Animals	*t*	GI	Ne
Whole Pedigree	47,884	1.95	10.87	204
Born between 2011 and 2021	6743	3.86	11.89	196

3.3. Effect of Genetic Selection on Inbreeding and Average Relatedness Coefficient of the Caballo de Deporte Español Population

The mean inbreeding coefficient (F) and average relatedness (AR) for the whole CDE population were 0.32% and 0.09%, respectively, and for the reference population they were 0.79% and 0.19%, respectively, whereas the population with animals showing an EBV over the population mean (EBV > 100) showed 0.27% and 0.19% for F and AR values, respectively. The hypothetical scenario where only CDE horses were allowed to be registered and reproduced, showed an F coefficient of 2.18% and an AR of 0.19%.

The evolution of the above parameters in all the sets considered and of the Ne computed via individual increase in inbreeding over the five periods analyzed is shown in Figure 2. Considering that in an ideal population with random matings, F would be half the AR, the larger than expected *F*-values would indicate an important subdivision and/or non-random mating pattern in the population.

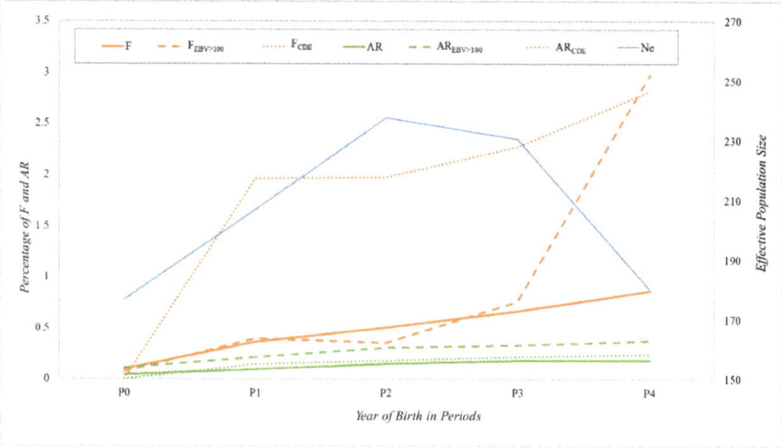

Figure 2. Evolution of average global inbreeding coefficient (F; orange solid line), for animals with breeding value over the mean for Show Jumping discipline ($F_{EBV > 100}$; orange dashed line) and for animals with both parents from CDE (F_{CDE}; orange dotted line). Evolution of average global relatedness coefficient (AR; green solid line), for animals with breeding value over the mean for *Show Jumping* discipline or ($AR_{EBV > 100}$; green dashed line) and for animals with both parents from CDE (AR_{CDE}; green dotted line). All values are in percentage. Evolution of effective population size computed via individual increase in inbreeding was also included (Ne; blue solid line). Where P0 is the period with animals born before 2000, P1 between 2000 and 2004, P2 between 2005 and 2009, P3 between 2010 and 2014, and P4 after 2015.

The results showed this subdivision in the three groups considered (total population, reference population and a hypothetical scenario with only CDE horses as reproducers), with a progressive increase over periods for the whole population, going from almost 0% F and AR values in P0 to 0.86% and 0.19%, respectively, in P4. For animals with EBV > 100 the F-values remained below 0.8% until P3, increasing considerably to almost 3.0% in P4, whereas the AR values remained below 0.4% over the four periods. Regarding the hypothetical scenario considered for the CDE population, the F-values increased considerably from almost 0.0% in P0 to almost 2.0% in P1 and P2 and continued increasing to 2.8% in P4. However, the AR values maintained a constant evolution with quite low values that went from 0.0% in P0 to a maximum of 0.25% in P4. In respect of Ne results, values in P0 and P4 were the lowest (around 180 animals), whereas P2 showed the highest Ne, with 237 horses.

3.4. Effects of Genetic Selection on the CDE Horse Breed

Genetic variances and heritability values from the genetic models used in the routine genetic evaluations of Show Jumping, Eventing and Dressage disciplines, are shown in Table 2. Heritability values were higher for Dressage than for Eventing or Show Jumping disciplines, ranging from 0.26 for Points obtained at the Dressage Reprise (P_R), to 0.32 for Points for Gallop (P_G) and Points for General Impression (P_{GI}). On the other hand, Show Jumping discipline showed the highest variance values for all the genetic model components.

Table 2. Estimation of variance components and heritability (h2) for the genetic evaluations computed for Show Jumping (SJ), Eventing (E) and Dressage (D) disciplines.

	Variables	Variance Genetic Components						h^2 ($\pm$SE)
		Animal	Rider	Rhi	Pe	Res	TV	
SJ	PS	93.22	-	193.43	55.58	1988.22	2330.44	0.04 ($\pm$0.0009)
	WTR	112.86	-	73.21	49.44	893.06	1128.57	0.10 ($\pm$0.0007)
E	PS_{SJ}	8.70	1.26	-	-	33.63	43.60	0.12 ($\pm$0.2635)
	FP_D	11.93	13.01	-	-	36.19	61.12	0.12 ($\pm$1.0982)
	PSc	41.83	30.04	-	-	281.05	352.91	0.12 ($\pm$2.9944)
D	P_R	4.09	5.53	-	6.28	9.38	25.29	0.26 ($\pm$0.0574)
	P_W	0.15	0.14	-	0.18	0.23	0.70	0.20 ($\pm$0.0002)
	P_T	0.14	0.11	-	0.19	0.21	0.65	0.31 ($\pm$0.0085)
	P_G	0.14	0.11	-	0.18	0.19	0.62	0.32 ($\pm$0.0011)
	P_S	0.13	0.11	-	0.19	0.21	0.64	0.30 ($\pm$0.1520)
	P_{GI}	0.14	0.11	-	0.19	0.19	0.62	0.32 ($\pm$0.0009)

Where PS = penalty score transformed to a positive scale; WTR = weighted total ranking; PS_{SJ} = penalty score transformed to a positive scale of the Show Jumping exercise; FP_D = final points obtained in the Dressage exercise; PS_C = penalty score transformed to a positive scale of the Cross exercise; P_R = points obtained at the dressage reprise; P_W = points for walk; P_T = points for trot; P_G = points for gallop; P_S = points for submission; P_{GI} = points for general impression; Animal = additive genetic effect; Rider = rider effect; Rhi = rider-horse interaction effect; Pe = permanent environment effect; Res = residual effect; TV = total variance; and h2 = heritability.

In order to ascertain the genetic change of the CDE breed regarding the objective trait, the EBV of Show Jumping, Eventing and Dressage performances characteristics and its evolution through the five periods considered was included in Table 3 for the most influential founders and their descendants.

Firstly, the 12 founders with the highest contributions to the entire CDE populations were included. Horses were ordered according to their year of birth. In the 'Order' column, the position of each founder in the ranking of global contributions and in each particular period was presented, showing the evolution of the 'importance' of each founder along time.

Table 3. Evolution of the genetic contribution of the twelve most contributing founders of the Caballo de Deporte Español breed in period 0, ordered by their contribution for each of the periods P1 to P4 (in percentage), with their mean estimated breeding value (EBV) and the mean EBV of its descendant (EBV$_d$) globally and per period, calculated based on the genetic evaluation of 2020 for *Show Jumping* (SJ), *Eventing* (Ev) and *Dressage* (Dr) disciplines.

Founders			F201	F464	F224	F445	F333	F203	F416	F1910	F229	F429	F430	F968
Year of Birth			1939	1950	1941	1950	1947	1939	1949	1963	1941	1949	1949	1956
Sex			M	F	M	F	F	F	M	M	M	M	F	M
Breed			TH	OHB	OHB	SF	SF	TH	TH	HOLST	AH	PRE	PRE	PRE
Contribution (%)			0.45	0.44	0.44	0.4	0.28	0.25	0.25	0.24	0.23	0.17	0.12	0.1
Order			1	2	3	4	5	6	7	8	9	10	11	12
Nd			5223	6641	6640	4632	3265	3672	3672	3233	1860	1623	1448	663
Global	EBV	SJ	101.3	101.5	101.5	101.7	102.4	103.2	103.2	98.9	98.6	98.7	99.3	99.9
Global	EBV	Ev	108.4	106.1	106.1	105.1	109.8	107.4	107.4	107.8	108.2	107.6	108.9	111.1
Global	EBV	Dr	98.7	97.2	97.5	96.8	100	99.4	99.4	100.7	100.4	107.2	102.7	-
Global	EBV$_d$	SJ	101.3	101.2	101.2	101.4	101.1	102.3	102.3	102.5	97.3	98.9	99.2	98.8
Global	EBV$_d$	Ev	107.2	107.2	107.2	107.2	107.9	107.2	107.2	108.1	106.5	108.4	108.5	111
Global	EBV$_d$	Dr	99.9	98.2	98.2	98.2	100.2	99.9	99.9	100	99.1	103	103	97.7
P0		Order	4	1	2	3	17	8	7	45	6	5	10	9
P0	EBV$_d$	SJ	101.9	101.3	101.3	101.5	101.6	102.4	102.4	101.9	98.9	98.9	98.9	99.2
P0	EBV$_d$	Ev	107.7	106.6	106.6	106.4	109	107.2	107.2	107.9	106.4	108.8	108.8	111.1
P0	EBV$_d$	Dr	99.2	97.6	97.6	97.5	99.6	98.4	98.4	99.8	99.6	103	103	97
P1		Order	3	1	2	4	5	28	27	51	6	39	58	74
P1	EBV$_d$	SJ	100.2	100	100	100.4	100.2	101.3	101.3	101.8	96.1	99.2	99.3	96.3
P1	EBV$_d$	Ev	107.4	107.8	107.8	108	108.1	106.9	106.9	108.1	106.7	107	107.4	111
P1	EBV$_d$	Dr	98.5	97.6	97.6	97.6	99	99.4	99.4	99.8	97.1	103.7	103.7	98.2
P2		Order	1	2	3	4	6	8	7	5	16	27	55	90
P2	EBV$_d$	SJ	101.7	101.9	101.9	101.9	101.3	102.7	102.7	102.6	92.7	98.3	99.8	97.9
P2	EBV$_d$	Ev	107.9	107.9	107.9	107.9	108.3	107.9	107.9	108	109.8	108.1	108.1	108.3
P2	EBV$_d$	Dr	99.7	99.5	99.5	99.3	99.8	100.4	100.4	99.3	97.3	100.6	100.5	101.8
P3		Order	1	3	4	5	12	10	9	2	22	56	73	103
P3	EBV$_d$	SJ	102.3	102.6	102.6	102.4	102.5	102.6	102.6	102.9	95.1	99.6	100.4	99.9
P3	EBV$_d$	Ev	104.5	105.5	105.5	106.2	104.9	107	107	108.4	106.5	108	108	-
P3	EBV$_d$	Dr	104.5	104.9	104.9	104.7	104.5	104.2	104.2	101.6	97.5	104.6	104.6	103.6
P4		Order	1	3	4	2	18	16	17	5	109	101	129	461
P4	EBV$_d$	SJ	102.2	102.9	102.9	102.7	103.7	103.3	103.3	104.9	94.3	-	-	-
P4	EBV$_d$	Ev	99	102.5	102.5	102.3	103.1	100.7	100.7	-	-	-	-	-
P4	EBV$_d$	Dr	104.3	103.2	103.2	103.2	104.5	100.5	100.5	102	-	106.3	106.3	104.7

Order indicates the order of the founder according to the percentage of contribution in the period/group considered, Nd is the number of all descendants of the founder throughout all periods, P0 is period with animals born before 2000, P1 between 2000 and 2004, P2 between 2005 and 2009, P3 between 2010 and 2014, and P4 after 2015.

Results showed that the five most contributing founders for the whole CDE population were from TH (F201), OHB (F224 and F464) and SF (F333 and F445) breeds. They all showed an EBV for Show Jumping discipline above 101 (i.e., higher than the mean), and the mean EBVs of their descendants were between 101 and 102 (the superiority of these founders seems to be transmitted). In fact, there is a general upward trend through the periods for the EBVs of the descendants of the 12 founders listed, reaching up to 104. In respect of Eventing and Dressage disciplines, results showed that the ranking of individuals based on their EBVs was similar to the Show Jumping and Eventing performances but opposite to Dressage performance. These results are consistent with the negative genetic correlation found between Dressage and Eventing (-0.12 ± 0.0199).

When accounting for the evolution of the genetic contribution of founders and of their descendants' EBVs over the P0 to P4 periods, most founders decreased their genetic contribution to the population across the years, together with their mean descendants' EBVs, which remained below 100 from P0 to P4.

On the other hand, these founders showed mean accuracy values close to 40%, with maximum values above 70%.

The accumulated genetic contribution of all the founders from the CDE Stud Book, according to the breed group and the period, is represented in Figure 3, also including mean EBV for Show Jumping discipline for each breed's founders and for their descendants. Breeds are ordered according to their breed founder's accumulated genetic contribution. Only the results from the Show Jumping discipline were used for this analysis, as more data were available.

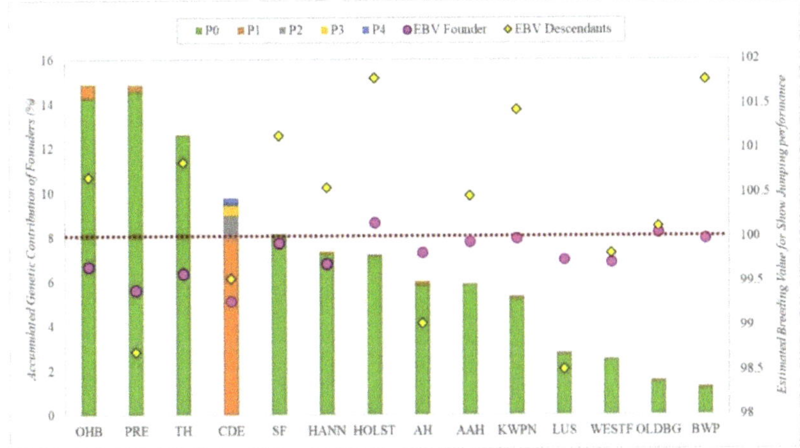

Figure 3. Genetic contribution of all founders classified by breed (each column), per period (different colors within columns), with their mean estimated breeding value (EBV Founder; pink circles) and the mean estimated breeding value of their descendants (EBV Descendants.; yellow rhombuses) for *Show Jumping* discipline, ordered according to contribution level of the founders. Where AAH is Anglo-Arab Horse, AH is Arab Horse, BWP is Belgian Warmblood, CDE is Caballo de Deporte Español, HANN is Hannoverian Horse, HOLST is Holsteiner Horse, KWPN is KWPN Horse, LUS is Lusitano Purebred, OLDBG is Oldenburg Horse, OHB is Other Horse Breeds, PRE is Pura Raza Española, TH is Thoroughbred, SF is Selle Français, WESTF is Westfalian Horse, P0 is period with animals born before 2000, P1 between 2000 and 2004, P2 between 2005 and 2009, P3 between 2010 and 2014, P4 after 2015. The red dots line indicate the mean estimated breeding value (EBV = 100).

OHB and PRE founders showed the highest genetic contribution, at almost 15%, whereas LUS, WESTF, OLDB and BWP founders showed the lowest, with accumulated contributions of less than 5%. In contrast, the mean EBV of the descendants was considerably higher than their founders' EBV for most breed groups, except for PRE, AH and LUS breed groups, in which we found the opposite outcome.

4. Discussion

Using composite breeds in domestic animals has been historically developed to benefit from the combinatory aptitude of this strategy in their progeny [19,20]. The amount of heterosis, as well as the particular outcome for any target trait, will depend on the proportion of genetic information coming from each of the breeds involved in the creation of the composite. With regard to the CDE population breed composition, preliminary analyses showed that the different breeds composing the CDE are still genetically different (there has been no homogenization of the breed). Notwithstanding, the genetic structure of the population has changed across generations, with CDE, PRE and OHB being the majority horse breeds in most periods. For CDE, the rising contributions (in percentage of the total number of horses of the CDE population) from P1 to P4 indicated a trend of breeders to prefer CDE stallions and/or mares as reproducers of the next generation. To fulfil the recommendation by Ref [3] of one or two breed crosses to account for optimal sport performance, another breed in addition to CDE should be used to obtain optimal results.

Regarding the available pedigree information, completeness level was slightly higher than that found in this breed in a study developed a few years after the Foundational Registry of the official Studbook was closed [1] and in a study from [21], comparing CDE results with other closed purebred populations. However, it was still less complete than other sport horse breeds such as the French Trotter [22] or the AH [23], where completeness values were above 50% as far back as the seventh generation while it reached this level in only the two most recent generations in CDE. This tendency was corroborated with GI and Ne values, being lower than those reported in the previous CDE paper [1]. These results also highlighted a better maintenance of the genetic diversity by year in this breed. These results, together with the lower t results than those reported in the previous CDE paper, could be due to foreign animals entering the studbook with some historic pedigree information (even if the amount of foreign horses was not large). Previous authors [24] reported similar values in the Brazilian Sport Horse.

When accounting for AR and F, the three sets of horses considered (whole CDE pedigree, animals with EBV > 100 and only CDE with CDE parents) showed greater F than expected under random mating from the corresponding AR values, thus indicating an important subdivision and/or non-random mating effect in these populations. A similar outcome was reported by Ref [1]. However, lower F and AR values has been observed in P4 (exclusive of the present study) than in the previous analyses. This could be due to the constant importation of foreign genetic material with limited information on the genealogy of the imported horses, so that, for F and AR calculations, these new foreign animals were considered as not inbred and not related to the rest of the studbook. Similar results were found in other sport horse breeds with open Stud Books [21,25]. Notwithstanding, of particular note is the large increase of mean *F*-value (mainly occurring in P4) for horses with EBV over the mean. This could indicate a trend in this last period to select CDE horses with higher EBV values according to Show Jumping performance genetic evaluations. These horses would tend to be more inbred by selecting parents coming from the same better performing ancestors. On the other hand, the simulation made of CDE with CDE parents' population, showed increased *F*-values through all periods, indicating that genetic variability would be considerably affected within a few years if the Stud Book policies become more restrictive, as already reported in [1]. Considering Ne, despite values in all periods being in accordance with other open sport horse breeds [21,24], the decreasing trend from P2 to P4 could be due to a loss of genetic variability within the population probably due to the selection efforts developed by breeders for Show Jumping, Eventing and/or Dressage performance, selecting only those animals that better fit their sporting purposes. Attention should be paid to this selection effect, in order not to lose critical genetic variability within this population in the future.

Results from the genetic contribution of the CDE founders, indicate that the five most contributing founders to CDE's genetic diversity were derived from TH, OHB and SF breed groups, all with EBVs for Show Jumping performance over the mean and transmitting also high EBVs to their descendants throughout all periods. This observation makes sense considering these breeds share similar performance goals with CDE, with the Show Jumping discipline being the main breeding selection objective for most of them [26–29]. These results also suggest that CDE breeders have been making selection based on EBV's results, as the higher the EBV of the founder (and of its descendants over periods), the higher genetic contribution it has had on the CDE population. This trend has been observed previously in other sport horse breeds [25,30–33]. On the other hand, we also found four founders from PRE and AH breed groups that, despite being within the list of the twelve most contributing founders, showed EBVs for Show Jumping below the mean and transmitting poor Show Jumping aptitude through all periods. However, when analyzing these founders for Eventing and Dressage results, they all show EBVs over the mean for both disciplines and good transmission to their progeny. This could be due to these founders being selected not for their genetic potential for Show Jumping performance but for Dressage, thus explaining the low EBV values for the former discipline. This different genetic selection was also supported by correlations found between disciplines, so that selecting for Dressage discipline would decrease performance in Eventing. Previous studies reported PRE as mainly bred for Dressage discipline [34,35], whereas AH were bred mainly for Endurance discipline [23], which shares some resilience and agility aptitudes with Eventing that are required for horses to compete successfully [36,37].

Furthermore, when comparing the heritability estimates for the three disciplines with previous studies, some differences were found. As regards to Show Jumping discipline, estimates were similar to those reported both in KWPN (0.11 [29]) and in a previous study for the CDE breed (0.047 to 0.085 [13]). When accounting for Eventing, heritability estimates in our study were higher than those reported in Great Britain's sport horse breeds (0.05, [32]) and of lower magnitude than previous estimates in the CDE horse (0.16, [3]). For Dressage discipline, our heritability values were higher than those reported in KWPN (0.11, [29]), and in British native horse breeds, Arabs and Warmbloods (0.110 to 0.152 [33]), but in the range of those reported for PRE horse breed (0.22 to 0.59, [34]). Differences in heritability estimates could be due to the different methods used for the estimation and the populations sampled, explaining the lower differences found with previous studies in national breeds.

On the other hand, when accounting for the genetic contribution of all founders (not just the highest performing), it seems that breeders tended to choose minority horse breeds (OHB) and PRE horses to form their CDEs. As regards the OHB group, the reason could be this breed incorporates several foreign sport horse breeds that have been selected for sport performance, particularly Show Jumping, before the CDE breed was created [27,29,37,38]. Whereas, for PRE, this could be due to its position as the main native national horse breed in Spain and geographically distributed all around the country [34,39], thus making it a very handy option for most CDE breeders. However, as reported previously, PRE horses transmit a better Dressage genetic potential than other breeds, thus PRE founder's breed groups EBV results were low, transmitting a poor genetic potential for Show Jumping to their descendants, as expected due to their main breeding orientation for Dressage.

Thus, it seems that a selection directed to different performance purposes has been made in CDE, with Eventing, Dressage and Show Jumping disciplines being selected at the same time. It has also to be noted that mean descendant's EBV showed higher values than their founder's for most breed groups, hence denoting genetic progress in the CDE breed. Therefore, selection has been effective in this population Although management has not been perfect, it seems that the selection of main breeding stock of the CDE breed was based on genetic results on sport performance of both selected stallions/mares and their descendants. Notwithstanding, despite descendants showing better EBVs than their CDE founders, there is still margin for improvement and the breeding association must make an effort to foster the performance of this breed by coordinating the management actions and encouraging the breeders to follow their recommendations for the selection and mating of the horses. Moreover, for future studies, if the levels of inbreeding increase, genetic models for sport performance in this breed could be improved by including pedigree inbreeding to account for inbreeding depression, as reported previously for other parameters [40]. However, for future studies, an introgression analysis using genotypic data would help reveal the real contributions of other breeds in the CDE pedigree.

5. Conclusions

Our results showed that, while a genetic progress can be observed in the CDE breed, with selection having an effect over this population, an improvement in population management is still needed to control the loss of genetic variability that could materialize if only

CDE horses with high genetic value are used as reproducers, without implementing any control on the rate of increase in inbreeding. Thus, an effort should be made by ANCADES to persevere with the task of coordination and education of their breeders to make them aware of what genetic evaluations mean and how to use this breeding tool adequately. Furthermore, an effort should be made to complete the pedigree depths of all breeds connected by relatives of CDE animals in order to improve the accuracy of the breeding evaluations and, hence, enhance the genetic progress of this breed.

Author Contributions: Conceptualization, M.V. and S.T.R.-R.; methodology, J.F., M.V., S.T.R.-R. and E.B.; software, E.B.; formal analysis, E.B.; investigation, E.B. and S.T.R.-R.; writing—original draft preparation, E.B.; writing—review and editing, J.F., M.V., S.T.R.-R. and E.B. All authors have read and agreed to the published version of the manuscript.

Funding: This research received no external funding.

Institutional Review Board Statement: Not applicable. No institutional animal care and use committee approval was needed due to no experimental procedures being conducted on animals in this study. All data used for the analyses were obtained from the pedigree information of the official CDE stud book. In respect of the information from routine genetic evaluations, data were obtained from official Show Jumping, Eventing and Dressage competitions, with no additional experimental procedures.

Informed Consent Statement: Not applicable.

Data Availability Statement: Restrictions apply to the availability of these data. Data were obtained from Asociación Nacional de Criadores de Caballo de Deporte Español (ANCADES) and are available from the corresponding author with the permission of ANCADES.

Acknowledgments: The authors wish to thank the Asociación Nacional de Criadores de Caballo de Deporte Español (ANCADES) for providing the data used in this study.

Conflicts of Interest: The authors declare no conflict of interest.

Appendix A

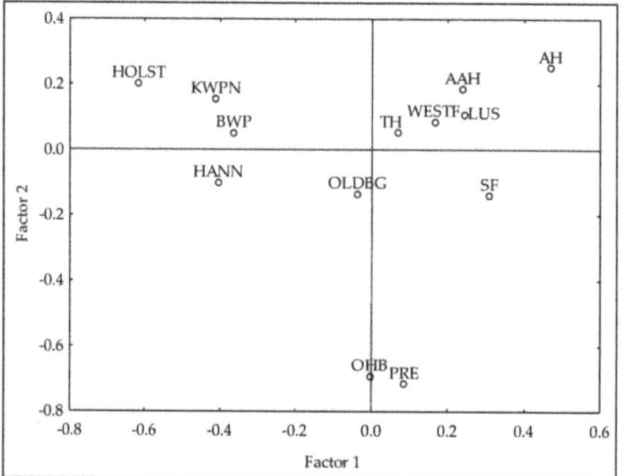

Figure A1. Principal Components Analysis with genealogical information from the horse breeds included in the Official Studbook of the Caballo de Deporte Español horse breed. Where AAH is Anglo-Arab Horse, AH is Arab Horse, BWP is Belgian Warmblood, HANN is Hannoverian Horse, HOLST is Holsteiner Horse, KWPN is KWPN Horse, LUS is Lusitano Purebred, OLDBG is Oldenburg Horse, OHB is Other Horse Breeds, PRE is Pura Raza Española, TH is Thoroughbred, SF is Selle Français and WESTF is Westfalian Horse.

References

1. Bartolomé, E.; Cervantes, I.; Valera, M.; Gutiérrez, J.P. Influence of foreign breeds on the genetic structure of the spanish sport horse population. *Livest. Sci.* **2011**, *142*, 70–79. [CrossRef]
2. Amuzu-Aweh, E.N.; Bijma, P.; Kinghorn, B.P.; Vereijken, A.; Visscher, J.; Van Arendonk, J.A.; Bovenhuis, H. Prediction of heterosis using genome-wide SNP-marker data: Application to egg production traits in white leghorn crosses. *Heredity* **2013**, *111*, 530–538. [CrossRef]
3. Cervantes, I.; Bartolomé, E.; Valera, M.; Gutiérrez, J.P.; Molina, A. Crossbreed genetic performance study in the eventing horse competition. *Anim. Prod. Sci.* **2015**, *56*, 1454–1462. [CrossRef]
4. BOE N° 7; Ministry of Agriculture, Fisheries and Feeding of Spain: Madrid, Spain, 2020.
5. Fernández, J.; Villanueva, B.; Toro, M.A. Optimum mating designs for exploiting dominance in genomic selection schemes for aquaculture species. *Genet. Sel. Evol.* **2021**, *53*, 14. [CrossRef] [PubMed]
6. Medugorac, I.; Veit-Kensch, C.E.; Ramljak, J.; Brka, M.; Marković, B.; Stojanović, S.; Bytyqi, H.; Kochoski, L.; Kume, K.; Grünenfelder, H.-P.; et al. Conservation priorities of genetic diversity in domesticated metapopulations: A Study in taurine cattle breeds. *Ecol. Evol.* **2011**, *1*, 408–420. [CrossRef] [PubMed]
7. Gutiérrez, J.P.; Goyache, F. A note on ENDOG: A computer program for analysing pedigree information. *J. Anim. Breed. Genet.* **2005**, *122*, 172–176. [CrossRef] [PubMed]
8. Maccluer, J.W.; Boyce, A.J.; Dyke, B.; Weitkamp, L.R.; Pfenning, D.W.; Parsons, C.J. Inbreeding and pedigree structure in standardbred horses. *J. Hered.* **1983**, *74*, 394–399. [CrossRef]
9. Boichard, D.; Maignel, L.; Verrier, É. The value of using probabilities of gene origin to measure genetic variability in a population. *Genet. Sel. Evol.* **1997**, *29*, 5–23. [CrossRef]
10. James, J.W. A note on selection differential and generation length when generations overlap. *Anim. Prod.* **1977**, *24*, 109–112. [CrossRef]
11. Groeneveld, E.; Westhuizen, B.v.d.; Maiwashe, A.; Voordewind, F.; Ferraz, J.B.S. POPREP: A generic report for population management. *Gen. Mol. Res.* **2009**, *8*, 1158–1178. [CrossRef] [PubMed]
12. Falconer, D.S.; Mackay, T.F.C. *Introduction to Quantitative Genetics*, 4th ed.; Addison Wesley Longman Limited: Essex, UK, 1996.
13. Bartolomé, E.; Menéndez-Buxadera, A.; Molina, A.; Valera, M. Plasticity effect of rider-horse interaction on genetic evaluations for show jumping discipline in sport horses. *J. Anim. Breed. Genet.* **2018**, *135*, 138–148. [CrossRef] [PubMed]
14. Groeneveld, E.; Kovac, M.; Mielenz, N. *VCE Users's Guide and Reference Manual, Version 6.0*; Institute of Farm Animal Genetics: Neustadt, Germany, 2010.
15. Malécot, G. *Les Mathématiques de l'Hérédite*; Masson et Cie: Paris, France, 1948.
16. Goyache, F.; Gutiérrez, J.P.; Fernández, I.; Gomez, E.; Alvarez, I.; Díez, J.; Royo, L.J. Using pedigree information to monitor genetic variability of endangered populations: The xalda sheep breed of asturias as an example. *J. Anim. Breed. Genet.* **2003**, *120*, 95–105. [CrossRef]
17. Gutiérrez, J.P.; Altarriba, J.; Díaz, C.; Quintanilla, R.; Cañón, J.; Piedrafita, J. Pedigree analysis of eight spanish beef cattle breeds. *Genet. Sel. Evol.* **2003**, *35*, 43–63. [CrossRef] [PubMed]
18. Lacy, R.C. Analysis of founder representation in pedigrees: Founder equivalents and founder genome equivalents. *Zoo Biol.* **1989**, *8*, 111–123. [CrossRef]
19. Zhang, S.; Bidanel, J.-P.; Burlot, T.; Legault, C.; Naveau, J. Genetic parameters and genetic trends in the chinese x european tiameslan composite pig line. I. genetic parameters. *Genet. Sel. Evol.* **2000**, *32*, 41–56. [CrossRef] [PubMed]
20. Tang, G.; Yang, R.; Xue, J.; Liu, T.; Zeng, Z.; Jiang, A.; Jiang, Y.; Li, M.; Zhu, L.; Bai, L.; et al. Optimising a crossbreeding production system using three specialised imported swine breeds in South-Western China. *Anim. Prod. Sci.* **2014**, *54*, 999–1007. [CrossRef]
21. Cervantes, I.; Gutiérrez, J.P.; Molina, A.; Goyache, F.; Valera, M. Genealogical analyses in open populations: The case of three Arab-derived Spanish horse breeds. *J. Anim. Breed. Genet.* **2009**, *126*, 335–347. [CrossRef] [PubMed]
22. Moureaux, S.; Verrier, É.; Ricard, A.; Mériaux, J. Genetic variability within French race and riding horse breeds from genealogical data and blood marker polymorphisms. *Genet. Sel. Evol.* **1996**, *28*, 83. [CrossRef]
23. Cervantes, I.; Molina, A.; Goyache, F.; Gutiérrez, J.P.; Valera, M. Population history and genetic variability in the Spanish Arab horse assessed via pedigree analysis. *Livest. Sci.* **2008**, *113*, 24–33. [CrossRef]
24. Medeiros, B.R.; Bertoli, C.D.; Garbade, P.; McManus, C.M. Brazilian sport horse: Pedigree analysis of the Brasileiro de hipismo breed. *Ital. J. Anim. Sci.* **2014**, *13*, 657–664. [CrossRef]
25. Hellsten, E.T.; Näsholm, A.; Jorjani, H.; Strandberg, E.; Philipsson, J. Influence of foreign stallions on the Swedish warmblood breed and its genetic evaluation. *Livest. Sci.* **2009**, *121*, 207–214. [CrossRef]
26. Tunnell, J.A.; Sanders, J.O.; Williams, J.D.; Potter, G.D. Pedigree analysis of four decades of quarter horse breeding. *J. Anim. Sci.* **1983**, *57*, 585–593. [CrossRef] [PubMed]
27. Koenen, E.P.C.; Aldridge, L.I.; Philipsson, J. An overview of breeding objectives for warmblood sport horses. *Livest. Prod. Sci.* **2004**, *88*, 77–84. [CrossRef]
28. Bokor, A.; Blouin, C.; Langlois, B. Possibility of Selecting racehorses on jumping ability based on their steeplechase race results in France, the United Kingdom and Ireland. *J. Anim. Breed. Genet.* **2007**, *124*, 124–132. [CrossRef] [PubMed]
29. Rovere, G.; Ducro, B.J.; van Arendonk, J.A.M.; Norberg, E.; Madsen, P. Analysis of competition performance in dressage and show jumping of Dutch warmblood horses. *J. Anim. Breed. Genet.* **2016**, *133*, 503–512. [CrossRef]

30. Hamann, H.; Distl, O. Genetic variability in Hanoverian warmblood horses using pedigree analysis. *J. Anim. Sci.* **2008**, *86*, 1503–1513. [CrossRef]

31. Solé, M.; Bartolomé, E.; José Sánchez, M.; Molina, A.; Valera, M. Predictability of adult show jumping ability from early information: Alternative selection strategies in the Spanish sport horse population. *Livest. Sci.* **2017**, *200*, 23–28. [CrossRef]

32. Kearsley, C.G.S.; Woolliams, J.a.; Coffey, M.P.; Brotherstone, S. Use of competition data for genetic evaluations of eventing horses in Britain: Analysis of the dressage, showjumping and cross country phases of eventing competition. *Livest. Sci.* **2008**, *118*, 72–81. [CrossRef]

33. Stewart, I.D.; Woolliams, J.A.; Brotherstone, S. Genetic evaluation of horses for performance in dressage competitions in Great Britain. *Livest. Sci.* **2010**, *128*, 36–45. [CrossRef]

34. Sánchez Guerrero, M.J.; Cervantes, I.; Valera, M.; Gutiérrez, J.P. Modelling genetic evaluation for dressage in Pura Raza Español horses with focus on the rider effect. *J. Anim. Breed. Genet.* **2014**, *131*, 395–402. [CrossRef] [PubMed]

35. Valera, M.; Molina, A.; Gutiérrez, J.P.; Gómez, J.; Goyache, F. Pedigree analysis in the andalusian horse: Population structure, genetic variability and influence of the carthusian strain. *Livest. Prod. Sci.* **2005**, *95*, 57–66. [CrossRef]

36. Raidal, S.L.; Love, D.N.; Bailey, G.D.; Rose, R.J. Effect of single bouts of moderate and high intensity exercise and training on equine peripheral blood neutrophil function. *Res. Vet. Sci.* **2000**, *68*, 141–146. [CrossRef] [PubMed]

37. Braam, Å.; Näsholm, A.; Roepstorff, L.; Philipsson, J. Genetic variation in durability of Swedish warmblood horses using competition results. *Livest. Sci.* **2011**, *142*, 181–187. [CrossRef]

38. Azcona, F.; Valera, M.; Molina, A.; Trigo, P.; Peral-García, P.; Solé, M.; Demyda-Peyrás, S. Impact of reproductive biotechnologies on genetic variability of Argentine polo horses. *Livest. Sci.* **2020**, *231*, 103848. [CrossRef]

39. Solé, M.; Valera, M.; Fernández, J. Genetic structure and connectivity analysis in a large domestic livestock meta-population: The case of the Pura Raza Español Horses. *J. Anim. Breed. Genet.* **2018**, *135*, 460–471. [CrossRef] [PubMed]

40. Perdomo-González, D.I.; Molina, A.; Sánchez-Guerrero, M.J.; Bartolomé, E.; Varona, L.; Valera, M. Genetic inbreeding depression load for fertility traits in Pura Raza Española mares. *J. Anim. Sci.* **2021**, *99*, 1–10. [CrossRef]

Article

Identification of Copy Number Variations in Four Horse Breed Populations in South Korea

Yong-Min Kim [1,†], Seok-Joo Ha [2,†], Ha-Seung Seong [1], Jae-Young Choi [3], Hee-Jung Baek [2], Byoung-Chul Yang [3], Jung-Woo Choi [2,*] and Nam-Young Kim [3,*]

1. Swine Science Division, National Institute of Animal Science, Rural Development Administration, Cheonan 31000, Republic of Korea
2. Department of Animal Science, College of Animal Life Sciences, Kangwon National University, Chuncheon 24341, Republic of Korea
3. Subtropical Livestock Research Institute, National Institute of Animal Science, RDA, Jeju 63242, Republic of Korea
* Correspondence: jungwoo.kor@gmail.com (J.-W.C.); rat1121@korea.kr (N.-Y.K.); Tel.: +82-33-250-8622 (J.-W.C.)
† These authors contributed equally to this work.

Citation: Kim, Y.-M.; Ha, S.-J.; Seong, H.-S.; Choi, J.-Y.; Baek, H.-J.; Yang, B.-C.; Choi, J.-W.; Kim, N.-Y. Identification of Copy Number Variations in Four Horse Breed Populations in South Korea. *Animals* 2022, 12, 3501. https://doi.org/10.3390/ani12243501

Academic Editors: María Dolores Gómez Ortiz and Isabel Cervantes

Received: 30 October 2022
Accepted: 8 December 2022
Published: 12 December 2022

Publisher's Note: MDPI stays neutral with regard to jurisdictional claims in published maps and institutional affiliations.

Simple Summary: The objective of this study is to detect copy number variations (CNVs) in four horse populations (Jeju horses, Thoroughbreds, Jeju riding horses, and Hanla horses) in South Korea. We found a total of 843 CNV regions (CNVRs) (164.3 Mb), which coincided with 7.2% of the reference horse genome. Overall, copy number losses were found more than gains and mixed CNVRs. A comparison of the CNVRs among the populations showed that a substantial number of CNVRs overlapped each other, while some CNVRs were found specifically in each population. We retrieved parts of CNVRs that overlapped with genes; these overlapping areas are potentially associated with traits of interest in horses. The Thoroughbred and crossbred populations had shared CNVRs overlapping with QTLs (Quantitative trait loci) that were associated with withers height and racing performance. Using gene ontology (GO) analysis, a total of 1884 functional genes were identified within the 577 CNVRs. GO analysis further showed that several of the genes are involved in the olfactory pathway and the nervous system.

Abstract: In this study, genome-wide CNVs were identified using a total of 469 horses from four horse populations (Jeju horses, Thoroughbreds, Jeju riding horses, and Hanla horses). We detected a total of 843 CNVRs throughout all autosomes: 281, 30, 301, and 310 CNVRs for Jeju horses, Thoroughbreds, Jeju riding horses, and Hanla horses, respectively. Of the total CNVRs, copy number losses were found to be the most abundant (48.99%), while gains and mixed CNVRs accounted for 41.04% and 9.96% of the total CNVRs, respectively. The length of the CNVRs ranged from 0.39 kb to 2.8 Mb, while approximately 7.2% of the reference horse genome assembly was covered by the total CNVRs. By comparing the CNVRs among the populations, we found a significant portion of the CNVRs (30.13%) overlapped; the highest number of shared CNVRs was between Hanla horses and Jeju riding horses. When compared with the horse CNVRs of previous studies, 26.8% of CNVRs were found to be uniquely detected in this study. The CNVRs were not randomly distributed throughout the genome; in particular, the *Equus caballus* autosome (ECA) 7 comprised the largest proportion of its genome (16.3%), while ECA 24 comprised the smallest (0.7%). Furthermore, functional analysis was applied to CNVRs that overlapped with genes (genic-CNVRs); these overlapping areas may be potentially associated with the olfactory pathway and nervous system. A racing performance QTL was detected in a CNVR of Thoroughbreds, Jeju riding horses, and Hanla horses, and the CNVR value was mixed for three breeds.

Keywords: CNV; structural variation; Jeju horse; thoroughbred; crossbred

1. Introduction

There is abundant evidence that phenotypic diversity can be in part attributed to genetic variation such as single nucleotide polymorphisms (SNPs) and copy number variations (CNVs) [1–3]. CNVs have typically been defined as DNA segments larger than 1 kb in length that have differing numbers of copies among individuals of a species. Recently, this definition has been modified to obtain a higher resolution (50 bp) via recent advances in high-density chip arrays and massively parallel sequencing technologies [4,5].

With the release of an equine reference genome assembly and subsequent massive SNP detection, various high-throughput SNP genotyping arrays were developed, enabling rapid genotyping at a reasonable price for diverse horse breeds [6–8]. Subsequently, this has led to much research aimed at locating genes associated with horse performance traits including speed index and durability, via genome-wide association studies [9,10]. However, there are still very limited numbers of publications available for horse CNVs, while they have been intensely investigated in human genomes.

In horse genomics, there are currently several published CNV studies that have used various technologies. Doan et al. (2012) published the first study to detect genome-wide CNVs in horses, using a whole-exome tiling array and an array-comparative genomic hybridization platform. Since the development of high-density SNP arrays for horses [7,11], CNVs could be also assessed throughout the genomes of various horse breeds, of which some studies investigated potential associations with traits of interest in horses, including body size and genetic risk factors for insect-bite hypersensitivity [12–15].

The Jeju horse (also known as the Jeju pony) was registered with the Food and Agricultural Organization of the United Nations as the only indigenous horse breed in South Korea. It is a small- to medium-sized breed (mean withers height: 116 cm) with a range of colors. The origin of the current Jeju horse remains to be further clarified, though it is known to have involved Mongolian horses [16,17]. Because of a remarkably reduced population, this breed was designated as a Korean natural monument (No. 347) in 1986. In an effort to utilize this native genetic resource in South Korea, the Jeju horse was crossed with Thoroughbreds. A racecourse was opened for horseback racing of Jeju horses and Jeju crossbreeds in 1990, and this led to an increase in the crossbreed population on Jeju Island. However, to our current knowledge, there is still no statutory regulation to clarify exact breed compositions of the crossbreeds, which are usually categorized as either riding horses for outdoor leisure or as racing horses, with respect to their genetic make-up (usually judged by how much Thoroughbred genetics contribute to the crossbreed). It has been reported that most of the crossbreeds used for horseback racing (usually referred to as Hanla horses) on Jeju Island are genetically more heavily influenced by Thoroughbreds than Jeju horses [18]; this is not so surprising, because Thoroughbreds are the most prominent breed for horse racing because of their agility and speed. Conversely, the crossbreed mostly used in riding, which tends to be smaller than the Hanla, is generally called the Jeju riding horse.

The main purpose of this study was to identify CNVs and investigate their patterns throughout the genome of four horse populations, including the Jeju, Thoroughbred, Jeju riding horse, and Hanla populations in the Republic of Korea. We used a total of 469 horse genotyping datasets derived from the Illumina GGP EquineSNP70 Genotyping Bead Chip array. Furthermore, functional enrichment analysis for CNVRs that are potentially associated with genes was performed.

2. Materials and Methods

2.1. Samples and Genotyping

We used a total of 469 individual genotypic datasets using the Illumina GGP Equine SNP70 Bead-chip array (Illumina, San Diego, CA, USA); all of the animals were obtained from the Subtropical Livestock Research Institute in the National Institute of Animal Science in the Republic of Korea. Of the total datasets, we newly genotyped in this study a total of 130 individuals that included Jeju horse (*n* = 4), Thoroughbred (*n* = 5), Jeju riding horse (*n* = 119), and Hanla horse (*n* = 2). We extracted genomic DNA from blood samples treated with EDTA. The samples' DNA concentration was adjusted to 50 ng/L. Furthermore, Thoroughbred (n = 134), Jeju riding horse (n = 63), and Hanla horse (n = 142) individual genotype data were included from a previous study [18]. Illumina GGP Equine SNP70 Beadchip, which includes approximately 65 k SNPs, was used for genotyping, and Illumina GenomeStudio software v.2.0 (Illumina, San Diego, CA, USA) was used to determine SNP genotypes.

2.2. Analysis of Population Genetic Structure

Principal Component Analysis (PCA) and admixture analysis were used to analyze the demographic structure of the horse populations on Jeju Island. The ggplot function in the R package [19] was used to show the relationships between the PC1 and PC2 coordinates after performing the "–pca" flag using PLINK v.1.9 [20] to produce eigenvectors and eigenvalues.

Furthermore, supervised ADMIXTURE (version 1.3) [21] was used to compute the proportions of ancestry (K) at K = 2 for crossbred populations (Hanla horse and Jeju riding horse) to confirm their admixed status. The result derived from ADMIXTURE was visualized using R plots.

2.3. CNV Detection

PennCNV v.1.0.5 (https://penncnv.openbioinformatics.org/en/latest/ (accessed on 1 August 2022)) [22], which is based on the Hidden Markov Model (HMM), was used to perform CNV calling. Four input files were used for running the PennCNV software: the signal intensity file (including the SNP Name, log R ratio (LRR), and B allele frequency (BAF)), the population frequency of the B allele (PFB) file, the SNP MAP file (includes the SNP Name, Chromosome, and Position) and the GC-Content file. SNP name, chromosomal position, BAF, and LRR data files were exported as a single file and divided using the PennCNV package's split option. The PFB file was created using the PennCNV "compile_pfb.pl" function and the GC-Content file was determined as the proportion of GC content both 1 MB upstream and downstream of the SNP locus in the reference genome (UCSC Genome Browser Downloads: https://hgdownload.soe.ucsc.edu/goldenPath/equCab3/bigZips/ (accessed on 30 July 2022). PennCNV was run according to the default criteria using the command line "detect_cnv.pl". Quality control was performed with "filter_cnv.pl" functions on the samples with a standard deviation of the Log Ratio (SD LRR) >0.3, BAF drift >0.01, and a wave factor (WF) >0.05. Further filtering was performed and CNVs with consecutive SNPs ≥3 and CNV length ≥10 kb were retained. Then "clean_cnv.pl" script was used to combine the adjacent CNVs that had a gap size of less than 20% of the total length of the CNVs. As a result, a total of 45 individuals were filtered (Thoroughbred (n = 19), Jeju riding horse (n = 25), Hanla (n = 1)), and 424 horses were included in the dataset.

2.4. CNVR Detection

After CNV detection, the CNVRs were detected by running HandyCNV [23]. "call_cnvr" generates CNV regions as the union of sets of CNVs that overlap by at least one base pair [1]. Each CNVR should have the same boundaries so that each individual can be classified as a diploid, CNVR-gain position (duplications), or CNVR-loss position (deletions). Gain and loss CNVRs that overlapped were combined into a single region to account for genomic regions where both events had happened (CNVR "mixed").

CNVRs have shared regions among each breed. The overlapped CNVRs were identified using BEDtools v2.17.0 (Quinlan laboratory, Salt Lake, UT, USA) [24]. We created one comprehensive CNVR table for further analysis. We calculated the sum of all CNVRs in a detected chromosome to determine the genomic percentage covered by CNVRs at the chromosomal level. The gene content of the CNVRs was evaluated using the reference genome EquCab 3.0 from the UCSC Genome Browser Gateway (UCSC Genomics Institute University of California, Santa Cruz, CA, USA).

2.5. Gene Contents and Functional Annotation in CNVRs

The NCBI GFF file was used to execute gene annotation to find genes that over lapped the detected CNVRs (https://www.ncbi.nlm.nih.gov/assembly/GCF_002863925 1/genome_assemblies_genome_gff (accessed on 30 July 2022)). Furthermore, the horse quantitative trait locus (QTL) information was annotated using the animal QTL database (https://www.animalgenome.org/cgi-bin/QTLdb/EC/index (accessed on 20 October 2022) on EC_3.0 database). PANTHER17.0 (Protein Analysis THrough Evolutionary Relation ships, version 17.0, http://www.pantherdb.org/ (accessed on 20 October 2022)) provided the functional enrichment and gene functional annotation of the CNVRs. Therefore, we conducted gene ontology (GO) analyses for the genes in CNVRs to offer insight into the functional enrichment of the CNVRs [25]. The threshold was set as false discovery rate (FDR) corrected p-value < 0.05 and genes were classified by their biological process molecular function, and cellular component.

3. Results and Discussion

3.1. Breed Compositions Used in this Study

In this study, we used four horse populations inhabiting Jeju Island. As aforemen tioned, the Jeju riding horse and Hanla are the crossbreeds of Thoroughbreds and Jeju horses, despite their different main uses (riding and racing for those breeds, respectively in the Korean horse industry. Namely, the Hanla horse is considered an independent breed in various previous studies [26,27]. In this study, it was reconfirmed via PCA and ADMIXTURE analyses that the Hanla horse forms an independent cluster (Figure S1a–c) However, in the case of the Jeju riding horse, it showed that the pattern was somewhat more irregular than that of the Hanla horse; this pattern was similarly observed in the ADMIXTURE analysis (Figure S1d). These results are not surprising, because there is no accurate documentation to enable such a distinction, particularly for the Jeju riding horse. Although it seems obvious that the Jeju riding horse has been less influenced by Thoroughbred genetics than the Hanla, it is difficult to clearly define the mixing ratio of Thoroughbred and Jeju horse. Therefore, further analysis is required to suggest a clear guideline to designate breed composition, particularly for the Jeju riding horse population

3.2. Identifying Genome-Wide CNVs

CNVs are structural variations that are a primary source of genetic diversity and phenotypic variation. As such, CNVs are recognized as significant for identifying genetic diversity among populations and in the evolution of breeds [28–30]. To profile genome wide CNVs, we first detected CNVs throughout 31 autosomes in each of the four horse populations on Jeju Island, including Jeju horses, Thoroughbreds, Jeju riding horses, and Hanla horses, using PennCNV v1.0.5 [22]. A total of 4482 CNVs were detected for the four horse breed populations; we found that copy number gains (2805 CNVs) were more abundant than losses (1677 CNVs). The higher frequency of gains compared with losses was observed for most of the horse populations except Thoroughbreds (Figure S2).

These results are similar to previous studies showing a high incidence of gains in horse populations. It has been suggested that gains in coding or enhancing sequences increase the genetic diversity of organisms, resulting in phenotypic variety and the putative potential to adapt in challenging environments [31]. For this reason, Jeju riding horses and Hanla horses are considered to be undergoing the process of removing mutations and adapting to the environment by artificial selection.

Of the four breeds, the Hanla showed the highest number of CNVs (1818 CNVs with 493 losses and 1325 gains) and Jeju horses had the lowest (41 CNVs with 11 losses and 30 gains). The difference in CNVs indicates that genetic variation from CNVs may con tribute to breed phenotypic diversity, but it may also result from the different demographic history and effective population sizes between breeds [32]. The very small number of CNVs observed in this study for Jeju horses may be a result of the small number of samples. The CNVs ranged in size from 11 kb in Thoroughbreds to 2816 kb in Jeju riding horses. The total length of CNVs ranged from 12 Mb in Jeju horses to 376 Mb in Hanla horses, which is consistent with the range of genome coverage from 0.51% to 16.46% for Jeju horses and Hanla horses, respectively (Table 1).

Table 1. Copy number variation (CNV) analysis of four horse breeds in South Korea.

Breed	Sample	CNV	CNL [1]	CNG [2]	Length Min [3]	Length Max [4]	Length Median [5]	Total Length in CNV	Length Average
Jeju horse	4	41	11	30	12,219	2,156,704	116,562	11,529,371	281,204.17
Thoroughbred	120	1340	716	624	10,509	2,563,439	155,989	374,113,603	279,189.26
Jeju riding horse	157	1283	457	826	10,718	2,816,442	147,819	358,366,183	279,318.93
Hanla horse	143	1818	493	1325	10,509	2,246,935	108,623	375,503,162	206,547.39

[1] CNL, CNV losses. [2] CNG, CNV gains. [3] Length Min, Minimum length of CNV. [4] Length Max, Maximum length of CNV. [5] Length Median, Median length of CNV.

3.3. Defining CNVRs

To detect CNVRs for each horse population, we merged CNVs that overlapped by at least one base pair. A total of 992 CNVRs were detected from all four populations. Hanla horses showed the highest number of CNVRs (310 CNVRs with 170 gains, 127 losses, and 13 mixed), and Jeju horses had the lowest number of CNVRs (30 CNVRs with 19 gains and 11 losses). The CNVR coverage of chromosomes is 0.24% (5.48 Mb) in Jeju horses, 3.2% (73.53 Mb) in Thoroughbreds, 3.47% (97.11 Mb) in Jeju riding horses, and 2.91% (66.32 Mb) in Hanla horses. Additionally, the longest CNVRs were identified in Jeju riding horses (2866 kb) and Hanla horses (2569 kb), while Thoroughbreds and Jeju horses had the least (2362 kb and 2156 kb, respectively) (Table 2). The CNVR distributions appear to be affected by the sample size; note that the small number of Jeju horses affected the very small number of CNVs detected and the CNVR distribution in the Jeju horse population.

Table 2. The basic statistic of copy number variation regions (CNVRs) of four horse breeds in South Korea.

Breed	CNVR	Gain	Loss	Mixed	Length Min [1]	Length Max [2]	Length Median [3]	Total Length in CNVR	CNVR Coverage [4]
Jeju horse	30	19	11	0	12,219	2,156,704	94,014	5,475,931	0.24%
Thoroughbred	281	97	175	9	10,509	2,563,439	163,969	73,526,600	3.22%
Jeju riding horse	301	164	117	20	10,718	2,865,712	161,474	79,107,681	3.47%
Hanla horse	310	170	127	13	10,509	2,569,249	134,033	66,321,742	2.91%

[1] Length Min: Minimum length of CNVR. [2] Length Max: Maximum length of CNVR. [3] Length Median: Median length of CNVR. [4] CNVR frequency in horse genome (Horse genome length/Total length in CNVR)

In each chromosome, the number of CNVRs was the highest on ECA1 in Thoroughbreds ($n = 25$), Jeju horses ($n = 5$), and Hanla horses ($n = 34$), and the number of CNVRs was the highest on ECA4 in Jeju riding horses. Conversely, the coverage of CNVRs on each chromosome was the highest on ECA12 (5.8% of the Jeju horse genome, 9.0% of the Jeju riding horse genome, and 8.4% of the Hanla horse genome) and on ECA7 (10.3% of the Thoroughbred genome) (Table S1, Figure 1).

As observed in prior studies with horse breeds, the largest shared CNVRs were discovered on ECA12 [33,34]. ECA12 displays the particular characteristic of being enriched with clusters of olfactory receptor genes, which is also observed in other mammalian genomes, and it has been proposed that this property affects the fight or flight response and temperament diversity in horses [35]. Therefore, the mild personality of Jeju horses, could be the result of selection for riding in a wide variety of environments.

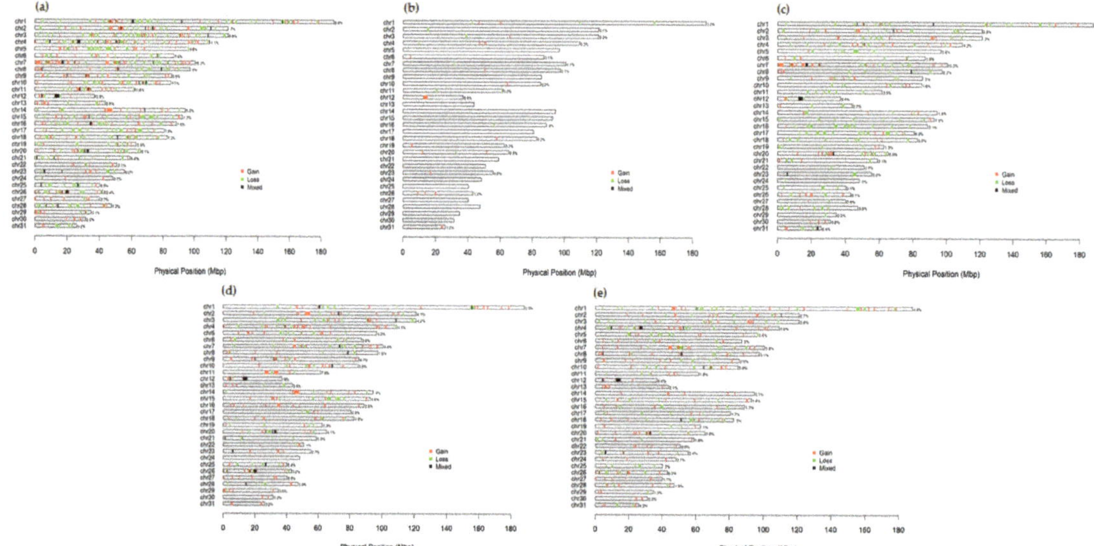

Figure 1. Plot of copy number variation regions (CNVR) on 31 horse (*Equus caballus*) autosomes. (**a**) Total CNVR distribution; (**b**) Jeju horse CNVR distribution; (**c**) Thoroughbred CNVR distribution; (**d**) Jeju riding horse CNVR distribution; (**e**) Hanla horse CNVR distribution. Red, green, and black represent gains, losses, and mixed CNVRs, respectively.

We assessed CNVRs that overlapped among the four populations (overlapping CNVRs). A total of 843 overlapping CNVRs was retrieved by considering CNVRs that overlapped by at least one base pair between populations. The overlapping CNVRs comprised 346 gains, 413 losses, and 84 mixed events (a mean length of 195 kb), and ranged from 0.4 kb to 2866 kb in size and covered 164,330 kb of the horse genome, which corresponds to 7.2% of the horse autosomes.

In each breed, CNVs and CNVRs showed more gain in the horse population (Tables 1 and 2). However, 843 overlapping CNVRs show more loss. Thoroughbred populations have long loss-CNVRs (more than 0.3 Mb, n = 60). Therefore, among overlapping CNVRs, there was an increase in loss-CNVRs (Figure 2).

The distribution of CNVRs across all autosomes varied considerably, with the highest number at 74 on ECA4 and the lowest at 4 on ECA24. The ratio of total estimated CNVR length per chromosome to the length of that chromosome varied from 16.3% for ECA7 to 0.7% for ECA24 (Figure 1).

Among the 843 CNVRs, 589 (70.6%) were specific CNVRs that did not overlap with other breeds and appeared only in each breed as follows: Jeju horses (n = 5), which had the fewest specific CNVRs, Thoroughbreds (n = 180), Jeju riding horses (n = 202), and Hanla horses (n = 202) (Figure 3, Table S2). Because Jeju riding horses and Hanla horses are crossbreeds produced by hybridizing Jeju horses and Thoroughbreds, we can expect these two breeds to exhibit significant characteristics in CNVRs. The majority of horse breeds were of recent origin and had undergone significant crossbreeding before closed breeds were founded, which has resulted in a high level of haplotype sharing [6,36], and thus fewer breed-specific CNVRs can be found than crossbred [37].

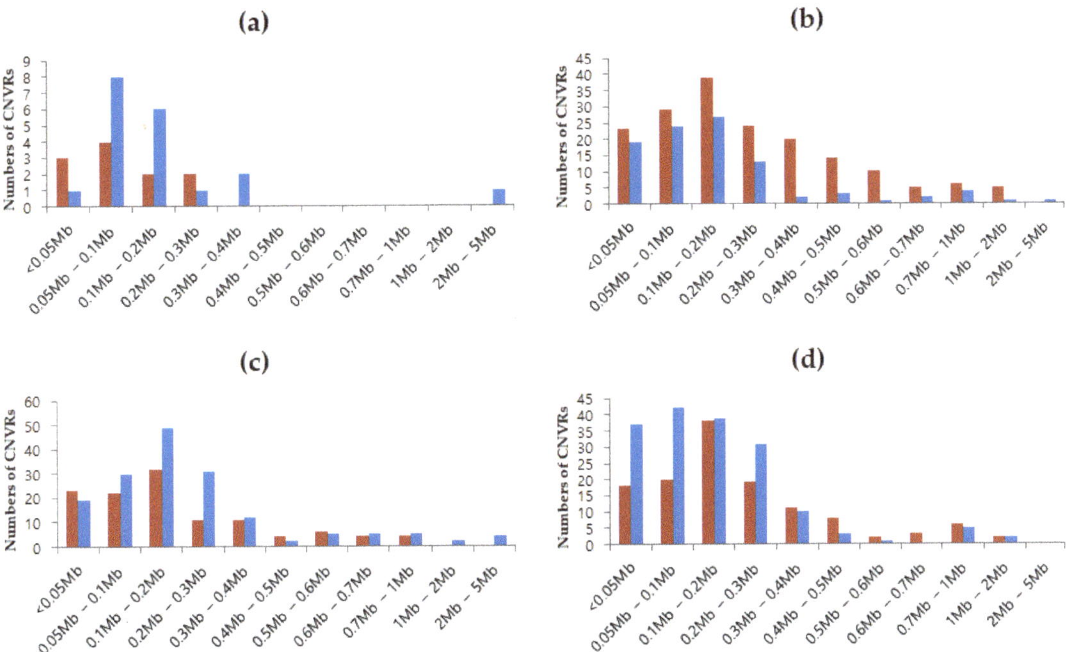

Figure 2. Gain and loss distribution by length of CNV region (CNVR) in four horse breeds in South Korea. (**a**) Jeju horse CNVR distribution; (**b**) Thoroughbred CNVR distribution; (**c**) Jeju riding horse CNVR distribution; (**d**) Hanla horse CNVR distribution. Red and blue represent losses and gains, respectively.

The breed-specific CNVR for each breed showed high losses for Jeju horses (80%), Thoroughbreds (82.78%), and Hanla horses (50.99%), while Jeju riding horses showed a gain of 49.01%. We confirmed that the CNVRs shared by the populations had a gain of 48.82%, confirming that CNVR gains are shared more among these populations than losses. The results of the present study were in agreement with the findings of Wang et al. (2014) [38] and Ghose et al. (2014) [34], who reported that losses prevailed over gains in most horse breeds.

When comparing CNVRs shared between breeds, the smallest number was found between Thoroughbreds and Jeju horses (16 CNVRs), and the largest number was found between Hanla and Jeju riding horses (168 CNVRs). Thoroughbreds shared 130 CNVRs with Jeju riding horses and 128 CNVRs with Hanla horses. In contrast, compared with Jeju horses, a relatively small number (22 CNVRs) were shared between Jeju horses and Jeju riding horses, and 21 CNVRs were shared between Jeju horses and Hanla horses (Figure 3). Jeju horses and Thoroughbreds are parental breeds, and they showed the lowest number of shared CNVRs because of their low genetic association. Conversely, Hanla and Jeju riding horses shared more CNVRs with Thoroughbreds than with Jeju horses. We conclude that this is because these breeds were produced with a heavier influence from Thoroughbreds than Jeju horses, with the aim of breeding horses for riding and racing purposes.

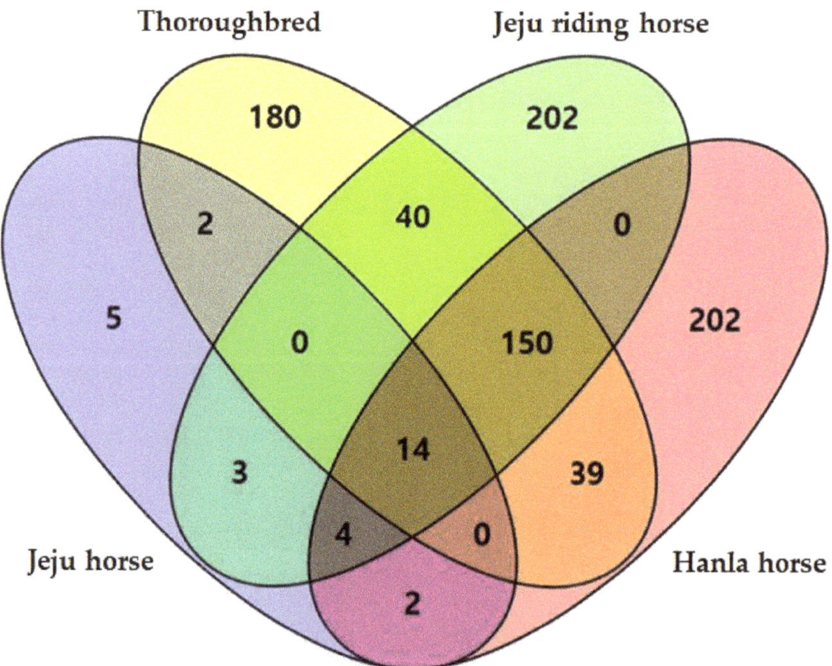

Figure 3. Number of overlapping copy number variation regions (CNVRs) between four breeds in South Korea. Purple is number of Jeju horse CNVR; Yellow is number of Thoroughbred CNVR; Green is number of Jeju riding horse CNVR; Red is number of Hanla horse CNVR, respectively.

3.4. Comparison of CNVRs to those Identified in Previous Studies

To characterize the CNVRs identified in this study in more detail, we compared them to those identified in ten previous studies that used various methods [13,15,33–35,38–42]. A total of 2519 CNVRs were identified across all 11 studies: 1844 (73.2%) CNVRs were identified in our study and the previous studies, and 675 (26.8%) CNVRs were identified only in this study (Table S3). The number of CNVRs found in this study was higher than that detected by Kader et al. (2016) (122 CNVRs) [41]. The number of CNVRs was smaller than that detected by Schurink et al. (2018) (5350 CNVRs), while the genome coverage rate of CNVRs was similar to Schurink et al.'s (2018) study [13].

When our results were compared individually with the results of the previous studies, the highest matching rate (approximately 33.99%) was with Solé et al. (2019). The lowest matching rate (approximately 4.22%) was with Wang et al.'s (2014) study (Table 3).

These matching rates may be the result of breed differences or the genome coverage ratio of the CNVRs in the various studies. Solé et al. (2019) [33] had the largest coverage ratio, which may have contributed to the high matching rate with our study; conversely, Wang et al. (2014) [38] had the smallest coverage ratio. With only a few exceptions, the higher the coverage, the higher the matching rate.

Table 3. Comparison of reported copy number variation regions (CNVRs) for horses (*Equus caballus*).

Study	Platform	Breed	Sample	CNV Analysis Algorithm	CNVR Count	CNVR Range (kb~Mb)	Genome Coverage %	Reference Genome	Overlapped CNVR Count with the Present Study	Overlapped CNVR Percentage with the Present Study
Doan et al. (2012)	Array CGH	15	16	ADM-2	775	0.2–3.5	3.7	EquCab 2.0	169	9.53%
Dupuis et al. (2013)	Illumina Equine 50 K SNP BeadChip	4	477	PennCNV	478	0.1–2.7	2.3	EquCab 2.0	160	8.90%
Ghosh et al. (2014)	Array CGH	16	38	Agilent Genomic Workbench	258	1–2.5	1.2	EquCab 2.0	97	7.44%
Wang et al. (2014)	Array CGH	6	6	segMNT	353	6.1–0.5	0.6	EquCab 2.0	105	4.22%
Kader et al. (2016)	Illumina Equine 70 K SNP BeadChip	3	96	PennCNV	122	0.2–2.2	0.8	EquCab 2.0	76	5.58%
Ghosh et al. (2016)	Array CGH	NA	63	Agilent Genomic Workbench	245	0.1–79.9	6.1	EquCab 2.0	87	7.30%
Schurink et al. (2018)	Axiom Equine Genotyping Array (670,796 SNPs)	1	222	PennCNV	5350	0.12–1.03	11.2	EquCab 2.0	658	18.41%
Solé et al. (2019)	Axiom Equine Genotyping Array (670,796 SNPs)	8	1755	Axiom® CNV summary	939	1–21.3	24.41	EquCab 2.0	360	33.99%
Wang et al. (2022)	Illumina Equine 70 K SNP BeadChip	10	282	PennCNV	495	1–2.3	1.8	EquCab 3.0	206	9.61%
Laceca et al. (2022)	670 K Affymetrix Axiom™ Equine Genotyping Array	1	654	PennCNV	1007	1–4.6	4.4	EquCab 3.0	163	8.87%
Present study	Illumina GGP Equine 70 K SNP BeadChip	4	469	PennCNV	843	0.4–2.9	7.2	EquCab 3.0	-	-

Comparing our CNVRs with other studies, we could confirm the CNVRs that overlapped with previous studies as well as discovered only in this study. The non-overlapped CNVRs with previous studies had unique genetic features of the Korean horse population used in this study. However, we consider that this is not an accurate reflection of actual breed differences, because there are differences (such as sample sizes, platform, and filtering criteria) in the conditions for detecting CNVRs between each study. Therefore, it is necessary to only compare studies that have the same conditions, such as platform and CNV and CNVR analysis algorithms.

3.5. QTLs Overlapping with CNVRs

We annotated previously reported horse QTLs to a total of 843 CNVRs, which were discovered throughout 31 autosomes of the four horse populations in this study (Table S2). We retrieved QTLs that overlapped with CNVRs, and we found several CNVRs that might be associated with phenotypic or economically desirable traits in horses. Because of the inheritance of a large proportion of CNVs, we have focused on CNVRs in two groups to observe CNVRs that might be derived from parental breeds (Jeju horses and Thoroughbreds): one with CNVRs shared between the Thoroughbred and crossbred populations (Jeju riding horses, Hanla horses, or both) and the other with CNVRs shared between the Jeju horse and crossbred populations.

Among the CNVRs shared between the Thoroughbred and crossbred populations (Jeju riding horses and Hanla horses), we identified 23 CNVRs (gains or losses) overlapping with QTLs that were previously reported to be associated with withers height. These regions were located on ECA7 (50.8–51.4 Mb; 73.5–73.6 Mb), ECA8 (51.5–51.6 Mb), ECA16 (34.1–34.9 Mb), and ECA26 (38.5–38.6 Mb). We also found one mixed-type CNVR (ECA20, 32.7–33.0 Mb) that overlapped with the QTL associated with racing performance. As the Thoroughbred is one of the main representative racing horse breeds, this CNVR in crossbred populations might be influenced by Thoroughbreds. For CNVRs shared between the Jeju horse and Jeju riding horse populations, we located one loss-type CNVR (ECA3, 38.8–39.0 Mb) shared between Jeju horses and Jeju riding horses that was associated with a QTL reported to have an association with white coat markings (Table S2).

3.6. Functional Annotation for CNVRs

Based on the NCBI (National Center for Biotechnology Information) annotation of the EquCab 3.0 genome, 1884 genes overlapped with 843 CNVRs (68.48%). We conducted functional annotation analysis on 1,884 genes using PANTHER. As a result, the most significantly enriched biological processes were included in three main categories: sensory perception of smell (raw *p*-value: 3.93×10^{-5}) and chemical stimulus (raw *p*-value: 4.91×10^{-5}) and nervous system processes (raw *p*-value: 4.42×10^{-5}) (Table 4).

Furthermore, when analyzing the function of genes present in CNVRs for each breed, GO with the highest enrichment in Jeju horses had many functions related to G-protein functions, such as "G-protein-coupled serotonin receptor signaling pathway", in the biological process. Furthermore, when analyzing the function of genes present in the CNVRs for each breed, Thoroughbreds had many functions related to olfactory senses such as "detection of chemical stimulus invalidated in sensory perception of smell" and "sense perception of smell" in the biological process. The Jeju riding horses had immunological and olfactory functions such as "antimicrobial humoral immune response mediated by antimicrobial peptide" and "detection of chemical stimulation in sensory perception of smell". Additionally, we identified olfactory-related functions in Hanla horses, such as "sense perception of smell, detection of chemical stimuli invalidated in sense perception of smell" (Table S4). Previous studies have shown that olfactory receptor and immune-related genes were located in CNVRs, and the results of our ontology study support this. In this study, additional genes related to the nervous system were annotated. In Jeju horses, we confirmed that G-protein-related genes were identified more than olfactory and immune functions, and in contrast, all varieties except Jeju horses showed a strong correlation with smell-related functions.

Table 4. Gene ontology (GO) terms (biological process) for annotated genes in copy number variation regions (CNVRs) of horses.

Category	ID	Term	REFLIST [1]	Upload [2]	Expected [3]	Fold Enrichment [4]	raw *p*-Value	FDR [5]
GO_BP	GO:0007608	sensory perception of smell	1039	114	75.35	1.51	3.93×10^{-5}	5.14×10^{-2}
GO_BP	GO:0050911	detection of chemical stimulus involved in sensory perception of smell	1021	112	74.04	1.51	4.46×10^{-5}	4.27×10^{-2}
GO_BP	GO:0050907	detection of chemical stimulus involved in sensory perception	1063	115	77.09	1.49	6.13×10^{-5}	4.90×10^{-2}
GO_BP	GO:0007606	sensory perception of chemical stimulus	1136	122	82.38	1.48	4.91×10^{-5}	4.41×10^{-2}
GO_BP	GO:0007600	sensory perception	1407	146	102.03	1.43	4.16×10^{-5}	4.60×10^{-2}
GO_BP	GO:0050877	nervous system process	1759	176	127.56	1.38	4.42×10^{-5}	4.54×10^{-2}
GO_BP	GO:0051716	cellular response to stimulus	5968	524	432.78	1.21	9.34×10^{-7}	2.69×10^{-3}
GO_BP	GO:0007165	signal transduction	4609	404	334.23	1.21	4.92×10^{-5}	4.16×10^{-2}
GO_BP	GO:0007154	cell communication	4952	434	359.11	1.21	1.95×10^{-5}	3.12×10^{-2}
GO_BP	GO:0023052	signaling	4860	425	352.43	1.21	3.02×10^{-5}	4.34×10^{-2}
GO_BP	GO:0032501	multicellular organismal process	5482	472	397.54	1.19	4.11×10^{-5}	4.93×10^{-2}
GO_BP	GO:0050896	response to stimulus	7140	613	517.77	1.18	9.50×10^{-7}	2.28×10^{-3}
GO_BP	GO:0050794	regulation of cellular process	10,163	833	736.99	1.13	2.09×10^{-6}	3.75×10^{-3}
GO_BP	GO:0065007	biological regulation	11,479	930	832.43	1.12	1.06×10^{-6}	2.18×10^{-3}

[1] The number of genes in the reference list that map to this particular annotation data category. [2] The number of genes in our uploaded list that map to this annotation data category. [3] The column contains the expected value, which is the number of genes expected in our list. [4] The column shows the fold enrichment of the genes observed in the uploaded list over the expected value (number in our list divided by the expected number). If it is greater than 1, it indicates that the category is overrepresented. Conversely, the category is underrepresented if it is less than 1. [5] The False Discovery Rate (FDR)-corrected values as calculated by the Benjamini-Hochberg procedure.

4. Conclusions

The characteristics of CNVs and CNV regions in the Korean horse populations of Jeju (Jeju horse, Thoroughbred, Jeju riding horse, and Hanla horse) were investigated using the Illumina GGP Equine SNP70 Beadchip in this study. We detected a total of 843 CNVRs throughout 31 autosomes, and these CNVRs covered approximately 7.2% of the Equine genome. The discovered CNVRs include both previously reported and novel CNVRs. These CNVRs overlapped with known horse QTLs such as those for withers height, racing performance, and white coat markings, and functional analyses revealed that genes associated with olfactory function and nervous response were highly expressed in CNVRs. The results derived from this study will extend understanding of the genetic composition of Korean horse populations and unique CNVRs throughout the horse genome; furthermore, it will provide resources for future studies on CNVs in horses.

Supplementary Materials: The following supporting information can be downloaded at: https://www.mdpi.com/article/10.3390/ani12243501/s1, Figure S1: PCA and admixture analysis results of four horse breeds; Figure S2: Gain and loss distribution by length of copy number variation (CNV) in four horse breeds in South Korea; Table S1: Distribution of copy number variation region (CNVR) in 31 horse (Equus caballus) autosomes; Table S2: Genome-wide CNVRs discovered in four horse populations; Table S3: Comparison of reported copy number variation regions (CNVRs) for horses (Equus caballus); Table S4: PANTHER gene ontology analysis of annotated genes in each horse breed

Author Contributions: Conceptualization, J.-W.C. and N.-Y.K.; methodology, J.-W.C., N.-Y.K., Y.-M.K. and H.-S.S.; sample collection, N.-Y.K., J.-Y.C. and B.-C.Y.; data curation and analysis, S.-J.H., H.-J.B., J.-W.C., N.-Y.K., Y.-M.K. and H.-S.S.; writing—original draft preparation, Y.-M.K. and S.-J.H.; writing–review and editing, Y.-M.K., S.-J.H., H.-S.S., J.-W.C. and N.-Y.K.; supervision, J.-W.C. and N.-Y.K.; project administration, J.-Y.C. and N.-Y.K.; funding acquisition, J.-W.C. and N.-Y.K. All authors have read and agreed to the published version of the manuscript.

Funding: This study was supported by National Research Foundation of Korea (Project no. NRF-2016R1D1A3B03934278). This work was also carried out with the support of "A study on characteristics related to the temperament in domestic riding horses (Project No. PJ01679701)", Rural Development Administration, Republic of Korea.

Institutional Review Board Statement: For sampling horse breeds used in this study, the study protocol and standard operating procedures were reviewed and approved by the National Institute of Animal Science's Institutional Animal Care and Use Committee (No. NIAS20222448).

Informed Consent Statement: Not applicable.

Data Availability Statement: The datasets of the current study are available from the corresponding author upon reasonable request and with permission of the National Institute of Animal Science RDA in the Republic of Korea.

Conflicts of Interest: The authors declare no conflict of interest.

References

1. Redon, R.; Ishikawa, S.; Fitch, K.R.; Feuk, L.; Perry, G.H.; Andrews, T.D.; Fiegler, H.; Shapero, M.H.; Carson, A.R.; Chen, W.; et al. Global variation in copy number in the human genome. *Nature* **2006**, *444*, 444–454. [CrossRef] [PubMed]
2. Stranger, B.E.; Forrest, M.S.; Dunning, M.; Ingle, C.E.; Beazley, C.; Thorne, N.; Redon, R.; Bird, C.P.; De Grassi, A.; Lee, C. Relative impact of nucleotide and copy number variation on gene expression phenotypes. *Science* **2007**, *315*, 848–853. [CrossRef] [PubMed]
3. Conrad, D.F.; Pinto, D.; Redon, R.; Feuk, L.; Gokcumen, O.; Zhang, Y.; Aerts, J.; Andrews, T.D.; Barnes, C.; Campbell, P.; et al. Origins and functional impact of copy number variation in the human genome. *Nature* **2010**, *464*, 704–712. [CrossRef] [PubMed]
4. MacDonald, J.R.; Ziman, R.; Yuen, R.K.; Feuk, L.; Scherer, S.W. The Database of Genomic Variants: A curated collection of structural variation in the human genome. *Nucleic Acids Res.* **2014**, *42*, D986–D992. [CrossRef]
5. Zarrei, M.; MacDonald, J.R.; Merico, D.; Scherer, S.W. A copy number variation map of the human genome. *Nat. Rev. Genet.* **2015**, *16*, 172–183. [CrossRef] [PubMed]
6. Wade, C.; Giulotto, E.; Sigurdsson, S.; Zoli, M.; Gnerre, S.; Imsland, F.; Lear, T.; Adelson, D.; Bailey, E.; Bellone, R. Genome sequence, comparative analysis, and population genetics of the domestic horse. *Science* **2009**, *326*, 865–867. [CrossRef]
7. McCue, M.E.; Bannasch, D.L.; Petersen, J.L.; Gurr, J.; Bailey, E.; Binns, M.M.; Distl, O.; Guérin, G.; Hasegawa, T.; Hill, E.W.; et al. A High Density SNP Array for the Domestic Horse and Extant Perissodactyla: Utility for Association Mapping, Genetic Diversity and Phylogeny Studies. *PLOS Genetics* **2012**, *8*, e1002451. [CrossRef]
8. Kalbfleisch, T.S.; Rice, E.S.; DePriest, M.S., Jr.; Walenz, B.P.; Hestand, M.S.; Vermeesch, J.R.; BL, O.C.; Fiddes, I.T.; Vershinina, A.O.; Saremi, N.F.; et al. Improved reference genome for the domestic horse increases assembly contiguity and composition. *Commun. Biol.* **2018**, *1*, 197. [CrossRef]
9. Meira, C.T.; Farah, M.M.; Fortes, M.R.; Moore, S.S.; Pereira, G.L.; Silva, J.A.I.V.; da Mota, M.D.S.; Curi, R.A. A genome-wide association study for morphometric traits in quarter horse. *J. Equine Vet. Sci.* **2014**, *34*, 1028–1031. [CrossRef]
10. McGivney, B.A.; Hernandez, B.; Katz, L.M.; MacHugh, D.E.; McGovern, S.P.; Parnell, A.C.; Wiencko, H.L.; Hill, E.W. A genomic prediction model for racecourse starts in the Thoroughbred horse. *Anim. Genet.* **2019**, *50*, 347–357. [CrossRef]
11. Schaefer, R.J.; Schubert, M.; Bailey, E.; Bannasch, D.L.; Barrey, E.; Bar-Gal, G.K.; Brem, G.; Brooks, S.A.; Distl, O.; Fries, R. Developing a 670k genotyping array to tag~ 2M SNPs across 24 horse breeds. *BMC Genomics* **2017**, *18*, 1–18. [CrossRef] [PubMed]
12. Metzger, J.; Schrimpf, R.; Philipp, U.; Distl, O. Expression levels of LCORL are associated with body size in horses. *PLoS ONE* **2013**, *8*, e56497. [CrossRef] [PubMed]
13. Schurink, A.; da Silva, V.H.; Velie, B.D.; Dibbits, B.W.; Crooijmans, R.; Franois, L.; Janssens, S.; Stinckens, A.; Blott, S.; Buys, N.; et al. Copy number variations in Friesian horses and genetic risk factors for insect bite hypersensitivity. *BMC Genet.* **2018**, *19*, 49. [CrossRef] [PubMed]
14. Corbi-Botto, C.M.; Morales-Durand, H.; Zappa, M.E.; Sadaba, S.A.; Peral-Garcia, P.; Giovambattista, G.; Diaz, S. Genomic structural diversity in Criollo Argentino horses: Analysis of copy number variations. *Gene* **2019**, *695*, 26–31. [CrossRef] [PubMed]
15. Laseca, N.; Molina, A.; Valera, M.; Antonini, A.; Demyda-Peyrás, S. Copy Number Variation (CNV): A New Genomic Insight in Horses. *Animals* **2022**, *12*, 1435. [CrossRef]

6. Jung, Y.-H.; Han, S.-H.; Shin, T.; Oh, M.-Y. Genetic characterization of horse bone excavated from the Kwakji archaeological site, Jeju, Korea. *Mol. Cells* **2002**, *14*, 224–230.
7. Cho, B.-W.; Jung, J.-H.; Kim, S.-W.; Kim, H.-S.; Lee, H.-K.; Cho, G.-J.; Song, K.-D. Establishment of genetic characteristics and individual identification system using microsatellite loci in Jeju native horse. *J. Life Sci.* **2007**, *17*, 1441–1446. [CrossRef]
8. Kim, N.Y.; Seong, H.-S.; Kim, D.C.; Park, N.G.; Yang, B.C.; Son, J.K.; Shin, S.M.; Woo, J.H.; Shin, M.C.; Yoo, J.H. Genome-wide analyses of the Jeju, Thoroughbred, and Jeju crossbred horse populations using the high density SNP array. *Genes Genomics* **2018**, *40*, 1249–1258. [CrossRef]
9. Wickham, H.; Chang, W.; Wickham, M.H. Package 'ggplot2'. Create elegant data visualisations using the grammar of graphics. *Version* **2016**, *2*, 1–189. Available online: https://CRAN.R-project.org/package=ggplot2 (accessed on 5 August 2022).
0. Purcell, S.; Neale, B.; Todd-Brown, K.; Thomas, L.; Ferreira, M.A.; Bender, D.; Maller, J.; Sklar, P.; De Bakker, P.I.; Daly, M.J. PLINK: A tool set for whole-genome association and population-based linkage analyses. *Am. J. Hum. Genet.* **2007**, *81*, 559–575. [CrossRef]
1. Alexander, D.H.; Novembre, J.; Lange, K. Fast model-based estimation of ancestry in unrelated individuals. *Genome Res.* **2009**, *19*, 1655–1664. [CrossRef] [PubMed]
2. Wang, K.; Li, M.; Hadley, D.; Liu, R.; Glessner, J.; Grant, S.F.; Hakonarson, H.; Bucan, M. PennCNV: An integrated hidden Markov model designed for high-resolution copy number variation detection in whole-genome SNP genotyping data. *Genome Res.* **2007**, *17*, 1665–1674. [CrossRef] [PubMed]
3. Zhou, J.; Liu, L.; Lopdell, T.J.; Garrick, D.J.; Shi, Y. HandyCNV: Standardized Summary, Annotation, Comparison, and Visualization of Copy Number Variant, Copy Number Variation Region, and Runs of Homozygosity. *Front. Genet.* **2021**, *12*, 731355. [CrossRef] [PubMed]
4. Quinlan, A.R.; Hall, I.M. BEDTools: A flexible suite of utilities for comparing genomic features. *Bioinformatics* **2010**, *26*, 841–842. [CrossRef]
5. Mi, H.; Thomas, P. PANTHER pathway: An ontology-based pathway database coupled with data analysis tools. *Methods Mol. Biol.* **2009**, *563*, 123–140. [CrossRef]
6. Seo, J.-H.; Park, K.-D.; Lee, H.-K.; Kong, H.-S. Genetic diversity of Halla horses using microsatellite markers. *J. Anim. Sci. Technol.* **2016**, *58*, 1–5. [CrossRef]
7. Jung, J.S.; Seong, J.; Lee, G.H.; Kim, Y.; An, J.H.; Yun, J.H.; Kong, H.S. Genetic diversity and relationship of Halla horse based on polymorphisms in microsatellites. *J.Anim. Reprod. Biotechnol.* **2021**, *36*, 76–81. [CrossRef]
8. Liu, G.E.; Hou, Y.; Zhu, B.; Cardone, M.F.; Jiang, L.; Cellamare, A.; Mitra, A.; Alexander, L.J.; Coutinho, L.L.; Dell'Aquila, M.E.; et al. Analysis of copy number variations among diverse cattle breeds. *Genome Res.* **2010**, *20*, 693–703. [CrossRef]
9. Pezer, Z.; Harr, B.; Teschke, M.; Babiker, H.; Tautz, D. Divergence patterns of genic copy number variation in natural populations of the house mouse (Mus musculus domesticus) reveal three conserved genes with major population-specific expansions. *Genome Res.* **2015**, *25*, 1114–1124. [CrossRef]
0. Di Gerlando, R.; Mastrangelo, S.; Tolone, M.; Rizzuto, I.; Sutera, A.M.; Moscarelli, A.; Portolano, B.; Sardina, M.T. Identification of Copy Number Variations and Genetic Diversity in Italian Insular Sheep Breeds. *Animals* **2022**, *12*, 217. [CrossRef]
1. Bekaert, M.; Conant, G.C. Gene duplication and phenotypic changes in the evolution of mammalian metabolic networks. *PLoS ONE* **2014**, *9*, e87115. [CrossRef] [PubMed]
2. Upadhyay, M.; da Silva, V.H.; Megens, H.-J.; Visker, M.H.; Ajmone-Marsan, P.; Bâlteanu, V.A.; Dunner, S.; Garcia, J.F.; Ginja, C.; Kantanen, J. Distribution and functionality of copy number variation across European cattle populations. *Front. Genet.* **2017**, *8*, 108. [CrossRef] [PubMed]
3. Sole, M.; Ablondi, M.; Binzer-Panchal, A.; Velie, B.D.; Hollfelder, N.; Buys, N.; Ducro, B.J.; Francois, L.; Janssens, S.; Schurink, A.; et al. Inter- and intra-breed genome-wide copy number diversity in a large cohort of European equine breeds. *BMC Genomics* **2019**, *20*, 759. [CrossRef]
4. Ghosh, S.; Qu, Z.; Das, P.J.; Fang, E.; Juras, R.; Cothran, E.G.; McDonell, S.; Kenney, D.G.; Lear, T.L.; Adelson, D.L.; et al. Copy Number Variation in the Horse Genome. *PLOS Genetics* **2014**, *10*, e1004712. [CrossRef] [PubMed]
5. Doan, R.; Cohen, N.; Harrington, J.; Veazey, K.; Juras, R.; Cothran, G.; McCue, M.E.; Skow, L.; Dindot, S.V. Identification of copy number variants in horses. *Genome Res.* **2012**, *22*, 899–907. [CrossRef] [PubMed]
6. Petersen, J.L.; Mickelson, J.R.; Rendahl, A.K.; Valberg, S.J.; Andersson, L.S.; Axelsson, J.; Bailey, E.; Bannasch, D.; Binns, M.M.; Borges, A.S. Genome-wide analysis reveals selection for important traits in domestic horse breeds. *PLoS Genetics* **2013**, *9*, e1003211. [CrossRef]
7. Nicholas, T.J.; Baker, C.; Eichler, E.E.; Akey, J.M. A high-resolution integrated map of copy number polymorphisms within and between breeds of the modern domesticated dog. *BMC Genomics* **2011**, *12*, 1–10. [CrossRef]
8. Wang, W.; Wang, S.; Hou, C.; Xing, Y.; Cao, J.; Wu, K.; Liu, C.; Zhang, D.; Zhang, L.; Zhang, Y.; et al. Genome-Wide Detection of Copy Number Variations among Diverse Horse Breeds by Array CGH. *PLoS ONE* **2014**, *9*, e86860. [CrossRef]
9. Dupuis, M.C.; Zhang, Z.; Durkin, K.; Charlier, C.; Lekeux, P.; Georges, M. Detection of copy number variants in the horse genome and examination of their association with recurrent laryngeal neuropathy. *Anim. Genet.* **2013**, *44*, 206–208. [CrossRef]
0. Ghosh, S.; Das, P.J.; McQueen, C.M.; Gerber, V.; Swiderski, C.E.; Lavoie, J.P.; Chowdhary, B.P.; Raudsepp, T. Analysis of genomic copy number variation in equine recurrent airway obstruction (heaves). *Anim. Genet.* **2016**, *47*, 334–344. [CrossRef]

41. Kader, A.; Liu, X.; Dong, K.; Song, S.; Pan, J.; Yang, M.; Chen, X.; He, X.; Jiang, L.; Ma, Y. Identification of copy number variation in three Chinese horse breeds using 70K single nucleotide polymorphism BeadChip array. *Anim. Genet.* **2016**, *47*, 560–56 [CrossRef] [PubMed]
42. Wang, M.; Liu, Y.; Bi, X.; Ma, H.; Zeng, G.; Guo, J.; Guo, M.; Ling, Y.; Zhao, C. Genome-Wide Detection of Copy Number Variant in Chinese Indigenous Horse Breeds and Verification of CNV-Overlapped Genes Related to Heat Adaptation of the Jinjiang Hors *Genes* **2022**, *13*, 603. [CrossRef] [PubMed]

Article

Copy Number Variation (CNV): A New Genomic Insight in Horses

Nora Laseca [1], Antonio Molina [1], Mercedes Valera [2], Alicia Antonini [3] and Sebastián Demyda-Peyrás [3,4,*]

1 Departamento of Genética, Universidad de Córdoba, Edificio Gregor Mendel, CN-IV KM396, 14071 Córdoba, Spain; ge2lagan@uco.es (N.L.); ge1moala@uco.es (A.M.)
2 Departamento de Agronomía, ETSIA, Universidad de Sevilla, Ctra Utrera Km 1, 41013 Sevilla, Spain; mvalera@us.es
3 Departamento de Producción Animal, Facultad de Ciencias Veterinarias, Universidad Nacional de La Plata, La Plata 1900, Argentina; antonini@fcv.unlp.edu.ar
4 Consejo Nacional de Investigaciones Científicas y Técnicas (CONICET), La Plata 1900, Argentina
* Correspondence: sdemyda@fcv.unlp.edu.ar

Simple Summary: This study aimed to contribute to our knowledge of CNVs, a type of genomic marker in equines, by producing, for the first time, a fine-scale characterization of the CNV regions (CNVRs) in the Pura Raza Española horse breed. We found not only the existence of a unique pattern of genomic regions enriched in CNVs in the PRE in comparison with the data available from other breeds but also the incidence of CNVs across the entire genome. Since these regions could affect the structure and dose of the genes involved, we also performed a gene ontology analysis which revealed that most of the genes overlapping in CNVRs were related to the olfactory pathways and immune response.

Abstract: Copy number variations (CNVs) are a new-fangled source of genetic variation that can explain changes in the phenotypes in complex traits and diseases. In recent years, their study has increased in many livestock populations. However, the study and characterization of CNVs in equines is still very limited. Our study aimed to investigate the distribution pattern of CNVs, characterize CNV regions (CNVRs), and identify the biological pathways affected by CNVRs in the Pura Raza Española (PRE) breed. To achieve this, we analyzed high-density SNP genotyping data (670,804 markers) from a large cohort of 654 PRE horses. In total, we identified 19,902 CNV segments and 1007 CNV regions in the whole population. The length of the CNVs ranged from 1.024 kb to 4.55 Mb, while the percentage of the genome covered by CNVs was 4.4%. Interestingly, duplications were more abundant than deletions and mixed CNVRs. In addition, the distribution of CNVs across the chromosomes was not uniform, with ECA12 being the chromosome with the largest percentage of its genome covered (19.2%), while the highest numbers of CNVs were found in ECA20, ECA12, and ECA1. Our results showed that 71.4% of CNVRs contained genes involved in olfactory transduction, olfactory receptor activity, and immune response. Finally, 39.1% of the CNVs detected in our study were unique when compared with CNVRs identified in previous studies. To the best of our knowledge, this is the first attempt to reveal and characterize the CNV landscape in PRE horses, and it contributes to our knowledge of CNVs in equines, thus facilitating the understanding of genetic and phenotypic variations in the species. However, further research is still needed to confirm if the CNVs observed in the PRE are also linked to variations in the specific phenotypical differences in the breed.

Keywords: copy number variation regions; functional clustering; SNP genotyping array; horse breed

Citation: Laseca, N.; Molina, A.; Valera, M.; Antonini, A.; Demyda-Peyrás, S. Copy Number Variation (CNV): A New Genomic Insight in Horses. *Animals* **2022**, *12*, 435. https://doi.org/10.3390/ani12111435

Academic Editor: Maria Teresa Sardina

Received: 3 May 2022
Accepted: 30 May 2022
Published: 2 June 2022

Publisher's Note: MDPI stays neutral with regard to jurisdictional claims in published maps and institutional affiliations.

1. Introduction

Copy number variations (CNVs) are defined as a change in the DNA sequence compared to a reference assembly due to the loss (deletions) or gain (insertions and duplications) of nucleotides bases. CNVs, which usually range from one kilo-base (kb) to several mega-bases (Mb) [1], were associated in livestock animals with changes in the phenotypic expression of simple traits (such as the presence or absence of horns, [2]), and also disease susceptibility and genetic disorders [3]. In addition, recent studies carried out on wildlife and livestock species have pointed to CNVs as a major source of genetic and phenotypic variation among individuals [4–6]. For this reason, increasing our knowledge of the existence and function of CNVs in livestock, particularly related to complex traits and environmental adaptability, contributes to a greater genetic improvement of economic and production traits and animal health [7]. Currently, CNVs can be detected using a range of different platforms, including array comparative genome hybridization (aCGH) [8], single nucleotide polymorphism (SNP) arrays [9], and next-generation sequencing (NGS) [10]. Particularly, the availability of SNP array data in large livestock populations genotyped for genomic breeding purposes has led to a considerable improvement in the characterization of the CNV landscape in some livestock species [11,12].

Studies on CNV diversity are relatively novel and scarce in horses. The first report was published by Doan, et al. [13], which suggested that CNVs are common in the horse genome and may modulate the biological processes underlying different traits. Thereafter, several studies have reported the association of CNVs with diseases [9,14–16], chromosomal abnormalities [17,18], and phenotypic traits [19–22], as well as reported CNV regions overlapping with several genes associated with the reproductive system [23] or adaptability to high temperature and humidity [24]. However, the largest study assessing the CNV landscape in several European horse breeds was recently published by Sole, et al. [12], where they identified CNV regions overlapping with QTLs previously associated with changes in fertility, coat color, conformation, and temperament. Although these studies provide a basis to understand the role of CNVs in equine biology, the current information is still insufficient for the efficient discovery of variants affecting the phenotype, and even more, its association with the phenotypes of complex traits.

The Pura Raza Española (PRE) breed, also known as the Andalusian breed, is the most important and widespread horse breed in Spain, with more than 250,000 active individuals. Although 23.3% of its census is distributed over 62 different countries around the world [25], the breeding program is managed worldwide by the Real Asociación Nacional de Criadores de Caballos de Pura Raza Española (ANCCE) from Spain. The PRE horse was recognized as an official breed in the 15th century [26] being considered as a horse of great beauty with a noble temperament and a great capacity for learning, which explains its success in certain competitions such as dressage, despite being originally selected as a saddle horse. Although a few studies analyzing the genomic landscape of the breed were recently published [17,25,26], to the best of our knowledge, there are no previous studies analyzing CNV variability in this breed.

The aim of this study was therefore to investigate for the first time the distribution pattern of CNVs, characterize CNV regions, and identify biological pathways affected by CNVRs in the PRE horse breed. To achieve this, we analyzed a large cohort of animals genotyped using high-density SNP genotyping technologies. In addition, we attempted to compare the CNV structure of this population with the genome-wide CNVRs identified in other horse breeds, and finally, compare the CNVRs found in different horse breeds.

2. Materials and Methods

2.1. Sampling, Genomic DNA Isolation, Genotyping, and Quality Control

We selected 805 living individuals from 373 PRE herds showing the present diversity of the population for single nucleotide polymorphism (SNP) genotyping using a range of criteria, including sample availability and low average relatedness among individuals.

Genomic DNA was isolated from blood or hair samples using DNeasy Blood & Tissue Kit (Qiagen, Hilden, Germany), following the manufacturer's protocol. The DNA was checked for quality and quantity by Nanodrop™ spectrophotometry (ThermoFisher scientific, Madrid, Spain) and gel electrophoresis. All the individuals were genotyped using the 670 K Affymetrix Axiom™ Equine Genotyping Array (ThermoFisher), including uniformly distributed 670,804 markers [27]. Firstly, Axiom Analysis Suite 5.0 software was used to process and filter genotypes based on DishQC (DQC) and call rate (CR) parameters (DQC $\geq$ 0.82, sample CR $\geq$ 0.95, and SNP CR $\geq$ 97), following the *Best Genotyping Practices Workflow* procedure. Only the samples and SNPs which passed the quality control were kept for the following analyses. The final dataset included 552,965 SNPs located on autosomal chromosomes.

2.2. CNV Data Analysis

CNV calling was performed using PennCNV v.1.0.5 software [28], based on an integrated hidden Markov model which incorporates multiple sources of information, including relative signal intensities (log R ratio, LRR) and minor allele frequencies (B allele frequency, BAF) per SNP, the distance between adjacent SNPs, the populational frequency of the B allele (PFB), and the GC content of the genomic regions in which each marker is located.

For this, we first extracted the LRR and BAF values per individual and marker from the raw genotyping files (CEL) using the Axiom™ CNV summary tool [29]. Thereafter, we compiled a PFB file by averaging the BAF values of each marker in the whole population, using the compile pfb script included in PennCNV. In addition, we estimated the percentage of GC content in the genomic region surrounding each marker position ($\pm$500 kb) using a self-made R script and FASTA information of the EquCab3 horse genome assembly [30] which is employed by the PennCNV algorithm to limit the effect of genomic waves produced by high GC content (according to Diskin, et al. [31]). Finally, we performed individual-based CNV calling using the *-test* option of PennCNV, with the *-gcmodel* and *-pfb* corrections.

CNV filtering and QC analysis were performed using the default PennCNV parameters (standard deviation for LRR $\leq$ 0.35, BAF drift < 0.01, and waviness factor $\leq$ 0.05). Only CNVs larger than 1 kb including at least five consecutive SNPs located on autosomal chromosomes were retained for further analysis since PennCNV calls for the sex chromosomes are unreliable and difficult to interpret (according to the software developer [28]). Finally, to determine the maximum number of CNVs that can be present in an animal for the analysis to be reliable (as proposed by Drobik-Czwarno, et al. [32]), we performed an outlier detection procedure assuming a two-tailed distribution, which determined the exclusion of all the individuals with more than 56 CNV calls (n = 151). This procedure was performed since PennCNV software tends to overestimate CNV fragments in individuals in which genotyping quality is not optimal. The final dataset included 654 horses.

2.3. Determination of CNV Regions and Gene Annotation

Individual CNV calls overlapping in at least one base pair in at least two animals were concatenated into CNV regions (CNVRs) using HandyCNV software [33]. The CNVRs were classified as gains (duplications), losses (deletions), or mixed CNVRs, in which both deletions and duplications were observed. In addition, we estimated the genomic percentage covered by CNVRs at a chromosome level as the sum of all the CNVRs in a given chromosome in relation to its total length.

Finally, the gene content of the CNVRs was assessed based on EquCab 3.0 as the reference genome using Ensembl Biomart [34]. Functional analysis, including Gene Ontology (GO) and Kyoto Encyclopedia of Genes and Genomes (KEGG) pathways, was established by using DAVID Bioinformatics v6.8 [35] and Uniprot online resources [36]. Finally, all the preliminary findings were confirmed by performing an extensive review of the available literature in public databases.

2.4. Comparison of CNVRs with Previous Studies

To determine the existence of differences in the CNV landscape among breeds and populations, we compared our results with eleven previous CNVR reports focused on

CNVR characterization in horses. For this, we combined all the CNVRs reported in thos studies to generate a large consensus CNVR list which was compared with our finding using HandyCNV software.

3. Results and Discussion

3.1. Detection of Genome-Wide CNVs

Advances in the identification of CNVs with a biological function are increasing since they affect genomic sequences which have been associated with a vital role in the regulation of gene functions by altering gene structure, dosage, and expression (causative variants), thus explaining a large part of the phenotypic variation in several traits and species [1], including the horse [37,38]. However, our knowledge of CNVs that contribute to complex traits and diseases in horses is still limited. In this study, we performed a populational analysis description of the CNV landscape by using high-density genotypes in a large cohort of 654 Pura Raza Español horses, identifying 19,902 segments located across the 31 autosomes. The average number of CNVs per individual was 30.43, ranging from 1.024 kb to 4.55 Mb, with an average of 96.07 kb (Table 1 and Figure 1a). The distribution of the CNVs of each chromosome is shown in Figure 1b; ECA12 and ECA20 being the chromosomes with the highest number of CNVs, with nearly ten times more CNVs than the average values. ECA12 is one of the smallest chromosomes in the horse, and therefore the increased CNV density detected may suggest that small chromosomes tend to retain a higher number of CNVs. However, ECA29, ECA30, and ECA31 are even smaller than ECA 12, without showing any sign of such an increase in CNV density (in fact, ECA31 was the chromosome showing the lower CNV density). Similarly, ECA20 was the densest chromosome in terms of SNPs analyzed with almost twice the SNPs per Mb than the rest of the genome ($\approx$507 vs. $\approx$235, respectively), in agreement with the recent findings reported by Rafter, et al. [39] which demonstrated that more CNVs could be detected using high-density than medium-density arrays in cattle. However, a similar SNP density was observed in BTA6 ($\approx$472, Table S1), in which the number of CNVs detected was similar, or even lower, than that observed in the rest of the genome. However, results of BTA12 and BTA20 are consistent with the previous findings reported by several authors in different breeds, in which the prevalence of CNVs in those chromosomes was also high [9,12,14,18,22,40]. Although we did not find a conclusive cause that may explain this peculiarity, it is logical to assume that these regions may carry genes involved in pathways in which a mutation or a change in the genome can provide an evolutive advantage in the species, or at least that they do not have a CNV, in which a duplication or deletion is incompatible with life.

Table 1. Summary of CNVs identified in Pura Raza Española breed.

CNV Type	CNVs *n*	Average Length (bp)	Min Length (bp)	Max Length (bp)
Homozygous deletion	3291	55,077	1150	1,098,544
Heterozygous deletion	3624	107,377	1063	3,209,464
Heterozygous duplication	12,886	103,367	1024	4,552,372
Homozygous duplication	101	95,655	2700	897,981

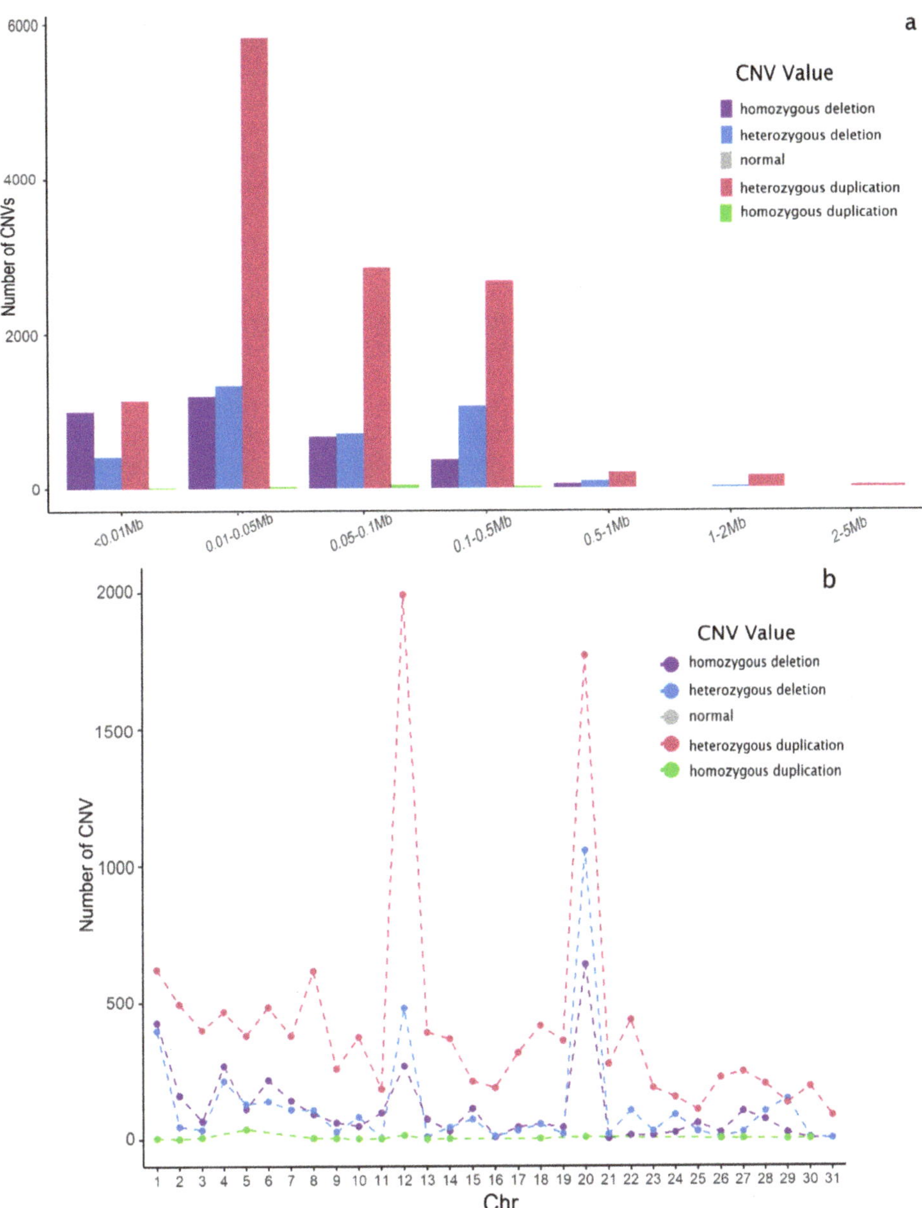

Figure 1. (**a**) Length distribution of CNVs identified; (**b**) chromosomal distribution of CNVs. Number of CNVs present on each chromosome. Purple (homozygous deletion), blue (heterozygous deletion), grey (normal), pink (heterozygous duplication), and green (homozygous duplication).

Although the number of deletions and duplications detected differed among chromosomes (Table S1), an interesting finding was the fact that the number of duplications (12,987) exceeded the number of deletions (6915) in most of them, with the sole exception of ECA1, ECA4, and ECA29, in which the opposite pattern was observed. These results are similar to other studies [9,12–14,16,24], which reported more gains than losses in several horse populations. Although CNVs can be a source of wide variability at the same locus, duplications are more likely to occur in large CNVs than deletions, since they are more tolerated by the genome since no loss of genetic material occurs [40,41]. This might be since duplications are kept for a long time since the deleted regions involving codifying or regulatory regions tend to be purged across the generations due to the existence of "purifying" selection [42]. For the same reason, it was proposed that duplications located in coding or enhancing sequences could increase the genetic diversity in the organisms thus contributing to phenotypic variation and the putative ability to thrive in adverse environments [43,44]. However, the loss of genetic material due to deletions might play a significant role in the genetics of complex traits, even though this has not been directly observed in several gene mapping studies [42]. Still, it is worth mentioning that opposite results have also been reported in horses [19,22,40], although in those studies, results may be explained by the use of different sequencing and genotyping platforms (medium-density arrays) during CNV detection, as well as by the scarce number of individuals analyzed, as demonstrated by Pawlina-Tyszko, et al. [16], Metzger, et al. [19], and Kader, et al. [20]. In this context, Di Gerlando, et al. [5] and Rafter, Gormley, Parnell, Kearney and Berry [39] demonstrated the importance of array density as a factor affecting the discovery of CNVs in sheep and cattle, high-density SNP arrays being associated with better resolution and sensitivity in the CNV detection. On the contrary, Sole, et al. [12] obtained similar results than us in terms of duplication/deletion ratios by analyzing nearly 1800 horses using the same HD SNP array employed in the present study. It is therefore important to mention that comparisons between studies involving CNV detection using different methodologies and algorithms should be made with caution.

3.2. Determination of CNV Regions

The overlapping CNVs in at least one base pair in at least two samples allowed us to detect 1007 CNVRs, including 694 gains, 139 losses, and 174 mixed regions (Table 2 and Table S2), covering 99.79 Mb (representing 4.4% of the genome). Among these, 109 (31 duplications, 19 deletions, and 59 mixed) were present in at least 5% of the PRE population analyzed. These results are higher than those obtained in several studies reported to date in other equine breeds, which range from 0.6 to 3.7% [13,14,20–24], but lower than those obtained by Schurink, et al. [9] and Metzger, et al. [19] using PennCNV software. In addition, although several CNVRs were observed in all the chromosomes (Figure 2), the coverage of CNVRs in each chromosome varied from 1.9% in ECA14 and ECA16 to 19.2% in ECA12. These results in terms of variability agree with most of the previous studies carried out on horses [9,12–14,16,19–24]. However, several of them reported that ECA12 is particularly enriched in CNVRs, including a cluster of genes associated with the development of olfactory receptors (ORs) [13,18,19]. In all of them, it has been suggested that these genes have undergone selection, or have even increased the number of copies, through CNV gain and loss processes during the domestication of the horse. However, it is worth mentioning that the overrepresentation of these OR genes in CNV regions is not only present in horses but also humans [45], cattle [46,47], pigs [48], and sheep [49].

Table 2. Summary of regions CNVs in Pura Raza Española breed.

CNVR Type	CNVRs *n*	Average Length (bp)	Min Length (bp)	Max Length (bp)	Total Length (bp)
Gains	694	78,815	1458	1,169,661	54,697,934
Losses	139	44,882	1063	635,203	6,238,703
Mixed	174	223,309	5458	4,921,979	38,855,746

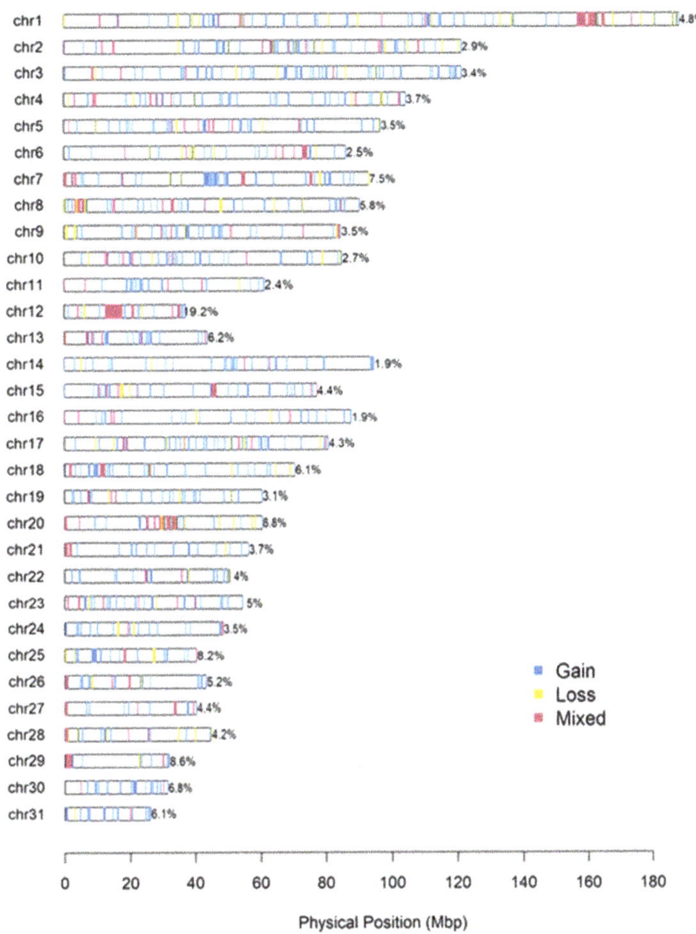

Figure 2. Map of CNVRs in the 31 equine autosome chromosomes. Blue, yellow, and red represent gain, loss, and mixed, respectively.

3.3. Comparative Analysis with Previously Known CNVRs

Differences in CNVRs have been successfully employed as a method of differentiation of horse breeds [12], since they allow us to find variants and regions which may be associated with its historical evolution, adaptability to a specific environment, or an improvement in the phenotype in traits of interest, such as the grey coat [50]. For this reason, we compared CNVRs identified in this study with those previously reported in eleven different reports [9,12–14,16,19–21,23,24] by generating a consensus list including all the different CNVRs reported previously (n = 8292; Table S3). Our analysis showed that 61% (614) of the CNVRs which we detected in PRE overlapped with some of these, whereas near 40% were described for the first time in this study. In terms of variability, this large difference detected among breeds supports the hypothesis that CNVRs can be used as a reliable genetic footprint to discriminate among breeds, as proposed by Sole, Ablondi, Binzer-Panchal, Velie, Hollfelder, Buys, Ducro, Francois, Janssens, Schurink, Viklund, Eriksson, Isaksson, Kultima, Mikko and Lindgren [12]. However, it must be taken into account that the detection methodology, the number of samples analyzed, the genotyping platform, and the criteria for searching for CNVRs employed in each study were different, and therefore, further research is still in need for a proper validation in the species. However, it was a highly interesting finding that most of the CNVR losses (95.7%), in contrast to only 63.6% of the CNVR gains, overlapped with previous CNVR reports, although it also may be explained by the fact that in most of the studies, the number of losses was higher than the number of gains, conversely to our findings. On the other hand, as we mentioned before, our analysis included 393 newly described CNVRs. This percentage agrees with Sole et al. [12], who reported a proportion of unique CNVs ranging from 30–50% by comparing Draught and Warmblood horses, which was lower than that reported by Schurink, et al. [9] in Friesian horses (58%). Interestingly, 22 of the unique regions detected in this study were quite common within the PRE population, being present in at least 5% of the horses analyzed. These findings also agree with Sole, et al. [12], who demonstrated the existing differences among the frequencies of common CNVRs between breeds.

Finally, we reviewed the CNVRs found in a total of 38 different breeds of horses in previous studies (Table S4). As we expected, the number of individuals analyzed was positively correlated with the number of CNVRs found (r^2 = 0.45). However, this value suggests the existence of a racial component since the correlation was moderate. For example, previous CNVR reports ranged from 1 in Brandenburger (2 individuals analyzed) to 5350 in Friesian (222 individuals analyzed), the breeds with more CNVR per individual being the Friesian, Vlaams paard, German draft, and Ardenner horses. In contrast, Warmbloods and Swedish warmblood were the breeds with the fewest CNVRs per individual, without taking into account the breeds in which the CNVRs were analyzed by consensus between methodologies [19]. Paradoxically, the Friesian breed was analyzed in two studies [9,12] and the number of CNVRs found was very different, probably as a result of the different methodologies used. Although most of the CNVs studies performed in the different breeds were performed using PennCNV, several different approaches and software (such as CNVRuler [51] or BEDTools [52]) were used to overlap the CNVs into CNV regions. Drawing conclusions from this comparison is extremely complex because the differences observed could be attributed to all the factors affecting the CNV call, but also in this case, to the use of different methodologies applied for CNVR discovery.

3.4. Functional Annotations of CNV Regions

CNVRs are involved in modulating the gene function in multiple ways, including changing gene structure, altering gene regulation, and exposing recessive alleles. For this reason, we performed functional analysis on the genes affected by a CNVR to understand its potential effects on biological processes in horses. Interestingly, 71.4% of the CNVRs contained genes (2105) (Table S5), among which 77.86% were protein-coding genes, 15.53% were long noncoding RNAs (lncRNAs), 2.94% were pseudogenes, 1.66% were microRNAs, 1.05% were small nuclear RNAs (snRNAs), and 0.76% were small nucleolar RNAs (snoRNAs). These results are similar to other studies in horses, indicating that a high percentage of CNVRs cover genomic regions involved in genes. For instance, Ghosh, et al. [23] found that 82% of the CNVRs identified concatenated with one or more genes. Similarly, other studies have reported that 80% [15], 79.3% [12], 69.7% [20], and 49.2% [9] of the CNVRs found involved genes.

Functional annotation analysis revealed that the most significantly enriched biological processes were included in three main categories: olfactory receptor activity (p Benjamini = 5.3 $\times$ 10^{-135}), G-protein coupled receptor activity (p Benjamini = 3.8 $\times$ 10^{-107}) and immune response (p Benjamini = 9.9 $\times$ 10^{-20}) (Table S6). As expected, the KEGG pathway analyses indicated that olfactory transduction was the most affected pathway (p Benjamini = 2.7 $\times$ 10^{-80}), with 311 genes involved. These olfactory system genes are essential for detecting, encoding, and processing chemo-stimuli that can carry information that is important for survival, social interactions, reproduction, and adaptation to the animal environment [53]. In this context, Palouzier-Paulignan, et al. [54] suggested that olfactory receptors are also involved in mammalian appetite regulation and feeding efficiency, and therefore, may be related to food intake. Here, our results agree with Hughes, et al. [55], who proposed that this large family of genes has undergone extensive expansion and contraction through duplication and pseudogenization, giving rise to new functionalities. In addition, several genes related to G-protein coupled receptor activity (358), immune response (66), steroid hormone biosynthesis (14), secondary metabolite biosynthesis, transport, and catabolism (10), and ovum development (5) were located within or nearby CNVRs, which revealed six different functional-term clusters that were significantly enriched (enrichment score higher than 4.76; Table 3). The ontology analysis that we performed agree with previous CNV studies in horses [9,14,19,22] that have identified the olfactory receptors and immunity-related genes as CNV hotspots. Furthermore, Young, et al. [45] provided evidence that OR enrichment in CNVs is not due to positive selection but to the frequent appearance of these genes in segmentally duplicated regions, and that the purifying selection against CNVs is lower in OR-containing regions than in regions containing essential genes [41].

Table 3. Significantly enriched annotation clusters and functional terms.

Functional Cluster (Enrichment Score)	Category	Term	Genes	p Value	P Benjamini
	GOTERM_MF_DIRECT	Olfactory receptor activity	355	6.9×10^{-138}	5.3×10^{-135}
	INTERPRO	Olfactory receptor	355	2.1×10^{-131}	3.1×10^{-128}
	GOTERM_MF_DIRECT	G-protein coupled receptor activity	358	9.8×10^{-110}	3.8×10^{-107}
	UP_SEQ_FEATURE	DOMAIN:G_PROTEIN_RECEP_F1_2	362	5.6×10^{-99}	3.7×10^{-96}
	UP_KW_BIOLOGICAL_PROCESS	Olfaction	282	5.0×10^{-97}	4.2×10^{-95}
Cluster 1	INTERPRO	GPCR, rhodopsin-like, 7TM	362	9.0×10^{-97}	6.8×10^{-94}
(94.64)	INTERPRO	G protein-coupled receptor, rhodopsin-like	348	4.5×10^{-93}	2.3×10^{-90}
	UP_KW_BIOLOGICAL_PROCESS	Sensory transduction	284	5.6×10^{-93}	2.4×10^{-91}
	UP_KW_MOLECULAR_FUNCTION	Transducer	368	4.4×10^{-84}	3.0×10^{-82}
	KEGG_PATHWAY	Olfactory transduction	311	8.6×10^{-83}	2.7×10^{-80}
	UP_KW_MOLECULAR_FUNCTION	Receptor	372	2.2×10^{-65}	7.4×10^{-64}
	UP_KW_CELLULAR_COMPONENT	Cell membrane	311	2.8×10^{-54}	1.2×10^{-52}
Cluster 2	UP_SEQ_FEATURE	DOMAIN:Ig-like	192	8.8×10^{-44}	2.9×10^{-41}
(40.74)	INTERPRO	Immunoglobulin-like domain	192	3.1×10^{-42}	1.2×10^{-39}
	INTERPRO	Immunoglobulin-like fold	233	2.2×10^{-38}	5.6×10^{-36}
Cluster 3	GOTERM_CC_DIRECT	Integral component of membrane	662	3.7×10^{-38}	1.8×10^{-35}
(30.68)	UP_SEQ_FEATURE	TRANSMEM:Helical	697	2.9×10^{-31}	6.3×10^{-29}
	UP_KW_CELLULAR_COMPONENT	Membrane	730	8.5×10^{-25}	1.8×10^{-23}
	GOTERM_BP_DIRECT	Phagocytosis, recognition	21	4.5×10^{-11}	2.6×10^{-8}
	GOTERM_BP_DIRECT	Phagocytosis, engulfment	21	2.2×10^{-10}	9.9×10^{-8}
	GOTERM_BP_DIRECT	Positive regulation of B cell activation	19	6.8×10^{-10}	2.2×10^{-7}
Cluster 4	GOTERM_CC_DIRECT	Immunoglobulin complex, circulating	19	2.9×10^{-9}	4.6×10^{-7}
(8.14)	GOTERM_BP_DIRECT	B cell receptor signaling pathway	22	1.2×10^{-8}	3.4×10^{-6}
	GOTERM_MF_DIRECT	Immunoglobulin receptor binding	19	1.4×10^{-8}	2.6×10^{-6}
	GOTERM_MF_DIRECT	Antigen binding	20	1.9×10^{-8}	2.9×10^{-6}
	GOTERM_BP_DIRECT	Complement activation, classical pathway	20	2.0×10^{-8}	5.2×10^{-6}
	GOTERM_BP_DIRECT	Defense response to bacterium	24	4.3×10^{-5}	4.5×10^{-3}
Cluster 5	UP_SEQ_FEATURE	DOMAIN:BPI1	12	1.1×10^{-7}	1.5×10^{-5}
(6.54)	INTERPRO	Lipid-binding serum glycoprotein, N-terminal	12	4.6×10^{-7}	5.3×10^{-5}
	INTERPRO	Bactericidal permeability-increasing protein, alpha/beta domain	12	4.6×10^{-7}	5.3×10^{-5}
Cluster 6	KEGG_PATHWAY	Graft-versus-host disease	22	9.1×10^{-7}	9.4×10^{-5}
(4.76)	KEGG_PATHWAY	Type I diabetes mellitus	20	2.6×10^{-5}	1.3×10^{-3}
	KEGG_PATHWAY	Allograft rejection	17	2.2×10^{-4}	$8.6 \times 10^{-3-}$

4. Conclusions

This study investigated for the first time the distribution pattern of CNVs and CNV regions in the Pura Raza Española horse breed. Our results revealed that a considerably large proportion of the genome (4.4%) was affected by CNVRs, although its distribution among the chromosomes was not uniform. Moreover, we found 394 CNVRs that had not yet been identified in different horse breeds, which may have contributed to the establishment of the PRE phenotype. Finally, functional annotation analysis of CNVRs revealed significant enrichment in genes related to olfactory transduction, olfactory receptor activity, and immune response, pointing to CNVs as hotspots for these genes. This study contributes to our knowledge of CNVs in the equine species and our understanding of genetic and phenotypic variations in the equine genome, but future research is needed to confirm if the observed CNVRs are also linked to phenotypical differences in complex traits.

Supplementary Materials: The following supporting information can be downloaded at: https://www.mdpi.com/article/10.3390/ani12111435/s1, Table S1: Number of CNVs (deletions and duplications) identified on the different chromosomes. Table S2: List of CNVRs found in the analyzed PRE horses. Table S3: Combination of all CNVRs found in previously published studies of horses. Table S4: Summary of the number of CNVRs identified in each breed of horse, previously published. Table S5: List of genes found and/or overlapping with CNV regions in PRE horse breed. Table S6: Functional annotation analysis performed using DAVID Database.

Author Contributions: Conceptualization, N.L., A.M., M.V. and S.D.-P.; methodology, N.L. and S.D.-P.; sample collection: A.M. and M.V.; genomic data, N.L. and S.D.-P.; data curation, N.L., A.A. and S.D.-P.; writing—original draft preparation, N.L., A.A., A.M. and S.D.-P.; writing—review and editing definitive version, N.L., A.M., M.V. and S.D.-P.; obtaining funding for analysis, A.M., M.V. and S.D.-P.; project coordination: A.M. and S.D.-P. All authors have read and agreed to the published version of the manuscript.

Funding: This research was funded by the AGL-2017-84217-P Research project from Ministerio de Economía, Industria y Competitividad of the Spanish Government and the PICT2018-0227 from the Agencia Nacional de Promoción Cientifica y Tecnológica (ANPCyT, Argentina). Nora Laseca is a doctoral fellow (PRE 2018-083492). The APC was partially funded by Universidad de Córdoba.

Institutional Review Board Statement: Not applicable.

Informed Consent Statement: All the individuals analyzed in this study are involved in the PRE breeding program which acknowledges the use of samples and data for scientific purpouses.

Data Availability Statement: The data that support the findings of this study are available from the Real Asociación Nacional de Criadores de Caballos de Pura Raza Española (ANCCE). Restrictions apply to the availability of these data, which were used under license for this study. The data are available from the authors with the permission of ANCCE. The data reviewed in this study were obtained from public databases.

Acknowledgments: The authors wish to thank the Asociación Nacional de Criadores de Caballos de Pura Raza Español (ANCCE) for their collaboration in this study and for providing the biological samples and genealogical data.

Conflicts of Interest: The authors declare no conflict of interest.

References

1. Zhang, F.; Gu, W.; Hurles, M.E.; Lupski, J.R. Copy Number Variation in Human Health, Disease, and Evolution. *Annu. Rev. Genom. Hum. Genet.* **2009**, *10*, 451–481. [CrossRef]
2. Guo, J.; Jiang, R.; Mao, A.; Liu, G.E.; Zhan, S.; Li, L.; Zhong, T.; Wang, L.; Cao, J.; Chen, Y.; et al. Genome-wide association study reveals 14 new SNPs and confirms two structural variants highly associated with the horned/polled phenotype in goats. *BMC Genom.* **2021**, *22*, 769. [CrossRef] [PubMed]
3. Bickhart, D.; Liu, G. The challenges and importance of structural variation detection in livestock. *Front. Genet.* **2014**, *5*, 37. [CrossRef]

4. Yang, L.; Niu, Q.; Zhang, T.; Zhao, G.; Zhu, B.; Chen, Y.; Zhang, L.; Gao, X.; Gao, H.; Liu, G.E.; et al. Genomic sequencing analysis reveals copy number variations and their associations with economically important traits in beef cattle. *Genomics* **2021**, *113*, 812–820. [CrossRef]

5. Di Gerlando, R.; Mastrangelo, S.; Tolone, M.; Rizzuto, I.; Sutera, A.M.; Moscarelli, A.; Portolano, B.; Sardina, M.T. Identification of Copy Number Variations and Genetic Diversity in Italian Insular Sheep Breeds. *Animals* **2022**, *12*, 217. [CrossRef]

6. Qiu, Y.; Ding, R.; Zhuang, Z.; Wu, J.; Yang, M.; Zhou, S.; Ye, Y.; Geng, Q.; Xu, Z.; Huang, S.; et al. Genome-wide detection of CNV regions and their potential association with growth and fatness traits in Duroc pigs. *BMC Genom.* **2021**, *22*, 332. [CrossRef]

7. Cao, X.-K.; Huang, Y.-Z.; Ma, Y.-L.; Cheng, J.; Qu, Z.-X.; Ma, Y.; Bai, Y.-Y.; Tian, F.; Lin, F.-P.; Ma, Y.-L.; et al. Integrating CNVs into meta-QTL identified GBP4 as positional candidate for adult cattle stature. *Funct. Integr. Genom.* **2018**, *18*, 559–567. [CrossRef]

8. Liu, M.; Fang, L.; Liu, S.; Pan, M.G.; Seroussi, E.; Cole, J.B.; Ma, L.; Chen, H.; Liu, G.E. Array CGH-based detection of CNV regions and their potential association with reproduction and other economic traits in Holsteins. *BMC Genom.* **2019**, *20*, 181. [CrossRef] [PubMed]

9. Schurink, A.; da Silva, V.H.; Velie, B.D.; Dibbits, B.W.; Crooijmans, R.P.M.A.; François, L.; Janssens, S.; Stinckens, A.; Blott, S.; Buys, N.; et al. Copy number variations in Friesian horses and genetic risk factors for insect bite hypersensitivity. *BMC Genet.* **2018**, *19*, 49. [CrossRef] [PubMed]

10. Moreno-Cabrera, J.M.; Del Valle, J.; Castellanos, E.; Feliubadaló, L.; Pineda, M.; Brunet, J.; Serra, E.; Capellà, G.; Lázaro, C.; Gel, B. Evaluation of CNV detection tools for NGS panel data in genetic diagnostics. *Eur. J. Hum. Genet. EJHG* **2020**, *28*, 1645–1655. [CrossRef] [PubMed]

11. Mielczarek, M.; Frąszczak, M.; Nicolazzi, E.; Williams, J.L.; Szyda, J. Landscape of copy number variations in Bos taurus: Individual and inter-breed variability. *BMC Genom.* **2018**, *19*, 410. [CrossRef] [PubMed]

12. Sole, M.; Ablondi, M.; Binzer-Panchal, A.; Velie, B.D.; Hollfelder, N.; Buys, N.; Ducro, B.J.; Francois, L.; Janssens, S.; Schurink, A.; et al. Inter- and intra-breed genome-wide copy number diversity in a large cohort of European equine breeds. *BMC Genom.* **2019**, *20*, 759. [CrossRef] [PubMed]

13. Doan, R.; Cohen, N.; Harrington, J.; Veazy, K.; Juras, R.; Cothran, G.; McCue, M.E.; Skow, L.; Dindot, S.V. Identification of copy number variants in horses. *Genome Res.* **2012**, *22*, 899–907. [CrossRef]

14. Dupuis, M.C.; Zhang, Z.; Durkin, K.; Charlier, C.; Lekeux, P.; Georges, M. Detection of copy number variants in the horse genome and examination of their association with recurrent laryngeal neuropathy. *Anim. Genet.* **2013**, *44*, 206–208. [CrossRef] [PubMed]

15. Ghosh, S.; Das, P.J.; McQueen, C.M.; Gerber, V.; Swiderski, C.E.; Lavoie, J.P.; Chowdhary, B.P.; Raudsepp, T. Analysis of genomic copy number variation in equine recurrent airway obstruction (heaves). *Anim. Genet.* **2016**, *47*, 334–344. [CrossRef]

16. Pawlina-Tyszko, K.; Gurgul, A.; Szmatoła, T.; Ropka-Molik, K.; Semik-Gurgul, E.; Klukowska-Rötzler, J.; Koch, C.; Mählmann, K.; Bugno-Poniewierska, M. Genomic landscape of copy number variation and copy neutral loss of heterozygosity events in equine sarcoids reveals increased instability of the sarcoid genome. *Biochimie* **2017**, *140*, 122–132. [CrossRef]

17. Demyda-Peyras, S.; Bugno-Poniewierska, M.; Pawlina, K.; Anaya, G.; Moreno-Millán, M. The use of molecular and cytogenetic methods as a valuable tool in the detection of chromosomal abnormalities in horses: A Case of sex chromosome chimerism in a Spanish Purebred colt. *Cytogenet. Genome Res.* **2013**, *141*, 277–283. [CrossRef]

18. Pirosanto, Y.; Laseca, N.; Valera, M.; Molina, A.; Moreno-Millán, M.; Bugno-Poniewierska, M.; Ross, P.; Azor, P.; Demyda-Peyrás, S. Screening and detection of chromosomal copy number alterations in the domestic horse using SNP-array genotyping data. *Anim. Genet.* **2021**, *52*, 431–439. [CrossRef]

19. Metzger, J.; Philipp, U.; Lopes, M.S.; da Camara Machado, A.; Felicetti, M.; Silvestrelli, M.; Distl, O. Analysis of copy number variants by three detection algorithms and their association with body size in horses. *BMC Genom.* **2013**, *14*, 487. [CrossRef]

20. Kader, A.; Liu, X.; Dong, K.; Song, S.; Pan, J.; Yang, M.; Chen, X.; He, X.; Jiang, L.; Ma, Y. Identification of copy number variations in three Chinese horse breeds using 70K single nucleotide polymorphism BeadChip array. *Anim. Genet.* **2016**, *47*, 560–569. [CrossRef]

21. Wang, W.; Wang, S.; Hou, C.; Xing, Y.; Cao, J.; Wu, K.; Liu, C.; Zhang, D.; Zhang, L.; Zhang, Y.; et al. Genome-wide detection of copy number variations among diverse horse breeds by array CGH. *PLoS ONE* **2014**, *9*, e86860. [CrossRef] [PubMed]

22. Corbi-Botto, C.M.; Morales-Durand, H.; Zappa, M.E.; Sadaba, S.A.; Peral-García, P.; Giovambattista, G.; Díaz, S. Genomic structural diversity in Criollo Argentino horses: Analysis of copy number variations. *Gene* **2019**, *695*, 26–31. [CrossRef] [PubMed]

23. Ghosh, S.; Qu, Z.; Das, P.J.; Fang, E.; Juras, R.; Cothran, E.G.; McDonell, S.; Kenney, D.G.; Lear, T.L.; Adelson, D.L.; et al. Copy Number Variation in the Horse Genome. *PLoS Genet.* **2014**, *10*, e1004712. [CrossRef] [PubMed]

24. Wang, M.; Liu, Y.; Bi, X.; Ma, H.; Zeng, G.; Guo, J.; Guo, M.; Ling, Y.; Zhao, C. Genome-Wide Detection of Copy Number Variants in Chinese Indigenous Horse Breeds and Verification of CNV-Overlapped Genes Related to Heat Adaptation of the Jinjiang Horse. *Genes* **2022**, *13*, 603. [CrossRef] [PubMed]

25. Solé, M.; Valera, M.; Fernández, J. Genetic structure and connectivity analysis in a large domestic livestock meta-population: The case of the Pura Raza Español horses. *J. Anim. Breed. Gen.* **2018**, *135*, 460–471. [CrossRef]

26. Poyato-Bonilla, J.; Laseca, N.; Demyda Peyrás, S.; Molina, A.; Valera, M. 500 years of breeding in the Carthusian Strain of Pura Raza Español horse: An evolutional analysis using genealogical and genomic data. *J. Anim. Breed. Gen.* **2021**, *139*, 84–99. [CrossRef]

27. Schaefer, R.J.; Schubert, M.; Bailey, E.; Bannasch, D.L.; Barrey, E.; Bar-Gal, G.K.; Brem, G.; Brooks, S.A.; Distl, O.; Fries, R.; et al. Developing a 670k genotyping array to tag ~2M SNPs across 24 horse breeds. *BMC Genom.* **2017**, *18*, 565. [CrossRef]

8. Wang, K.; Li, M.; Hadley, D.; Liu, R.; Glessner, J.; Grant, S.F.A.; Hakonarson, H.; Bucan, M. PennCNV: An integrated hidden Markov model designed for high-resolution copy number variation detection in whole-genome SNP genotyping data. *Genome Res.* **2007**, *17*, 1665–1674. [CrossRef]

9. Thermofisher. Axiom CNV Summary Tool User Manual. Available online: https://tools.thermofisher.com/content/sfs/manuals/axiom_cnv_summary_tool_usermanual.pdf (accessed on 15 August 2020).

0. Kalbfleisch, T.S.; Rice, E.S.; DePriest, M.S.; Walenz, B.P.; Hestand, M.S.; Vermeesch, J.R.; O'Connell, B.L.; Fiddes, I.T.; Vershinina, A.O.; Petersen, J.L.; et al. EquCab3, an Updated Reference Genome for the Domestic Horse. *bioRxiv* **2018**, 306928. [CrossRef]

1. Diskin, S.J.; Li, M.; Hou, C.; Yang, S.; Glessner, J.; Hakonarson, H.; Bucan, M.; Maris, J.M.; Wang, K. Adjustment of genomic waves in signal intensities from whole-genome SNP genotyping platforms. *Nucleic Acids Res.* **2008**, *36*, e126. [CrossRef]

2. Drobik-Czwarno, W.; Wolc, A.; Fulton, J.E.; Dekkers, J.C.M. Detection of copy number variations in brown and white layers based on genotyping panels with different densities. *Genet. Sel. Evol.* **2018**, *50*, 54. [CrossRef] [PubMed]

3. Zhou, J.; Liu, L.; Lopdell, T.J.; Garrick, D.J.; Shi, Y. HandyCNV: Standardized Summary, Annotation, Comparison, and Visualization of Copy Number Variant, Copy Number Variation Region, and Runs of Homozygosity. *Front. Genet.* **2021**, *12*, 731355. [CrossRef] [PubMed]

4. Aken, B.L.; Ayling, S.; Barrell, D.; Clarke, L.; Curwen, V.; Fairley, S.; Fernandez Banet, J.; Billis, K.; García Girón, C.; Hourlier, T.; et al. The Ensembl gene annotation system. *Database* **2016**, *2016*, baw093. [CrossRef]

5. Huang, D.W.; Sherman, B.T.; Lempicki, R.A. Systematic and integrative analysis of large gene lists using DAVID bioinformatics resources. *Nat. Protoc.* **2009**, *4*, 44–57. [CrossRef] [PubMed]

6. The UniProt, C. UniProt: The universal protein knowledgebase in 2021. *Nucleic Acids Res.* **2021**, *49*, D480–D489. [CrossRef]

7. Anaya, G.; Moreno-Millán, M.; Bugno-Poniewierska, M.; Pawlina, K.; Membrillo, A.; Molina, A.; Demyda-Peyrás, S. Sex reversal syndrome in the horse: Four new cases of feminization in individuals carrying a 64,XY SRY negative chromosomal complement. *Anim. Reprod. Sci.* **2014**, *151*, 22–27. [CrossRef]

8. Ghosh, S.; Davis, B.W.; Rosengren, M.; Jevit, M.J.; Castaneda, C.; Arnold, C.; Jaxheimer, J.; Love, C.C.; Varner, D.D.; Lindgren, G.; et al. Characterization of a Homozygous Deletion of Steroid Hormone Biosynthesis Genes in Horse Chromosome 29 as a Risk Factor for Disorders of Sex Development and Reproduction. *Genes* **2020**, *11*, 251. [CrossRef]

9. Rafter, P.; Gormley, I.C.; Parnell, A.C.; Kearney, J.F.; Berry, D.P. Concordance rate between copy number variants detected using either high- or medium-density single nucleotide polymorphism genotype panels and the potential of imputing copy number variants from flanking high density single nucleotide polymorphism haplotypes in cattle. *BMC Genom.* **2020**, *21*, 205. [CrossRef]

0. Locke, D.P.; Sharp, A.J.; McCarroll, S.A.; McGrath, S.D.; Newman, T.L.; Cheng, Z.; Schwartz, S.; Albertson, D.G.; Pinkel, D.; Altshuler, D.M.; et al. Linkage Disequilibrium and Heritability of Copy-Number Polymorphisms within Duplicated Regions of the Human Genome. *Am. J. Hum. Genet.* **2006**, *79*, 275–290. [CrossRef]

1. Redon, R.; Ishikawa, S.; Fitch, K.R.; Feuk, L.; Perry, G.H.; Andrews, T.D.; Fiegler, H.; Shapero, M.H.; Carson, A.R.; Chen, W.; et al. Global variation in copy number in the human genome. *Nature* **2006**, *444*, 444–454. [CrossRef]

2. Conrad, D.F.; Andrews, T.D.; Carter, N.P.; Hurles, M.E.; Pritchard, J.K. A high-resolution survey of deletion polymorphism in the human genome. *Nat. Genet.* **2006**, *38*, 75–81. [CrossRef] [PubMed]

3. Chain, F.J.J.; Feulner, P.G.D.; Panchal, M.; Eizaguirre, C.; Samonte, I.E.; Kalbe, M.; Lenz, T.L.; Stoll, M.; Bornberg-Bauer, E.; Milinski, M.; et al. Extensive Copy-Number Variation of Young Genes across Stickleback Populations. *PLoS Genet.* **2014**, *10*, e1004830. [CrossRef] [PubMed]

4. Bekaert, M.; Conant, G.C. Gene Duplication and Phenotypic Changes in the Evolution of Mammalian Metabolic Networks. *PLoS ONE* **2014**, *9*, e87115. [CrossRef] [PubMed]

5. Young, J.M.; Endicott, R.M.; Parghi, S.S.; Walker, M.; Kidd, J.M.; Trask, B.J. Extensive copy-number variation of the human olfactory receptor gene family. *Am. J. Hum. Genet.* **2008**, *83*, 228–242. [CrossRef]

6. Seroussi, E.; Glick, G.; Shirak, A.; Yakobson, E.; Weller, J.I.; Ezra, E.; Zeron, Y. Analysis of copy loss and gain variations in Holstein cattle autosomes using BeadChip SNPs. *BMC Genom.* **2010**, *11*, 673. [CrossRef]

7. Zhou, Y.; Connor, E.E.; Wiggans, G.R.; Lu, Y.; Tempelman, R.J.; Schroeder, S.G.; Chen, H.; Liu, G.E. Genome-wide copy number variant analysis reveals variants associated with 10 diverse production traits in Holstein cattle. *BMC Genom.* **2018**, *19*, 314. [CrossRef]

8. Paudel, Y.; Madsen, O.; Megens, H.-J.; Frantz, L.A.F.; Bosse, M.; Crooijmans, R.P.M.A.; Groenen, M.A.M. Copy number variation in the speciation of pigs: A possible prominent role for olfactory receptors. *BMC Genom.* **2015**, *16*, 330. [CrossRef]

9. Igoshin, A.V.; Deniskova, T.E.; Yurchenko, A.A.; Yudin, N.S.; Dotsev, A.V.; Selionova, M.I.; Zinovieva, N.A.; Larkin, D.M. Copy number variants in genomes of local sheep breeds from Russia. *Anim. Genet.* **2022**, *53*, 119–132. [CrossRef]

50. Rosenberg, N.A. distruct: A program for the graphical display of population structure. *Mol. Ecol. Notes* **2004**, *4*, 137–138. [CrossRef]

51. Kim, J.-H.; Hu, H.-J.; Yim, S.-H.; Bae, J.S.; Kim, S.-Y.; Chung, Y.-J. CNVRuler: A copy number variation-based case–control association analysis tool. *Bioinformatics* **2012**, *28*, 1790–1792. [CrossRef]

52. Quinlan, A.R.; Hall, I.M. BEDTools: A flexible suite of utilities for comparing genomic features. *Bioinformatics* **2010**, *26*, 841–842. [CrossRef] [PubMed]

53. Spehr, M.; Munger, S.D. Olfactory receptors: G protein-coupled receptors and beyond. *J. Neurochem.* **2009**, *109*, 1570–1583. [CrossRef] [PubMed]

54. Palouzier-Paulignan, B.; Lacroix, M.-C.; Aimé, P.; Baly, C.; Caillol, M.; Congar, P.; Julliard, A.K.; Tucker, K.; Fadool, D.A. Olfaction Under Metabolic Influences. *Chem. Senses* **2012**, *37*, 769–797. [CrossRef]
55. Hughes, G.M.; Boston, E.S.M.; Finarelli, J.A.; Murphy, W.J.; Higgins, D.G.; Teeling, E.C. The Birth and Death of Olfactory Receptor Gene Families in Mammalian Niche Adaptation. *Mol. Biol. Evol.* **2018**, *35*, 1390–1406. [CrossRef] [PubMed]

Article

Y Chromosome Haplotypes Enlighten Origin, Influence, and Breeding History of North African Barb Horses

Lara Radovic [1,2], Viktoria Remer [1], Carina Krcal [1], Doris Rigler [1], Gottfried Brem [1], Ahmed Rayane [3], Khadija Driss [4], Malak Benamar [5], Mohamed Machmoum [6], Mohammed Piro [6], Diana Krischke [7,8], Ines von Butler-Wemken [7] and Barbara Wallner [1,*]

[1] Institute of Animal Breeding and Genetics, University of Veterinary Medicine Vienna, 1210 Vienna, Austria
[2] Vienna Graduate School of Population Genetics, University of Veterinary Medicine Vienna, 1210 Vienna, Austria
[3] World Barb Horse Organization O.M.C.B., Organisation Mondiale Du Cheval Barbe, 148 Avenue de l'ALN Caroubier Hussein Dey Alger, Algiers 16000, Algeria
[4] Tunisian Horse Breeding Organization F.N.A.R.C., Foundation Nationale pour l'Amélioration de la Race Chevaline, Sidi Thabet 2020, Tunisia
[5] Royal Horse Breeding Association SOREC, Société Royale d'Encouragement Du Cheval, Haras Regional la Kasbah de Bouznika, B. P. 52, Bouznika 13100, Morocco
[6] Veterinary Genetic Laboratory, Hassan II, Agronomic and Veterinary Institute, B. P. 6202, Rabat 10101, Morocco
[7] Barb Horse Breeding Organization VFZB e. V., Verein der Freunde und Züchter des Berberpferdes e.V., Kirchgasse 11, 67718 Schmalenberg, Germany
[8] Department of Animal Breeding, University of Kassel, Nordbahnhofstr. 1a, 37213 Witzenhausen, Germany
* Correspondence: barbara.wallner@vetmeduni.ac.at; Tel.: +43-1-25077-5625

Citation: Radovic, L.; Remer, V.; Krcal, C.; Rigler, D.; Brem, G.; Rayane, A.; Driss, K.; Benamar, M.; Machmoum, M.; Piro, M.; et al. Y Chromosome Haplotypes Enlighten Origin, Influence, and Breeding History of North African Barb Horses. *Animals* **2022**, *12*, 2579. https://doi.org/10.3390/ani12192579

Academic Editors: Isabel Cervantes and María Dolores Gómez Ortiz

Received: 26 August 2022
Accepted: 23 September 2022
Published: 27 September 2022

Publisher's Note: MDPI stays neutral with regard to jurisdictional claims in published maps and institutional affiliations.

Simple Summary: Bred over centuries in the Maghreb region, on a corridor between the Arab and the Western world, the North African Barb horse has been touched by many influences in the course of history. The present study investigated the paternally inherited Y chromosome in today's Barbs and Arab-Barbs collected from North Africa and Europe, with the aim to link genetic patterns and narrative history. A broad Y chromosomal spectrum was observed, as well as regional disparities among populations. Y chromosomal patterns illustrated a tight connection of Barb horses with Arabians and several other breeds, including Thoroughbreds. Besides, results depict footprints of past migrations between North Africa and the Iberian Peninsula.

Abstract: In horses, demographic patterns are complex due to historical migrations and eventful breeding histories. Particularly puzzling is the ancestry of the North African horse, a founding horse breed, shaped by numerous influences throughout history. A genetic marker particularly suitable to investigate the paternal demographic history of populations is the non-recombining male-specific region of the Y chromosome (MSY). Using a recently established horse MSY haplotype (HT) topology and KASP™ genotyping, we illustrate MSY HT spectra of 119 Barb and Arab-Barb males, collected from the Maghreb region and European subpopulations. All detected HTs belonged to the Crown haplogroup, and the broad MSY spectrum reflects the wide variety of influential stallions throughout the breed's history. Distinct HTs and regional disparities were characterized and a remarkable number of early introduced lineages were observed. The data indicate recent refinement with Thoroughbred and Arabian patrilines, while 57% of the dataset supports historical migrations between North Africa and the Iberian Peninsula. In the Barb horse, we detected the HT linked to Godolphin Arabian, one of the Thoroughbred founders. Hence, we shed new light on the question of the ancestry of one Thoroughbred patriline. We show the strength of the horse Y chromosome as a genealogical tool, enlighten recent paternal history of North African horses, and set the foundation for future studies on the breed and the formation of conservation breeding programs.

Keywords: North African horse; Barb; Arab-Barb; Y chromosome; haplotype

---

1. Introduction

The history and origin of the North African horse have been long debated [1]. Still there is no confirmation of horses inhabiting Africa, or evidence of domesticated horses roaming around the continent in early prehistoric time, but discussions about an "Equus Algericus" found near Tiaret (Algeria) still remain [1,2]. However, historical and archeological findings indicate that the introduction of the domesticated horse to North Africa was likely in the late second millennium BCE, via several routes following human migrations and conquests (e.g., through Strait of Gibraltar or Egypt) [3–5].

The origin stories of the North African Barb horse lead off the Barbary coast in the Maghreb region (today's Algeria, Tunisia, Morocco), hence the name "Barb". Foremost Numidian horses and their crosses are especially discussed as founders of the breed [6,7]. Complex patterns of human and horse migrations in the North African region peaked around the 7th century, concurrent with the Muslim conquests [8,9]. Later, during the occupation of the Iberian Peninsula by the Moors, from the early 8th to the late 15th century, migrations between North Africa and Iberian Peninsula were frequently ongoing [3,8,10,11], and the influence of the North African horses onto Iberian stocks was substantial [12,13].

Numerous myths exist on the multilayer history of the Barb horse, for example phenotypic traits relate the discussion about the progenitors to Mongolian horses, as well as the rare light-colored (cream-gene) and piebald (sabino) horses, corresponding to the Turkoman and the Akhal Teke breed [1,14]. Barbs had a prominent role as war horses and for breeding in Europe [1,12]. Notably, Barb horses were used in the Punic wars (264–146 BCE) that were fought between Romans and Carthage, and later exported to Europe by Carthaginian conquests [14]. Likewise, more heavy horses were introduced to the Maghreb region first by Romans (from 146 BCE) and later in the 17th century by Louis XIV [7]. However, after the 18th century, breeding declined dramatically because Barb horses were no longer used for the military cavalry, due to the shift of military tactics that began in the 19th century [1,15]. More recently, from the end of the 19th century onwards, cross-breeding of North African and coldblooded horses from France resulted in the "Breton-Barb". In addition, crosses of the Barb horse with Thoroughbreds, Anglo-Arabs, and French Trotters in North Africa were reported [1,12,15]. Above all, systematic cross-breeding with Arabian horses founded the "Arab-Barb" breed in the Maghreb region. In the 20th century during both world wars, French colonial cavalry and later also under Rommel´s regime, captured Barb horses and this contributed to their diffusion throughout Europe [16]. Moreover, from 1965 onwards, the African horse sickness significantly reduced North African Barb horse populations and prevented horse export to Europe for over ten years from Algeria [17], and from Morocco during 1987–1991 [18].

In 1987, the "Organisation Mondiale du Cheval Barbe (OMCB)" was founded to preserve the purebred Barb horse and its cross populations ("derivates"), especially the Arab-Barb horse [19]. The OMCB is nowadays recognized as a competent authority for setting up the breeding programs. Breed registries were only recently established for Barbs and Arab-Barbs in the Maghreb region (1886 in Algeria, 1896 Tunisia, and 1914 Morocco) [1,14]. Since then, the studbooks remained open so that phenotypically classified horses can be entered retrospectively, even if no known ancestry can be proven (defined as "*Inscription à Titre Initial*", "ITI") [20]. Additionally, European registries are established (in France in 1989, Germany 1992, Switzerland 1993, and in Belgium from 1992–2017) and their studbooks are closed. Barb horses and the Arab-Barb horses are separated in different studbooks or studbook sections according to the OMCB stud-book regulations. The stud-book section for Arab-Barbs is still open for Arab/Barb crosses as well as crosses of Arab-Barbs with either Barbs or Arabs. All over, studying ancestry and breeding histories in North African horses via pedigree documentation is limited.

The census population size in the Maghreb countries is about 5500 for Barbs and 180,000 for Arab-Barbs [21,22]. Out of those, 1800 Barbs and 26,000 Arab-Barbs are registered in studbooks. In contrast, the European subpopulation constitutes about 2800 Barbs and 4000 Arab-Barbs, out of which 520 and 440 horses (Barbs and Arab-Barbs, respectively) are registered for breeding in the OMCB recognized studbooks. They produce about 160 foals per year [21,22]. The breeding programs for Barbs and Arab-Barbs are mainly based on characteristic phenotypic traits, robustness, and behavior rather than uniform breeding goals. Today, these horses are used for "Fantasia" (also known as *"Tbourida"* in Morocco and *"Mchef"* in Tunisia) a traditional equestrian war game dating back to the 16th century, as well as for agricultural work, carriage, riding, dressage, and equestrian art, as well as racing (only Arab-Barbs) in North Africa [1,12,19]. In Europe, they are used as leisure horses, for endurance-riding, historical dressage, jumping, and working equitation [1,16].

According to the diverse use and breeding areas, the North African Barb and Arab-Barb horse populations are characterized by broad phenotypic variation [1,22,23]. Within the Arab-Barbs, this strongly depends on the percentage of Arabian ancestry [24,25]. Investigation of blood group markers, protein, and DNA polymorphisms in North African subpopulations showed a pronounced genetic variation within the Barbs and the Arab-Barbs. Private alleles and high levels of heterozygosity were noted, however, no significant genetic differentiation was observed between Barb and Arab-Barb populations [26–29]. Likewise, apparent phenotypic differences distinguish the purebred Barb horse from the Arabian horse [1,23,25,30,31]. Microsatellite analysis showed similarities between the Arab-Barb and Arabian horses and a clear genetic separation of both breeds from Thoroughbreds [27–29]. The maternally inherited mitochondrial DNA showed close genetic relationships between Iberian breeds and Barb horses [11,32]. Nevertheless, the relationship between the North African Barb and the Arab horse has been continuously debated, till today [33].

A prominent genetic marker for inferring the ancestry of populations is the non-recombining, male-specific region of the Y chromosome (MSY). The MSY is inherited exclusively from the father to his sons and thus MSY haplotypes (HTs) mirror the paternal lineages in a population. MSY analysis is best established in humans where it is widely used in population genetics, genealogical research, and forensics [34–36]. In domestic horses, the MSY was long excluded from population genetic studies due to the lack of informative sequence polymorphism (reviewed in [37]). Nevertheless, a stable MSY HTs topology based on slowly evolving biallelic markers was constructed by mapping next generation sequencing (NGS) data to a 6.5 Mb horse MSY draft reference [38]. The MSY HTs of domestic horses are clearly distinct from those in the extant Przewalski's horses. The most pronounced MSY signature among domestic horses is the ~2000-year-old "Crown" haplogroup (HG), recounting various breeds from Central and South Europe, East Asia, North and South America [38,39]. It was proposed that the dominance of the Crown HG is a hallmark of the recent breeding influence of stallions of Oriental origin [38,40]. The crown topology supports the hypothesis [41,42] that only a limited number of stallions contribute to today´s horse population. Only some Asian horses [43,44] and Northern European breeds (e.g., [45]) seemed to be unaffected by the recent Oriental introgression, and thus kept their autochthonous HTs outside the Crown ("Non Crown"). Within the Crown, three HGs were defined (H, A, and T) and the HT signatures of three English Thoroughbred founders [38], as well as Arabian patrilines [39] were recently successfully delineated.

In horses, MSY analysis can unmask patrilines that contributed to a breed; thus, impart motifs of their male demography, and shed light on complex breeding histories. In this study, we investigated MSY HTs in North African Barb horses with the aim to link Y-chromosomal patterns to narratively known historical events. We hypothesize that the long-lasting input of foreign blood and complex migrations in the Maghreb region will be mirrored in their MSY HT spectrum. In addition, due to indigenous origin, regional and less intensive selection strategies [1], we might detect the preservation of autochthonous HTs in some North African horses' patrilines.

2. Materials and Methods

2.1. Sample Set

Biological samples were collected from 119 males, of Barbs (n = 84) and Arab-Barbs (n = 35) in Morocco, Algeria, Tunisia, and the European subpopulations. To ensure that many patrilines were represented in the dataset, pedigree information (available for 86 horses), provided by breeding authorities and associations, was considered in the sampling strategy as previously described [39]. Hence, oversampling of relatives was averted from the dataset by keeping six males per foundation sire at maximum. Additionally, we included 33 randomly sampled horses without pedigree information (10 European and 23 North African samples) to complement and capture population variation beyond documented patrilines. The dataset including individual male tail line information for ancestors born prior to 1990 is given in a string format in Table S1.

2.2. MSY Genotyping

We inferred MSY haplotype spectrum of 119 samples according to the previously reported horse Y phylogeny [38,39]. For genotyping, we created a downscaled HT structure based on 65 selected HT-determining variants as markers (61 SNVs, 3 short Indels, and 1 microsatellite, see Supplementary Table S2). The resulting tree served as the backbone and samples were placed onto branches of the tree via MSY marker screening.

For variant screening, genomic DNA was isolated from hair roots or blood with the nexttec® DNA Isolation Kit. The DNA was then diluted with TE buffer to the uniform concentration of 5 ng/μL. Genotyping of variants was performed using competitive allele specific PCR SNV genotyping assays (KASP™, lgcgroup.com (accessed on 2 July 2021)) following the standard protocol on a CFX96 Touch™ Real-Time PCR Detection System. Samples with known allelic state were included as positive controls, while DNA from females and non-template controls were used as negative controls. Information on variants (coordinates on LipY764, alleles, and flanking regions) are published in [38,39].

Genotyping of the amplicon length of the tetranucleotide microsatellite fBVB (GATA14 /GATA15) was performed on an ABI 3130xl Genetic Analyzer, as previously described [38]. In synopsis, for the fragment analysis, one PCR primer was tagged with FAM fluorescent dye (fwd_FAM: ACAACCTAAGTGTCTGTGAATGA; rev: CCCAATAATATTCCACT GCGTGT, expected amplicon length 204 bp). PCR was carried out in a 20 μL reaction volume containing 0.4 μM of each primer. The reaction temperature was increased to 95 °C for 5 min for initial DNA denaturation, followed by 35 cycles of 30 s at 95 °C, 40 s at 58 °C annealing temperature and 40 s at 72 °C, and a final extension step of 30 min at 72 °C. Finally, GeneMarker® was used to size the alleles relative to the internal size standard.

Genotyping was conducted in a consecutive manner by first testing the Crown determining variant rAX. If a sample carried the derived C-allele for this variant, allocation of the sample into main Crown HGs H, A, or T was conducted by testing markers fYR, rW, and rA. Each sample was then typed for the markers determining the substructure of the HG it clusters into. We then merged the genotyping information of all tested variants and imputed the allelic state of markers that were not tested or detected in the sample set according to the previously published HT structure [38,39] (see Figure 1 and Supplementary Table S2). We generated a median-joining HT network with program Network 10.2 [46] and redrew it as a HT frequency plot (Figure 1) in Canva Pro (https://www.canva.com (accessed on 29 June 2022)). Pie charts were drawn and scaled to the respective number of samples with RStudio version 4.0.3. [47].

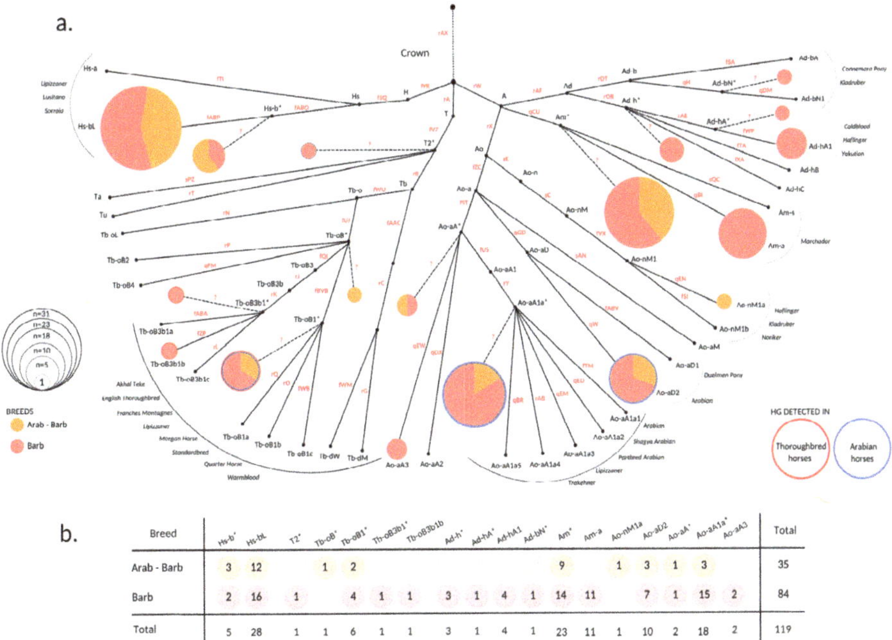

Breed	Hs-b*	Hs-bL	T2*	Tb-oB*	Tb-oB1*	Tb-oB3b1*	Tb-oB3b1b	Ad-h*	Ad-hA*	Ad-hA1	Ad-bN*	Am*	Am-a	Ao-nM1a	Ao-aD2	Ao-aA*	Ao-aA1a*	Ao-aA3	Total
Arab - Barb	3	12		1	2							9		1	3	1	3		35
Barb	2	16	1		4	1	1	3	1	4	1	14	11		7	1	15	2	84
Total	5	28	1	1	6	1	1	3	1	4	1	23	11	1	10	2	18	2	119

Figure 1. MSY haplotype spectra of North African horses. (**a**) HT frequency plot based on the MSY tree after [38,39]. HT determining variants used to construct the downscaled tree for genotyping are denoted on branches in red. Additional information is given in Supplement, Table S2. Clustering of 119 North African horses based on genotyping result is illustrated as pies. Pie radiuses are scaled to the number of allocated individuals and colors of the portions correspond to different breeds. HG names are labeled accordingly. HTs located on internal nodes are denoted with an asterisk (*) and trailed with dashed lines that originate from corresponding internal nodes. Unascertained variants that would determine * HTs are denoted with question marks (?). HTs framed with blue and/or red borders denote that they were detected previously in Arabian (blue border) and Thoroughbred (red border) horses [38,39]. Non-colored points express HTs that were not detected in the North African sample set. Gray list on the sides of the network indicates the breeds the HTs were previously reported [38–40]; (**b**) Number of individuals that allocate within detected HTs. Sample information details are given in Supplement Table S1.

3. Results

To investigate the MSY HT spectra of North African horses, 119 males representing 84 Barbs and 35 Arab-Barbs were genotyped. The results showed that all samples allocated into the Crown HG. In total, we distinguished 18 HTs and all three previously defined Crown HGs (A, H, and T) were represented in our sample set. The broad Crown MSY HT spectra was comparable in Barbs and Arab-Barbs (Figure 1). This is in contrast to patterns in other today's breeds [38,39] that showed distinct clustering on the tree. Remarkably, only half of the males analyzed carried defined HTs, whereas 61 males got placed at internal nodes of the backbone topology (See Figure 1 and Table S2). The samples allocated at inner nodes are marked with an asterisk (*) in their HT identifier and distinguished with dashed lines in Figure 1. For instance, the sample that allocates into Tb-oB* HT carried the derived allele for the fUJ marker and was placed onto the branch Tb-oB, but it carried the ancestral allele at the markers determining subsequent HTs in our backbone tree (rP, qFM, fQI, and fBVB). The inner node clustering of samples occurs when the HT of the horse is not represented by the tree due to ascertainment bias, and only the HG and the branching point could be determined.

More than half (56%, n = 67) of the analyzed individuals are distributed acros
two HGs, Am (n = 34) and Hs-b (n = 33), respectively (see Figure 1). Other than North
African horse, these HGs were so far only detected in some South American and Iberian
breeds [35,36]. Besides, we observed grouping of 28 (24%) males into Ao-aA1a* and Ao
aD2 HTs. Those HTs were designated recently as signatures for Arabian horses [39]. The
arrangement of the internal branching points in the strictly hierarchical MSY HT tree
topology reflects the emergence of the mutations over time. Hence, the HTs Ao-aA* (n = 2
and Ao-aA3 (n = 2) can be interpreted as hints to earlier introduced lines of presumably
Arabian origin, that evolved and are still preserved in the North African Barb horse. We
further aggregated ten males in the Tb HG. Among those, two males clustered onto early
branching points (T2* and Tb-oB*) and six were allocated in the HT Tb-oB1*. This HT
was previously reported in Akhal Teke, Turkoman, Thoroughbreds, as well as Arabian
horses [38–40]. The Tb-oB1* in North African horses can be explained as the recent influ
ence of stallions from that region. Noteworthy, we detected Tb-oB3b1*, the HT basal to
the HTs detected in the progeny of the Thoroughbred´s founder sire 'Godolphin Arabian
which are (Tb-oB3b1a/b/c) [38], in a Barb breeding stallion from Morocco. We found the
signature of recent influence of Warmblood or Thoroughbred in a single horse from France
carrying Tb-oB3b1b, but did not observe the typical Thoroughbred and Trotter HGs Tb-dW
and Tb-dM [38,40]. Moreover, ten males carried HGs, which are today mainly found in
Coldbloods and European Ponies [39,40], namely Ad-h (8), Ad-b (1), and Ao-n (1). Here
we again observed well resolved HTs (for example Ad-hA1), as well as earlier branching
off HTs (Ad-bN*, Ad-h*).

Roughly half of our sample set was collected in Europe and the other half in Algeria
Morocco, and Tunisia (see Figure 2 and Table S1). The samples from Algeria and Morocco
clustered in 8 HTs each. The European samples clustered into 16 HTs. Seven HTs were
represented only within this population group in our sample set, noting that two HTs
Ad-hA* and Tb-oB*, were detected in ITI horses directly imported from, respectively
Algeria and Morocco. The broad HT spectrum detected in samples collected in Algeria
Morocco, and Europe was not corroborated by Tunisian data. All collected Barbs (n = 9
and Arab-Barbs (n = 2) from Tunisia and all males exported from Tunisia to Europe (see
below) allocated into HT Hs-bL.

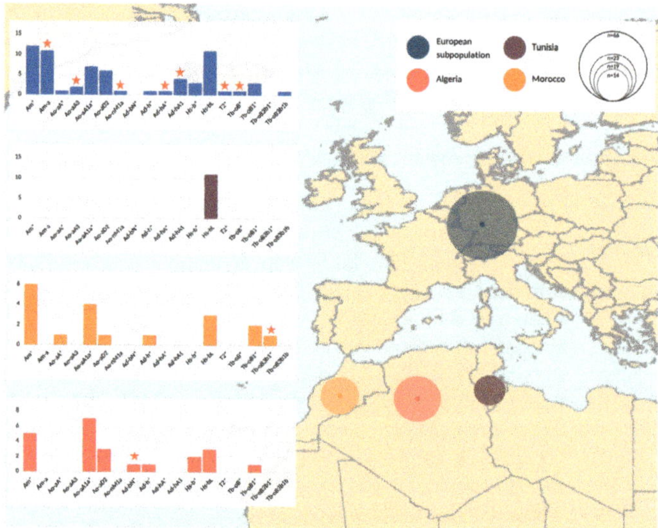

Figure 2. Geographical representation of MSY haplotypes. Populations analyzed are denoted with different

colors and circles on the map correspond to the sample size. Details are given in Supplement, Table S1. Summary information of genotyping results and regional differences are visualized with bar plots. The x axis on the bar plots corresponds to detected HTs, while the y axis indicates number of samples that correspond to each of the bars (HTs). The samples assigned to inner nodes are marked with an asterisk (*) in their HT identifier. Red stars indicate HGs that were found exclusively in the corresponding subpopulation (e.g., seven HGs denoted with red stars in the European subpopulation are found only among samples collected in European countries, and were not observed in samples from Maghreb countries).

Among the 66 European samples, nine were collected from horses imported from Algeria (4), Morocco (4), or Tunisia (1). Complementing pedigree information was available for another 56 European samples (see Supplementary Table S1). This documentation reveals that the majority, namely 50, of the European males also directly trace back paternally to Maghrebian stallions exported from Algeria, Morocco, and Tunisia to Europe during the last 35 years (see Figure 3 and Supplementary Table S1). Hence, only seven out of the 66 males in the European dataset could not be linked explicitly to a hitherto known Maghrebian line from documented records. Among those, five individuals descend from four stallions, who were inscripted as ITI in the course of the foundation of the French studbook in 1989. For one sample, we had no pedigree information, and for one founder, the country of origin was unknown (see Supplementary Table S1).

Overall, the full dataset (n = 119) included 33 individuals without pedigree information (10 European and 23 horses from Maghreb) and the HT pattern in horses with and without pedigree were comparable (Supplementary Table S1).

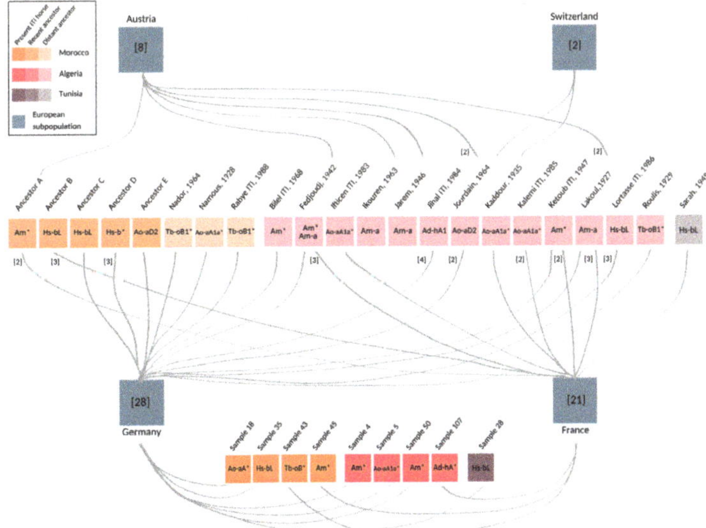

Figure 3. Maghrebian roots of European stallions. Fifty-nine European individuals, who were imported or their patrilines trace back to North Africa, are grouped based on their current registry (blue boxes). Number of horses included from Austria (n = 8), Switzerland (n = 2), Germany (n = 28), and France (n = 21) are denoted in square brackets. The paternal ancestors of the sampled individuals several generations back in time, as well as present individuals imported to France and Germany, are shown as colored boxes. The opacity of boxes indicates temporal layers whereas the brightest boxes on the bottom are present ITI horses, followed by recent ancestors born after 1990 (Ancestor A, B, C, D, and E), and lightest colored distant ancestors, in the middle. Name and year of the birth of ancestors is given for distant ancestors. MSY HTs, revealed from the European progeny, are shown within each stallion's box. HT identifiers attributed with asterisk (*) denote inner node clustering. The grey lines

connect the stallions with their descendants sampled in the respective European countries. Number in the brackets and adjacent to the left side of connection lines represent the number of descendant from each stallion found in European samples, if different from one. Pedigree details and full list o samples are given in Supplementary Table S1.

4. Discussion

The significant role of North Africa, as a transit route, during the Islamic conquest anc migratory movements between countries of the region [3,14], raised our interest on the Y chromosomal signature of North African Barb horses. While the MSY HT signatures of the Arabian and the Thoroughbred and their recent breeding influences are well described [38,39] the historically impactful North African horse remains enigmatic. We applied MSY haplotyping in a total of 84 Barbs and 35 Arab-Barbs, whereas half of our samples were collected in Europe and the other half in Algeria, Tunisia, and Morocco (see Figure 2 and Supplementary Table S1 and hypothesized that the MSY signature will mirror the variety of encountered influences. On the other hand, due to the documented indigenous origin and regional subgroups in North Africa, we expected partial representation of autochthonous patrilines.

The results of haplotyping indicate that no distantly related lineages were retainec in the collected sample set since all horses clustered within the Crown HG. In line with previously determined time to the most recent common ancestor [38], we can state tha the MSY of North African horses only reflects the last 1500 years of population history The sole detection of the Crown mirrors influences of Oriental stallions [40]. Interestingly we report a broad HT spectrum of North African horses across the Crown HGs (18 HTs) However, unlike other breeds (like Arabians and Thoroughbreds), for which it was possible to pin-point characteristic HGs and even discriminate discrete sublines with the use of pedigrees [38,39], the diffused HT distribution result in a tangled MSY footprint of North African horses. The observed preservation of a variety of HTs may be the consequence of less intensive selection on males and different breeding goals in North African regions Interestingly, MSY results were comparable in Barbs and Arab-Barbs. This verifies the inter-crossing and gene flow till today between the North African horse populations, as already depicted with autosomal genetic markers [27,28,48].

However, the broad HT spectrum was not supported from Tunisian samples (n = 11) where all nine Barbs and two Arab-Barbs were monomorphic, carrying a single HT (Hs-bL (Figure 2). This may demonstrate geographical disparities in breeding goals, supported by regional differences reported in the phenotype [1,23,30], as well as genetic spatial interpolation (e.g., [27]). In contrast, the analysis of microsatellites resulted in similarity of Moroccan and Tunisian Barb horse populations [29]. Regional differences are highlighted when we compare the HTs represented in Europe to the Maghreb region. Samples from European countries harbored seven HTs that were not represented in the samples collected in North Africa. Three of those patrilines were imports from North Africa after 2001 and four HTs trace back to the French ITI-inscriptions in 1989. Their private HTs may be explained with geographical separation of former exports to France. Additionally, we found two HTs each private for Moroccan and Algerian Barbs (Tb-oB3b1* and Ad-bN* respectively). Compared to Tunisia, we observe similar MSY patterns in Europe, Morocco and Algeria. One explanation for greater similarity of HTs among the latter three could be the tighter historical connection between those regions (export especially of ITI horses from Morocco and Algeria to Europe as seen in Figure 3). Nevertheless, we should interpret these findings with caution since it is possible that despite our efforts to collect a representative sample set from the Maghreb, the numbers of horses available from Tunisia was lower (n = 11). Hence, we could have underestimated HT diversity in that region.

All we see today is what is left throughout the time, and the MSY is a perfect tool to trace patrilines that shaped present populations. The relationship between the North African Barb and the Arabian horse has been continuously debated [33]. We noted a prominent clustering to Ao-aA1*, a HG previously detected in Arabian lines [39]. The detection of numerous Arabian HTs demonstrates the significant influence of Arabian stallion lines in Barbs and Arab-Barbs. A clear Arabian signature was visible in about a third of the analyzed samples. For the Arab-Barbs, the results are not surprising since the breed is based on Barbs refined with Arabians [49]. On the other hand, assignment of "purebred Barbs" to Arabian HGs may reflect, as hypothesized, recent historical migratory movements resulting in admixture, because the studbooks for the "purebred Barbs" are still open in North Africa and stallions without pedigrees are used for breeding.

Two third of the analyzed samples (85 North African horses) did not carry the Arabian signature HTs. Particularly interesting is that among those were 27 Arab-Barbs. In addition, we detected indications of recent upgrading with European Coldbloods in four males (Ad-hA1), which could be explained with the discussed influence of Coldblood stallions imported to North Africa [12]. Moreover, only a single individual carried an unambiguous sign of Warmblood or Thoroughbred male ancestry (Tb-oB3b1b) [38].

Barbs were used for upgrading and formation of many modern breeds [12,50]. There have also been reports on their contribution to Thoroughbreds, Anglo-Arabs, and French Trotters. Interestingly, North African horses´ HTs share branching points basal to the HTs observed in many todays Coldblood, British and European Ponies (Figure 2; detected in Ad-h, Ad-b, and Ao-n HTs) [39,40], which can be interpreted as the influence of the North African horses had on those breeds further back in time. Deeper investigation is needed to validate the proposed correlation.

A particularly remarkable finding was the observation of the HT basal to the HTs spread through the Godolphin Arabian sire line (Tb-oB3b1*) [38] in a Barb horse. There is still controversy about the ancestry of Godolphin Arabian, one of the foundation sires of the English Thoroughbred (exported from Tunisia to France in 1731). He is often referred to as Godolphin Barb due to his North African origin [51] and phenotypic marks different from the Arabian horse [1,12,13]. The MSY finding, namely detection of the basal Godolphin Barb HT in a Barb horse, again fuels the discussion on the origin of Godolphin Arabian, whether he was a Turkoman stallion with partial Arabian blood [52] or corroborates the hypothesis that the Tb-oB3b1 HG made its way into the Thoroughbred via the Barb horse [1,49].

When we look further back in time, from the Carthaginian civilization in the 1st century and Muslimic conquests in the 7th century to recurrent migrations with Iberian Peninsula (8th to 15th century), North Africa served as a main migratory route for many cultures [3,14]. Every culture that was present in the region could have left footprints in the horses' genomes, and this was depicted on the MSY. Notably, influence from the Middle East could be attributed to inner clustering of individuals to Ao-aA* and Tb-oB*, as well as allocation to Ao-aA3, Ao-aD2, Tb-oB1*, and T2* HGs. This grouping may indicate previously discussed influence of the ancestors of Arabian and Turkoman lineages on North African horses.

From the viewpoint of interactions between the North African regions and the Iberian Peninsula, previous research delineated homogenous mtDNA patterns within ancient [53] and modern [11,32] horse populations in Iberia and North Africa. Particularly, it is speculated that Barb and Iberian horses have a common origin [54]. A great number of North African horses that were analyzed [32] shared mtDNA HTs reported in South American and Iberian breeds. Accordingly, we note that two highly frequent HGs (Am and Hs-b) represented in our dataset also allocate Iberian and New World horse breeds, like Marchador (Am), Lusitano, and Sorraia (Hs-b) [39]. Iberian and New World breeds are not yet comprehensively studied for their MSY HTs, but the preliminary joint clustering could reflect the gene flow and recent shared ancestry of North African Barb and Iberian horses. However, to fully explain the assumed shared ancestry further back in time, as well as the magnitude of gene flow, and indices on New World horses´ ancestry, we should complement the dataset with additional Iberian and New World horse breeds in the future. Early separated populations, like the West African Barb, the Spanish Barb (USA), and South American breeds, as well as ancient DNA samples from the Maghreb, should enlighten another chapter in horse history. Additionally, basal allocation of samples in the tree topology and underrepresentation of private HTs (Figure 1) raises a discussion on technical limitation of our analysis. The MSY backbone topology was constructed based on the ascertainmen panel from [39], where five Barbs and one Arab-Barb were sequenced. However, it seem this is still insufficient, and more individuals need to be sequenced in order to clarify MS' signatures private for North African horses, in particular in HGs Hs-b, and Am.

Overall, North African horses retained the print of the "early Oriental influence starting with the Muslim conquests. With the observed broad HT spectrum, these horse could be a reservoir of genetic diversity—although their population is small. Furthe investigation of additional males, especially from the Maghreb regions, is needed to precise influential patrilines, as this is of particular practical interest for breeding. The MS' patterns should be considered together with autosomal markers, as well as mitochondria DNA, while constructing necessary conservation breeding programs, to preserve the Nortl African Barb horse.

5. Conclusions

Our study highlights the value of the Y chromosome analysis for horse populatior genetics and for the first time, enlightens recent paternal population history of the Nortl African Barb horses. Obtained MSY HT spectra point to, on the one hand, that stallions were probably wide-spread hundreds of years preceding the formation of modern horse breeds, and on the other hand, indicate the impact on historical migrations and recen upgrading. However, with our approach, it is at the moment not possible to pin-poin where and when the ancestors of North African Barbs came from, as well as the directior of gene flow. Future analysis on ancient DNA, as well as inclusion of more diverse Bart populations, are essential for dating of the origin of HGs, and exact inference of genetic influences. In addition, the ascertainment bias represented with HTs that are not fully resolved indicates that, even though the Crown is well described, there is still a lot lef to explore in future research. Finally, our findings enhanced our knowledge of paterna. ancestry of the breed and provided basis for future work and establishment of conservatior breeding programs.

Supplementary Materials: The following supporting information can be downloaded at: https //www.mdpi.com/article/10.3390/ani12192579/s1, Table S1: Sample set information; Table S2 Genotyping results; Table S3: Variant information.

Author Contributions: Conceptualization of the project was completed by L.R., V.R., I.v.B.-W. and B.W. Funding acquisition was performed by B.W. Methodology was determined by L.R. and B.W Data collection was performed by L.R., V.R., C.K. and D.R. Resources were provided by A.R., K.D. M.B., D.K., I.v.B.-W. and B.W. Data analysis was performed by L.R., V.R. and B.W. The original draf was written by L.R. and B.W. Writing—review and editing was done by L.R., V.R., D.R., G.B., A.R. K.D., M.B., M.M., M.P., D.K., I.v.B.-W. and B.W. Supervision of the project was performed by B.W. All authors have read and agreed to the published version of the manuscript.

Funding: This research received no external funding.

Institutional Review Board Statement: The study was discussed and authorized by the University of Veterinary Medicine Vienna´s institutional ethics and welfare committee in accordance with GSP guidelines and national legislation (ETK-10/05/2016). The study was carried out in compliance with the applicable guidelines specified in the preceding document.

Informed Consent Statement: The biological material (hair roots or blood samples) and permission of use were obtained from breeding associations and private horse owners. If pedigree data were available, informed consent was acceded. All samples used in the study are coded.

Data Availability Statement: Not applicable.

Acknowledgments: The authors thank all horse owners who contributed with samples. We thank GeneControl GmbH Grub, Germany for providing retained hair samples from the VFZB registered male horses, and especially Petra Jürgens (Germany), Jessica Pfeiffer (Austria), Veronika Leichtfried (Austria), Claudia Lazzarini (Switzerland), Caroline Duffeau (France), Claire Martin (France), and

Bénédicte Fournel (France). We also thank Royal Horse Breeding Association (SOREC, Morocco) for the support.; Open Access Funding by the University of Veterinary Medicine Vienna.

Conflicts of Interest: The authors declare no conflict of interest. The funders had no role in the design of the study; in the collection, analyses, or interpretation of the data; in the writing of the manuscript or in the decision to publish the results. The statements made herein are solely the responsibility of the authors.

References

1. Kadri, A. *The Barb, A Legendary Horse*; CPS Editions—Zaki Bouzid Editions: El Madania, Algiers, Algeria, 2009; ISBN 978-9961-771-11-2.
2. Chaid-Saoudi, Y. La préhistoire du cheval en afrique du nord. In *Le Cheval Barbe*; Organisation Mondiale du Cheval Barbe, Ed.; Caracole: Lausanne, Switzerland, 1989; pp. 35–39.
3. Gonzaga, P.G. *A History of the Horse: The Iberian Horse from Ice Age to Antiquity*; Allen, J.A. & Co. Ltd.: London, UK, 2003; Volume 1, pp. 173–231. ISBN 0851318673.
4. Blench, R.M. A survey of ethnographic and linguistic evidence for the history of livestock in Africa. In *The Archaeology of Africa: Ethnographic and Linguistic Evidence for the Prehistory of African Ruminant Livestock, Horses and Ponies*, 1st ed.; Shaw, T., Sinclair, P., Andah, B., Okpoko, A., Eds.; Routledge: London, UK, 1994; pp. 71–103. ISBN 9780203754245.
5. González-Fortes, G.; Tassi, F.; Trucchi, E.; Hanneberger, K.; Paijamans, J.L.A.; Díez-del-Molino, D.; Schroeder, H.; Susca, R.R.; Barroso-Ruíz, C.; Bermudez, F.J.; et al. A western route of prehistoric human migration from Africa into the Iberian Peninsula. *Proc. R. Soc. B Biol. Sci.* **2019**, *286*, 1895. [CrossRef]
6. Azzaroli, A. La Cavalerie Numide. In *Le Cheval Barbe*; Organisation Mondiale du Cheval Barbe, Ed.; Caracole: Lausanne, Switzerland, 1989; pp. 40–46.
7. Gaudois, M. Le cheval barbe et ses croisements. In *Le Cheval Barbe*; Organisation Mondiale du Cheval Barbe, Ed.; Caracole: Lausanne, Switzerland, 1989; pp. 85–93.
8. UNESCO International Scientific Committee. *General History of Africa III: Africa from the Seventh to the Eleventh Century*; Fasi, M.E., Hrbek, I., Eds.; UNESCO Publishing: Paris, France, 2000; Volume 3, ISBN 92-3-101-709-8.
9. Arauna, L.R.; Mendoza-Revilla, J.; Mas-Sandoval, A.; Izaabel, H.; Bekada, A.; Benhamamouch, S.; Fadhlaoui-Zid, K.; Zalloua, P.; Hellenthal, G.; Comas, D. Recent Historical Migrations Have Shaped the Gene Pool of Arabs and Berbers in North Africa. *Mol. Biol. Evol.* **2017**, *34*, 318–329. [CrossRef]
10. UNESCO International Scientific Committee. *General History of Africa, IV: Africa from the Twelfth to the Sixteenth Century*; Niane, D.T., Ed.; UNESCO Publishing: Paris, France, 2000; Volume 4, ISBN 92-3-101-710-1.
11. Royo, L.J.; Álvarez, I.; Beja-Pereira, A.; Molina, A.; Fernández, I.; Jordana, J.; Gómez, E.; Gutiérrez, J.P.; Goyache, F. The Origins of Iberian Horses Assessed via Mitochondrial DNA. *J. Hered.* **2005**, *96*, 663–669. [CrossRef] [PubMed]
12. Rayane, A. Le Barbe à l´origine de la création d´autres races. In *L´Histoire du Petit Cheval d´Afrique du Nord- Le Barbe dans sa Première Grandeur*; SATINFO Spa-Filiale Sonelgaz: Djasr Kasentina, Algeria, 2015; pp. 41–63. ISBN 978-9947-0-4308-0.
13. Barbie de Préaudeau, P. La fortune du cheval barbe en occident. In *Le Cheval Barbe*; Organisation Mondiale du Cheval Barbe, Ed.; Caracole: Lausanne, Switzerland, 1989; pp. 51–57.
14. Waldron Grutz, J. The Barb. *Saudi Aramco World* **2007**, *58*, 8–17. Available online: https://archive.aramcoworld.com/issue/2007 01/the.barb.htm (accessed on 23 December 2021).
15. Bogros, D. Les stud-books du cheval barbe: Histoire et perspective. In *Le Cheval Barbe*; Organisation Mondiale du Cheval Barbe, Ed.; Caracole: Lausanne, Switzerland, 1989; pp. 66–72.
16. Krischke, D. Selektion in kleinen Populationen am Beispiel des Berberpferdes. Master's Thesis, University of Göttingen, Göttingen, Germany, 2011.
17. Gourand, J.-L. Les sept vies du cheval barbe. In Proceedings of the Conférences Scientifiques, El Jadida, Morocco, 23 October 2010; pp. 8–23.
18. Dennis, S.J.; Meyers, A.E.; Hitzeroth, I.I.; Rybicki, E.P. African Horse Sickness: A Review of Current Understanding and Vaccine Development. *Viruses* **2019**, *11*, 844. [CrossRef] [PubMed]
19. Khaled, H. L'organisation mondiale du cheval barbe, l'assemblée constitutive. In *Le Cheval Barbe*; Organisation Mondiale du Cheval Barbe, Ed.; Caracole: Lausanne, Switzerland, 1989; pp. 162–169.
20. Seghrouchni, M. Protocoles de réglement de l´inscription à titre initial (ITI) au stud-book des chevaux Barbes et Arabe-Barbes. In Proceedings of the Conférences Scientifiques, El Jadida, Morocco, 24 October 2009; pp. 13–14.
21. Yaaraf, M.; El Kohen, M. Le cheval barbe situation actuelle et perspectives. In Proceedings of the Conférences Scientifiques, El Jadida, Morocco, 24 October 2009; pp. 7–12.
22. El Kohen, M. Réflexions sur le standard du cheval Barbe. In Proceedings of the Conférences Scientifiques, El Jadida, Morocco, 23 October 2010; pp. 24–29.
23. Mebarki, M.; Kaidi, M.; Benhenia, K. Morphometric description of Algerian Arab-Barb horse. *Rev. Méd. Vét.* **2018**, *169*, 180–184.
24. Guedaoura, S.; Cabaraux, J.-F.; Moumene, A.; Tahraoui, A.; Nicks, B. Evaluation morphometrique de chevaux de race barbe et derives en Algerie. *Ann. Méd. Vét* **2011**, *155*, 14–22. Available online: https://www.vfzb.de/app/download/10966262626/Barbe_Agerie+2011.pdf?t=1431942192 (accessed on 18 January 2022).

25. Boujenane, I.; Touati, I.; Machmoum, M. Mensurations corporelles des chevaux Arabe-Barbes au Maroc. *Rev. Méd. Vét.* **2008**, *159*, 144–149. Available online: https://www.vfzb.de/app/download/10966264626/AB_Maroc.pdf?t=1431942192 (accessed on 18 January 2022).

26. Ouragh, L.; Mériaux, J.-C.; Braun, J.-P. Genetic blood markers in Arabian, Barb and Arab-Barb horses in Morocco. *Anim. Genet.* **1994**, *25*, 45–47. [CrossRef] [PubMed]

27. Berber, N.; Gaouar, S.; Leroy, G.; Kdidi, S.; Tabet Aouel, N.; Säidi Mehtar, N. Molecular characterization and differentiation of five horse breeds raised in Algeria using polymorphic microsatellite markers. *J. Anim. Breed. Genet.* **2014**, *131*, 387–394. [CrossRef]

28. Jemmali, B.; Haddad, M.M.; Barhoumi, N.; Tounsi, S.; Lasfer, F.; Trabelsi, A.; Ben Aoun, B.; Gritli, I.; Ezzar, S.; Younes, A.B.; et al. Genetic diversity in Tunisian horse breeds. *Arch. Anim. Breed.* **2017**, *60*, 153–160. [CrossRef]

29. Piro, M.; Alyakine, H.; Ezzaouia, M.; Lasfar, F.; Ahmed, H.O.; Ouragh, L. Analyse génétique et relations phylogénétiques du cheval Barbe par l' utilisation des microsatellites. *Rev. Mar. Sci. Agron. Vét.* **2019**, *7*, 149–157.

30. Chabchoub, A.; Landolsi, F.; Jary, Y. Étude des paramètres morphologiques de chevaux Barbes de Tunisie. *Rev. Méd. Vét.* **2004**, *155*, 31–37. Available online: https://www.vfzb.de/app/download/10966227726/Barbes_Tunesie_RMV.pdf?t=1431942192 (accessed on 19 January 2022).

31. Rahal, K.; Guedioura, A.; Oumouna, M. Paramètres morphométriques du cheval barbe de Chaouchaoua, Algérie. *Rev. Méd. Vét.* **2009**, *160*, 586–589. Available online: https://www.vfzb.de/app/download/10966235026/Barbe_Tiaret.pdf?t=1431942192 (accessed on 16 January 2022).

32. Jansen, T.; Forster, P.; Levine, M.A.; Oelke, H.; Hurles, M.; Renfrew, C.; Weber, J.; Olek, K. Mitochondrial DNA and the origins of the domestic horse. *Proc. Natl. Acad. Sci. USA* **2002**, *99*, 10905–10910. [CrossRef]

33. Tobey, E. The Palio Horse: Object of trade and diplomacy. In *The Culture of the Horse: Status, Discipline, and Identity in the Early Modern World*; Raber, K., Tucker, T.J., Eds.; Palgrave Macmillan: New York, NY, USA, 2016; pp. 67–68. ISBN 1403966214.

34. Jobling, M.A.; Tyler-Smith, C. Human Y-chromosome variation in the genome-sequencing era. *Nat. Rev. Genet.* **2017**, *18*, 485–497. [CrossRef]

35. Kayser, M. Forensic use of Y-chromosome DNA: A general overview. *Hum. Genet.* **2017**, *136*, 621–635. [CrossRef]

36. Underhill, P.A.; Kivisild, T. Use of y chromosome and mitochondrial DNA population structure in tracing human migrations. *Annu. Rev. Genet.* **2007**, *41*, 539–564. [CrossRef]

37. Raudsepp, T.; Finno, C.J.; Bellone, R.R.; Petersen, J.L. Ten years of the horse reference genome: Insights into equine biology, domestication and population dynamics in the post-genome era. *Anim. Genet.* **2019**, *50*, 569–597. [CrossRef]

38. Felkel, S.; Vogl, C.; Rigler, D.; Dobretsberger, V.; Chowdhary, B.P.; Distl, O.; Fries, R.; Jagannathan, V.; Janečka, J.E.; Leeb, T.; et al. The horse Y chromosome as an informative marker for tracing sire lines. *Sci. Rep.* **2019**, *9*, 6095. [CrossRef]

39. Remer, V.; Bozlak, E.; Felkel, S.; Radovic, L.; Rigler, D.; Grilz-Seger, G.; Stefaniuk-Szmukier, M.; Bugno-Poniewierska, M.; Brooks, S.; Miller, D.C.; et al. Y-Chromosomal Insights into Breeding History and Sire Line Genealogies of Arabian Horses. *Genes* **2022**, *13*, 229. [CrossRef]

40. Wallner, B.; Palmieri, N.; Vogl, C.; Rigler, D.; Bozlak, E.; Druml, T.; Jagannathan, V.; Leeb, T.; Fries, R.; Tetens, J.; et al. Y Chromosome Uncovers the Recent Oriental Origin of Modern Stallions. *Curr. Biol.* **2017**, *27*, 2029–2035.e5. [CrossRef]

41. Lindgren, G.; Backström, N.; Swinburne, J.; Hellborg, L.; Einarsson, A.; Sandberg, K.; Cothran, G.; Villà, C.; Binns, M.; Ellegren, H. Limited number of patrilines in horse domestication. *Nat. Genet.* **2004**, *36*, 335–336. [CrossRef]

42. Wutke, S.; Sandoval-Castellanos, E.; Benecke, N.; Döhle, H.-J.; Friederich, S.; Gonzalez, J.; Hofreiter, M.; Lõugas, L.; Magnell, O.; Malaspinas, A.S.; et al. Decline of genetic diversity in ancient domestic stallions in Europe. *Sci. Adv.* **2018**, *4*, eaap9691. [CrossRef]

43. Felkel, S.; Vogl, C.; Rigler, D.; Jagannathan, V.; Leeb, T.; Fries, R.; Neuditschko, M.; Rieder, S.; Velie, B.; Lindgren, G.; et al. Asian horses deepen the MSY phylogeny. *Anim. Genet.* **2018**, *49*, 90–93. [CrossRef]

44. Han, H.; Wallner, B.; Rigler, D.; MacHugh, D.E.; Manglai, D.; Hill, E.W. Chinese Mongolian horses may retain early domestic male genetic lineages yet to be discovered. *Anim. Genet.* **2019**, *50*, 399–402. [CrossRef]

45. Castaneda, C.; Juras, R.; Khanshour, A.M.; Randlaht, I.; Wallner, B.; Rigler, D.; Lindgren, G.; Raudsepp, T.; Cothran, E.G. Population genetic analysis of the Estonian native horse suggests diverse and distinct genetics, ancient origin and contribution from unique patrilines. *Genes* **2019**, *10*, 629. [CrossRef]

46. Bandelt, H.J.; Forster, P.; Röhl, A. Median-joining networks for inferring intraspecific phylogenies. *Mol. Biol. Evol.* **1999**, *16*, 37–48. [CrossRef]

47. RStudio Team. *RStudio: Integrated Development for R. RStudio*; PBC: Boston, MA, USA, 2020; Available online: http://www.rstudio.com/ (accessed on 27 April 2022).

48. Jemmali, B.; Haddad, M.M.; Lasfer, F.; Ben Aoun, B.; Ezzar, S.; Kribi, S.; Ouled Ahmed, H.; Ezzaouia, M.H.; Rekik, B. Investigation de la diversité génétique des races Barbe et Arabe Barbe en Tunisie. *J. New Sci. Agric. Biotechnol.* **2015**, *21*, 948–956. Available online: https://www.jnsciences.org/agri-biotech/29-volume-21/99-investigation-de-la-diversite-genetique-des-races-barbe-et-arabe-barbe-en-tunisie.html (accessed on 11 January 2022).

49. Hendricks, B. *International Encyclopedia of Horse Breeds*; University of Oklahoma Press: Norman, OK, USA, 2007; ISBN 9780806138848.

50. Goldstein, H. *The Earth Walkers: Horses & Humans—Our Journey Together on Planet Earth*; Balboa Press: Bloomington, IN, USA, 2019; ISBN 1982229241.

51. Hunter, A. *American Classic Pedigrees (1914–2002)*; Blood-Horse Publications: Lexington, KY, USA, 2003; pp. 9–10.

2. Mackay-Smith, A. *Speed and the Thoroughbred: The Complete History, XXVII*; Lanham, M.D., Ed.; Derrydale Press: New York, NY, USA, 2000; p. 193. ISBN 1586670409.
3. Matoso Silva, R.; Bower, M.; Detry, C.; Valenzuela, S.; Fernández-Rodríguez, C.; Davis, S.; Riquelme Cantal, J.A.; Álvarez-Lao, D.; Arruda, A.M.; Viegas, C.; et al. Tracing the History of the horse in Iberia and North Africa through ancient DNA. In *Chromatography and DNA Analysis in Archaeology*; Oliviera, C., Morais, R., Morillo Cerdán, A., Eds.; NPRINT: Esposende, Portugal, 2015; pp. 217–227. ISBN 978-989-99468-1-1.
4. De la Vega Sedano, J.F. Analogies et différences entre le Barbe et l´Espagnol. In *Le Cheval Barbe*; Organisation Mondiale du Cheval Barbe, Ed.; Caracole: Lausanne, Switzerland, 1989; pp. 94–101.

Article

A *KIT* Variant Associated with Increased White Spotting Epistatic to *MC1R* Genotype in Horses (*Equus caballus*)

Laura Patterson Rosa [1,*], Katie Martin [1], Micaela Vierra [1], Erica Lundquist [1], Gabriel Foster [1], Samantha A. Brooks [2] and Christa Lafayette [1,*]

[1] Etalon, Inc., Menlo Park, CA 94025, USA; khoefs@etalondx.com (K.M.); mvierra@etalondx.com (M.V.); elundquist@etalondx.com (E.L.); gfoster@etalondx.com (G.F.)

[2] Department of Animal Science, UF Genetics Institute, University of Florida, Gainesville, FL 32610, USA; samantha.brooks@ufl.edu

* Correspondence: laura.patterson@sulross.edu (L.P.R.); clafayette@etalondx.com (C.L.); Tel.: +1-650-380-2995 (C.L.)

Citation: Patterson Rosa, L.; Martin, K.; Vierra, M.; Lundquist, E.; Foster, G.; Brooks, S.A.; Lafayette, C. A *KIT* Variant Associated with Increased White Spotting Epistatic to *MC1R* Genotype in Horses (*Equus caballus*). *Animals* **2022**, *12*, 1958. https://doi.org/10.3390/ani12151958

Academic Editors: Isabel Cervantes and María Dolores Gómez Ortiz

Received: 21 June 2022
Accepted: 1 August 2022
Published: 2 August 2022

Publisher's Note: MDPI stays neutral with regard to jurisdictional claims in published maps and institutional affiliations.

Simple Summary: Although over 40 genetic variants are known to influence white spotting and markings on the horse, many still have unknown genetic causes. Furthermore, some seem to be influenced by pigmentation (base coat color) genes. We investigated two horses demonstrating a heritable white spotting pattern of no known genetic cause. Through sequencing of the coding region of candidate genes, we identified a mutation in the *KIT proto-oncogene, receptor tyrosine kinase* (*KIT*) gene changing the coded amino acid, predicted to be deleterious to protein function. We further evaluated this variant in a population of 147 horses, characterized using photographs scored by three independent observers using a standardized Average Grade of White score. The *KIT* mutation is significantly associated with a quantitative increase in white pattern ($p = 3.3 \times 10^{-12}$) and demonstrates an influence of the *MC1R Extension* locus. We also report a complete link between the previously reported *KIT W19* allele and this mutation. We propose to name this mutation *W34* following established nomenclature. Given the quantitative effect on white markings and *MC1R* influence, genetic testing for this allele can be of value for horse owners that desire to select for white patterns.

Abstract: Over 40 identified genetic variants contribute to white spotting in the horse. White markings and spotting are under selection for their impact on the economic value of an equine, yet many phenotypes have an unknown genetic basis. Previous studies also demonstrate an interaction between *MC1R* and *ASIP* pigmentation loci and white spotting associated with *KIT* and *MITF*. We investigated two stallions presenting with a white spotting phenotype of unknown cause. Exon sequencing of the *KIT* and *MITF* candidate genes identified a missense variant in *KIT* (rs1140732842, NC_009146.3:g.79566881T>C, p.T391A) predicted by SIFT and PROVEAN as not tolerated/deleterious. Three independent observers generated an Average Grade of White (AGW) phenotype score for 147 individuals based on photographs. The *KIT* variant demonstrates a significant QTL association to AGW ($p = 3.3 \times 10^{-12}$). Association with the *MC1R Extension* locus demonstrated that, although not in LD, *MC1R e/e* (chestnut) individuals had higher AGW scores than *MC1R E/-* individuals ($p = 3.09 \times 10^{-17}$). We also report complete linkage of the previously reported *KIT W19* allele to this missense variant. We propose to term this variant *W34*, following the standardized nomenclature for white spotting variants within the equine *KIT* gene, and report its epistatic interaction with *MC1R*.

Keywords: dominant white; white pattern; Arabian horse; American Quarter Horse; American Paint Horse; chestnut

---

1. Introduction

The easily observed expression of novel mutations causing white spotting phenotypes provides straightforward targets in studies of equine genetic variation. Forty-four genetic variants, most of which are located in the *KIT proto-oncogene, receptor tyrosine kinase* (*KIT*) and the *melanocyte inducing transcription factor* (*MITF*) genes, are implicated in white spotting and depigmentation phenotypes in the horse (*W1-W28, W30-W33* and *SB1* on *KIT*; *SW1, SW3, SW5, SW6* and *SW7* on *MITF*; *SW2, SW4, LWO, TO* and *GR* on other genomic locations) [1–16]. Phenotypes for white spotting alleles vary from white markings on the legs and extremities, as observed with *KIT* ^{W20}, to large white patches on the body, legs and head or completely white phenotypes in a few other mutant *KIT, MITF* or multiallelic individuals. Occasionally, carriers of well-documented white spotting alleles may present no to minimal white markings uncharacteristic of the spotting variant, a condition sometimes named "crypto" for its cryptic expression [17].

Loci likely altering the eumelanin vs. pheomelanin pigment proportion are also associated with the extent of depigmentation observed in the horse. The "chestnut" coat color, caused by a *Melanocortin 1 Receptor* (*MC1R*, termed the *"Extension"* locus, symbol *"E"* or *"e"*) loss-of-function mutation (*e/e*) and resulting in predominantly pheomelanin pigmentation, displays greater *KIT*-associated white markings [18–20]. Comparatively, "black" and "bay" coat colors, possessing the dominant eumelanin-producing functional *MC1R* (*E/-*) [20], demonstrate greater *MITF*-associated white markings [18]. Interactions between *KIT* and *MC1R* could be due to linkage, as both genes are located on *Equus caballus Autosome 3* (ECA3), although separated by ~42 Mbps and a centromere [21–23]. Alternatively, the *e/e* genotype may decrease melanocyte quantity or migration, intensifying the white spotting QTL effect of *KIT* mutations [18].

White markings and spotting phenotypes are varying selection in some horse breeds, depending on breeding goals and registry requirements. For the American Paint Horse, a white spotting phenotype enables registration in the "Regular" registry rather than the "Solid Paint-Bred" registry, significantly impacting the economic value of the horse [8,24]. The American Quarter Horse Association did not previously allow registration of horses with white spotting phenotypes (a rule that changed in 2004), yet statements noting white spotting as "undesirable" and "uncharacteristic" are maintained in the regulations (AQHA Official Handbook, 67th Edition, 2019). Therefore, white markings or spotting phenotypes with unknown/novel associated genetic loci are of commercial interest for the equine industry, encouraging further studies on these variants.

We describe the investigation of a white spotting phenotype (extended white markings on limbs and the ventral thoracic region), yet negative for all published white variants (*W1-W28, W30-W33, SW1-SW7, LWO, SB1* and *TO*) [1–16]. Using a quantitative white score phenotype, we tested the association of this allele along with the *MC1R Extension* and *ASIP Agouti* loci in 41 cases and 94 breed-matched controls lacking published white variants. We report the significant association of a missense *KIT* polymorphism with a quantitative increase in white spotting on the coat, as well as the epistatic interaction of this allele with the *MC1R* loss-of-function genotype (chestnut). We also report the complete linkage of the previously published *KIT* c.1322A>G; p.(Y441C) (*W19*) allele with this variant in 12 genotyped individuals.

2. Materials and Methods

2.1. Horses and Putative Candidate Variant Inspection

An Arabian and a Mangalarga stallion, submitted to Etalon Diagnostics (Menlo Park, CA, USA) for commercial genotyping services, demonstrated a notable white spotting phenotype but no spotting or depigmentation alleles at any of the 44 known loci (*GR, W1-W33, SW1-SW7, LWO, SB1* and *TO*) [1–16]. Both individuals exhibited white markings extending past the distal part of the carpal and tarsal joints, as well as facial white markings extending from the forehead to the upper and lower lips. Additionally, the Arabian showed a large white spot on the ventral region of the body. The Mangalarga stallion was also reported to produce similar white phenotypes on its offspring (Figure 1). Given the likely heritable phenotype, we pursued further evaluation of coding regions of candidate genes *KIT* and *MITF* (known to cause phenotypically similar white spotting patterns in the horse) using previously described methods of targeted exon sequencing and alignment [13].

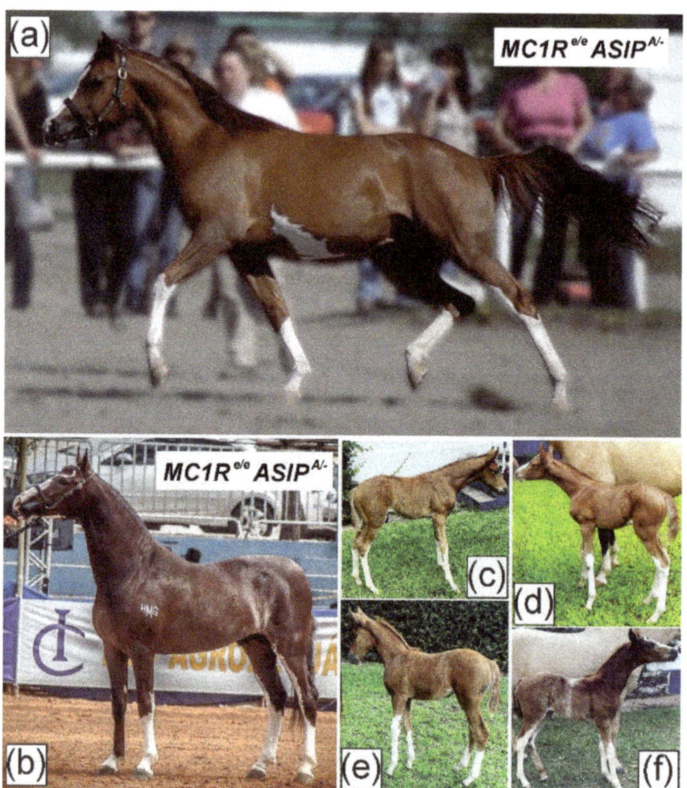

Figure 1. The subject stallions (**a**,**b**), demonstrating spotting phenotype extending past the distal part of the carpal and tarsal joints, as well as a facial white marking extending from the forehead to the upper and bottom lips along homozygote *MC1R e/e* phenotype, (**a**) also possess ventral white markings extending past the ribcage; (**c**–**f**) the respective offspring of the Mangalarga (**b**) stallion demonstrating the heritable phenotype.

Putative candidate polymorphisms were further evaluated by predicting functional impacts using the PROVEAN [25] and SIFT [26] webtools, using the NCBI Equus Caballus annotation release 103. To test for associations between novel variants and white spotting phenotypes, while controlling for confounding effects originating from other known white pattern variants, 41 unrelated individuals carrying only the candidate variant, as well as 94 negative control individuals that did not possess other known white spotting or depigmentation alleles (of 1431 previously genotyped by Etalon Diagnostics [27] representing the general horse population) having submitted photographs, were selected for further phenotyping (n = 135). Following genotypic selection for the novel candidate variants, we observed that all *W19* individuals (n = 12, four times the number of individuals in the original *W19* publication) in the Etalon Diagnostics genotyped population possessed at least one allele of the candidate *KIT* variant; thus, we performed a second analysis including this allele (n = 147).

2.2. Linkage Disequilibrium Analysis with Dominant White 19 and MC1R

Due to the co-localization of loci on the *Equus caballus* autosome 3 (ECA3), we calculated the linkage disequilibrium (LD) between candidate *KIT* loci, *W19* and the *MC1R* polymorphism using Haploview V4.1 (Broad Institute, MIT and Harvard), in the 147 phenotyped individuals.

2.3. Phenotyping and Statistical Analysis

Three blinded observers (LPR, KM and EM) scored the white phenotypes based on anonymized photographs of individual horses, following a previously published procedure for white pattern scoring [19]. In short, photos submitted by the owners were reviewed by KM and EM, selecting for 1 or 2 photographs that best represented the individual, where all 4 legs, the front of the head and any ventral/lateral white were visible for scoring purposes. Aside from the 12 compound *KIT W19* individuals, only two horses (EdX1476 and EdX4739) showed uneven body markings; these were scored based on the side with the highest amount of white. We then scored the amounts of white on the head, legs and body for each individual horse, which were then combined in a total score for each observer, summed, then averaged by the number of observers ($n = 3$), generating the "Average Grade of White" (AGW) value [19]. We modified the original score, awarding one point for white within a square comprising the ventral thorax and ribcage, and one point if white patterns were observed outside of the delimited region, to better quantify body markings (Figure 2). The modified scheme graded white from a minimum score of 0 to a maximum of 38, with a minimal effect on the original score maximum value of 36 [19]. Correlations between observer scores were evaluated using Pearson's pairwise multivariate correlation on SAS JMP Pro V15 (SAS Institute, Cary, NC, USA).

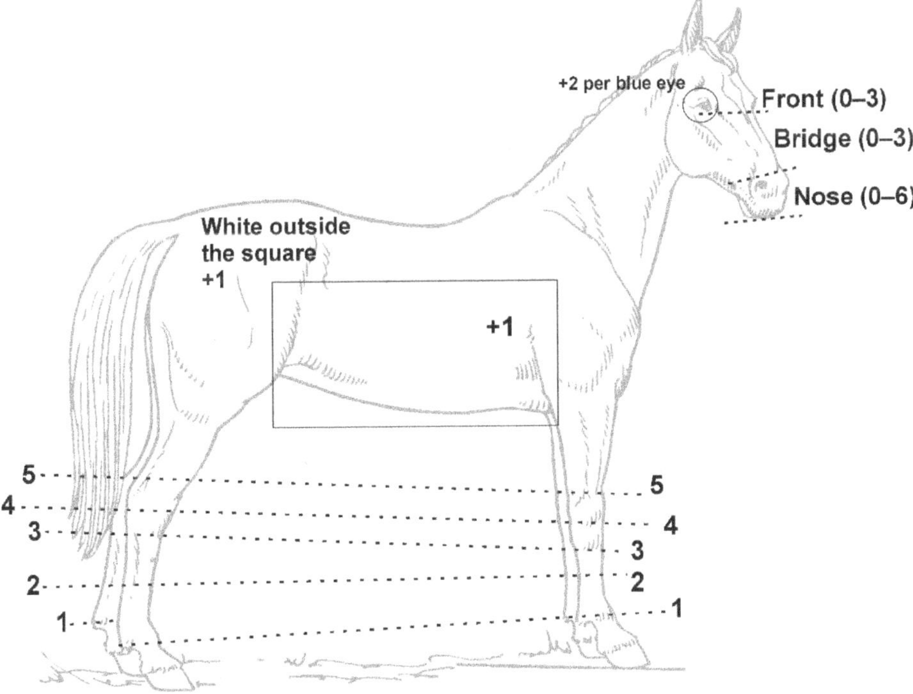

Figure 2. Average Grade of White phenotyping system as modified from that published by Rieder et al. [19]. Dashed lines demonstrate anatomical locations and limits for each score. Leg scores range from 0 (no white) to 5 (white above the respective dotted line).

A multiple linear regression modeling the effects of candidate loci, coat color (*MC1R* and *ASIP* genotype) and *KIT W19* on AGW was done using SAS JMP Pro V15 (SAS Institute, Cary, NC, USA). We independently evaluated the genotype impact of *MC1R*, *ASIP* and the candidate mutation with (n = 147 horses) and without (n = 135 horses) the presence of *W19* on the AGW (Table 1 and Table S2). Genotypes for all variants were obtained through the Etalon Diagnostics (Menlo Park, CA, USA) commercial testing.

Table 1. Effect sizes of *MC1R Extension*, *ASIP Agouti* and *KIT rs1140732842* loci on AGW (n = 135 horses) using binomial regression.

	AICc	*MC1R Extension*	*ASIP Agouti*	*KIT rs1140732842*
AGW	918.99			
AGW + *MC1R Extension*	903.95	17.99		
AGW + *ASIP Agouti*	915.54		5.57	
AGW + *KIT rs1140732842*	869.13			32.51
AGW + *MC1R Extension* + *KIT rs1140732842*	843.23	30.26		40.51
AGW + *MC1R Extension* + *ASIP Agouti*	904.51	13.51	1.54	
AGW + *MC1R Extension* + *ASIP Agouti* + *KIT rs1140732842*	845.29	28.14	0.13	39.09
p-value (full model)		4.71×10^{-7}	0.7199	5.09×10^{-14}

p-values are given for the full model incorporating AGW and genetic effects.

3. Results

3.1. Variant Analysis Suggests a Candidate in KIT Influenced by MC1R

Exon sequencing of both stallions identified homozygosity for two non-synonymous variants, NC_009159.3:g.21551234C>G in *MITF* and NC_009146.3:g.79566881T>C in *KIT*, respectively recorded as *rs1148371483* and *rs1140732842* on the Ensembl Variation Annotation release 104 [28]. We did not observe an association between the *MITF* variant and the AGW phenotype in the 135 horses (p = 0.4256; *W19* individuals excluded). Functional predictions also support that the *MITF* variant, a glycine to alanine change, is not a likely candidate as SIFT and PROVEAN predicted its effect as neutral (score = 0.23, SIFT; score = −0.356, PROVEAN) [29].

The presence of the alternate allele at *rs1140732842* is associated with a substantial effect on the AGW quantitative phenotype (F (2, 132) = 32.51, (ANOVA) p = 3.3 × 10^{-12}) (Table 1). The *KIT* variant *rs1140732842* is predicted by the SIFT method as not tolerated (score = 0.03), and as deleterious by PROVEAN (score = −3.363). This variant substitutes an uncharged polar threonine to a nonpolar alanine in the *KIT* protein structure (p.T391A). The *MC1R* genotype alone also has a small yet significant effect on AGW (F (1, 133) = 17.99 (ANOVA) p = 4.12 × 10^{-5}), with *e/e* (chestnut) individuals demonstrating higher AGW scores on average (mean score of 7.63) than *MC1R E/-* (eumelanin dominant) individuals (mean score of 2.54) (Figure 3). The *ASIP Agouti* locus has no significant effect on the AGW phenotype (Table 1). The *KIT rs1140732842* variant appears to have an incomplete penetrance autosomal dominant, or additive mode of inheritance for AGW, that could be cryptic in *MC1R E/-* individuals. When evaluating the effect of *W19* in AGW (Table S2, N = 147) the best-fitting model included all four loci (AICc = 919.73). Notably, the AGW scores from the three independent observers were highly correlated (r(147) > 0.9846, p < 4.38 × 10^{-87}), demonstrating the repeatability of the AGW scoring methodology in this study.

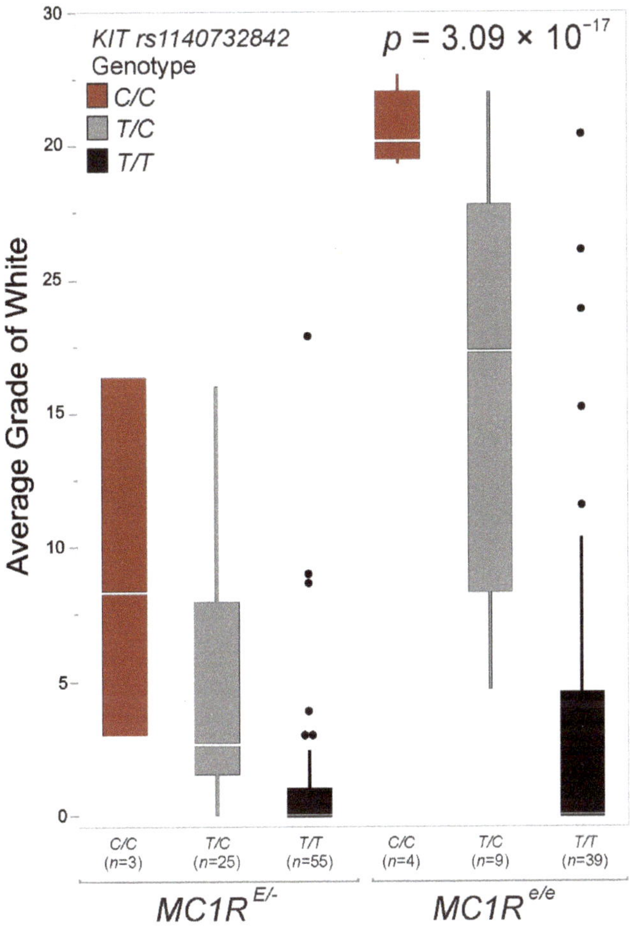

Figure 3. Genotype distribution of the *rs1140732842* polymorphism in 135 individuals by Average Grade of White and *MC1R* genotype (black/chestnut).

3.2. Linkage Disequilibrium between KIT and MC1R

We did not observe linkage disequilibrium between *MC1R* alleles and the *KIT rs1140732842* variant in the 147 horses ($r^2 = 0.0001$, $D' = 0.065$, LOD = 0.06). Similarly, Brooks et al. [24] demonstrated that the *KIT* ^{W20} allele (exon 14) was not in linkage disequilibrium with the *MC1R Extension* locus in a cohort of 364 American Paint Horses [25]. However, the *KIT W19* mutant allele (exon 8, 13.1 Kb apart) is in perfect linkage with the *rs1140732842* (exon 7) C variant ($r^2 = 0.17$, $D' = 1$, LOD = 6.49). Three horses demonstrated compound genotypes (EdX4400 and EdX1926: *KIT rs1140732842 C/C W19/KIT+* and EdX2927: *KIT rs1140732842 C/C W19/W19*), indicating that the *KIT W19* may have appeared after the *rs1140732842* mutant variant, as we observed heterozygosity of the *W19* allele in the presence of homozygosity of the *KIT rs1140732842 C* allele, yet not the opposite (Figure 4, File S2).

Figure 4. Phenotypic examples of the *rs1140732842* (*W34*) and compound *W19* allele action in different *MC1R Extension* and *ASIP Agouti* loci, as well as respective Average Grade of White (AGW) scores.

4. Discussion

Based on VGNC:19433 and NP_001157338.1 annotations as wild-type models, the computational analysis of the protein change in folding free energy upon mutation predicts that the p.T391A variant is destabilizing (Supplementary File S1) [30]. The amino acid threonine is highly conserved (Genomic Evolutionary Rate Profiling (GERP) score = 3.33) in this position in 91 eutherian mammals, including humans and mice [28], which could explain the observed low (1.51%) minor allele frequency of the *rs1140732842* variant in 1431 genotyped horses.

The effects of black or chestnut base coat colors on white spotting patterns were previously observed in the Arabian [31,32], the Franches-Montagnes horse [18,19] and the American Paint Horse [24]. While there is no linkage between the *MC1R Extension* locus and the candidate *KIT* variant in our cohort, the epistatic effect might be explained by other biological mechanisms. It is possible that the *MC1R e/e* genotype negatively affects the proliferation and differentiation of melanocytes, as observed in the murine recessive yellow (*Mc1r^e*) model [33]. Lower activity of the *MC1R* receptor, as is likely to result from the loss-of-function variant, promotes pheomelanin production [34]. As *KIT* is also involved in melanocyte pigmentation and development, the combined effect of deleterious alleles at both loci likely promotes a higher likelihood of failed melanocyte migration or maturation, resulting in unpigmented skin devoid of melanocytes [35]. Furthermore, exon screening cannot rule out the possibility that *rs1140732842* is tagging a haplotype bearing a non-coding regulatory change in the *KIT* gene.

Individuals possessing at least one *rs1140732842* alternate allele Ⓒ included the Arabian and its crosses, as well as Warmblood, Rocky Mountain Horse, American Quarter Horse, American Paint Horse, Appaloosa, Mustang, Mangalarga, Mangalarga Marchador and Morgan breeds (Table S1). Notably, the three original individuals in the *KIT W19* publication are also recorded as compound *rs1140732842* heterozygotes [5]. The *W19* allele seems to further increase the AGW ($p = 1.88 \times 10^{-20}$) and, given the AICc results, has some effect on the *ASIP* locus that we could not properly access in our study due to the small sample size and confounding effect of the *rs1140732842* variant. Due to the observed linkage, further evaluation of the *W19* allele's phenotypic effects alone is suggested, along with a possible effect of base coat color, including respective genotypes at *ASIP* and *MC1R* as suggested by the model.

5. Conclusions

We report a white spotting QTL associated with the *KIT* variant *rs1140732842* and modified by the presence of the *MC1R* loss-of-function pheomelanin genotype in the horse, as well as the observed linkage of *KIT W19* to this variant in our population and in the original publication. We propose to designate this polymorphism as *W34*, following the standardized nomenclature for white spotting variants within the *KIT* gene. Given the *KIT rs1140732842* alternate allele's significant association and QTL effect on white markings and its *MC1R* epistatic influence, genetic testing for this variant can be of value for horse owners that desire to select for quantitative white phenotypes.

Supplementary Materials: The following supporting information can be downloaded at: https://www.mdpi.com/article/10.3390/ani12151958/s1, Table S1: Genotypes and AGW phenotypic information for the 147 individuals; Table S2: Effect sizes of *MC1R Extension*, *ASIP Agouti*, *KIT W19* and *KIT rs1140732842* loci on AGW (n = 147 horses) using binomial regression; File S1: DUET—Protein Stability Change Upon Mutation prediction results for the *p.T391A* variant impact on protein stability; File S2: Graphical representation of the Linkage Disequilibrium heatmap between *MC1R*, *KIT W19* and *rs1140732842* (*W34*), as well as respective allele frequency and haplotype demonstration of the *W19* C genotype (r^2 = 0.17, D' = 1, LOD = 6.49) to *W34* C genotype in 147 horses.

Author Contributions: Conceptualization, L.P.R., K.M., S.A.B. and C.L.; methodology, L.P.R. and S.A.B.; software, L.P.R. and G.F.; validation, L.P.R., K.M., G.F. and S.A.B.; formal analysis, L.P.R.; investigation, L.P.R. and K.M.; resources, E.L. and C.L.; data curation, L.P.R., K.M., M.V. and G.F.; writing—original draft preparation, L.P.R.; writing—review and editing, L.P.R., K.M., M.V., E.L., G.F., S.A.B. and C.L.; visualization, L.P.R.; supervision, K.M., E.L., S.A.B. and C.L.; project administration, K.M., E.L. and C.L.; funding acquisition, E.L., K.M. and C.L. All authors have read and agreed to the published version of the manuscript.

Funding: This research received no external funding.

Institutional Review Board Statement: Not applicable.

Informed Consent Statement: Informed consent was obtained from all owners of the animal subjects involved in the study.

Data Availability Statement: Not applicable.

Acknowledgments: The authors would like to thank Deedra Holloway, José Roberto Petrucci, Emelia Wolf, Nancy Groenenboom, Natalie DiBerardinis, Roberta Nugent, Tammie Patrick, Sherry Bumbarger, Laetitia Lhéritier, Stephanie Wind, Steve Hugus, Janet Macdonald, Honey Belshe, Brandi Fulgo and João Antônio Rodrigues Caldas, as well as all the owners of subject horses, who provided samples, photos and background information for the *W34* study, and Erik Martin (EM), for his collaboration in this study.

Conflicts of Interest: All authors are affiliated with Etalon Diagnostics, which offers testing for white pattern variants.

References

1. Haase, B.; Brooks, S.A.; Schlumbaum, A.; Azor, P.J.; Bailey, E.; Alaeddine, F.; Mevissen, M.; Burger, D.; Poncet, P.-A.; Rieder, S. Allelic heterogeneity at the equine KIT locus in dominant white (W) horses. *PLoS Genet.* **2007**, *3*, e195. [CrossRef] [PubMed]
2. Haase, B.; Brooks, S.A.; Tozaki, T.; Burger, D.; Poncet, P.A.; Rieder, S.; Hasegawa, T.; Penedo, C.; Leeb, T. Seven novel KIT mutations in horses with white coat colour phenotypes. *Anim. Genet.* **2009**, *40*, 623–629. [CrossRef] [PubMed]
3. Haase, B.; Rieder, S.; Tozaki, T.; Hasegawa, T.; Penedo, M.C.T.; Jude, R.; Leeb, T. Five novel KIT mutations in horses with white coat colour phenotypes. *Anim. Genet.* **2011**, *42*, 337–339. [CrossRef]

4. Hauswirth, R.; Haase, B.; Blatter, M.; Brooks, S.A.; Burger, D.; Drögemüller, C.; Gerber, V.; Henke, D.; Janda, J.; Jude, R. Mutation in MITF and PAX3 cause "splashed white" and other white spotting phenotypes in horses. *PLoS Genet.* **2012**, *8*, e100265. [CrossRef]

5. Hauswirth, R.; Jude, R.; Haase, B.; Bellone, R.R.; Archer, S.; Holl, H.; Brooks, S.A.; Tozaki, T.; Penedo, M.C.T.; Rieder, S. Novel variants in the KIT and PAX 3 genes in horses with white-spotted coat colour phenotypes. *Anim. Genet.* **2013**, *44*, 763–76. [CrossRef]

6. Haase, B.; Jagannathan, V.; Rieder, S.; Leeb, T. A novel KIT variant in an Icelandic horse with white-spotted coat colour. *Anim. Genet.* **2015**, *46*, 466. [CrossRef] [PubMed]

7. Brooks, S.A.; Bailey, E. Exon skipping in the KIT gene causes a Sabino spotting pattern in horses. *Mamm. Genome* **2005**, *16*, 893–90. [CrossRef]

8. Brooks, S.A.; Lear, T.L.; Adelson, D.L.; Bailey, E. A chromosome inversion near the KIT gene and the Tobiano spotting pattern in horses. *Cytogenet. Genome Res.* **2008**, *119*, 225–230. [CrossRef]

9. Holl, H.; Brooks, S.; Bailey, E. De novo mutation of KIT discovered as a result of a non-hereditary white coat colour pattern. *Anim. Genet.* **2010**, *41*, 196–198. [CrossRef]

10. Dürig, N.; Jude, R.; Holl, H.; Brooks, S.A.; Lafayette, C.; Jagannathan, V.; Leeb, T. Whole genome sequencing reveals a novel deletion variant in the KIT gene in horses with white spotted coat colour phenotypes. *Anim. Genet.* **2017**, *48*, 483–485. [CrossRef]

11. Holl, H.M.; Brooks, S.A.; Carpenter, M.L.; Bustamante, C.D.; Lafayette, C. A novel splice mutation within equine KIT and the W15 allele in the homozygous state lead to all white coat color phenotypes. *Anim. Genet.* **2017**, *48*, 497–498. [CrossRef] [PubMed]

12. Martin, K.; Vierra, M.; Foster, G.; Brooks, S.A.; Lafayette, C. De novo mutation of KIT causes extensive coat white patterning in a family of Berber horses. *Anim. Genet.* **2020**, *52*, 135–137. [CrossRef] [PubMed]

13. Patterson Rosa, L.; Martin, K.; Vierra, M.; Foster, G.; Lundquist, E.; Brooks, S.A.; Lafayette, C. Two variants of KIT causing white patterning in Stock-type horses. *J. Hered.* **2021**, *112*, 447–451. [CrossRef] [PubMed]

14. Metallinos, D.L.; Bowling, A.T.; Rine, J. A missense mutation in the endothelin-B receptor gene is associated with Lethal White Foal Syndrome: An equine version of Hirschsprung disease. *Mamm. Genome* **1998**, *9*, 426–431. [CrossRef] [PubMed]

15. Esdaile, E.; Till, B.; Kallenberg, A.; Fremeux, M.; Bickel, L.; Bellone, R.R. A de novo missense mutation in KIT is responsible for dominant white spotting phenotype in a Standardbred horse. *Anim. Genet.* **2022**, *53*, 534–537. [CrossRef] [PubMed]

16. Patterson Rosa, L.; Martin, K.; Vierra, M.; Foster, G.; Brooks, S.A.; Lafayette, C. Non-frameshift deletion on MITF is associated with a novel splashed white spotting pattern in horses (*Equus caballus*). *Anim. Genet.* **2022**, *53*, 538–540. [CrossRef]

17. Stachurska, A.; Jansen, P. Crypto-tobiano horses in Hucul breed. *Czech J. Anim. Sci.* **2015**, *60*, 1–9. [CrossRef]

18. Haase, B.; Signer-Hasler, H.; Binns, M.M.; Obexer-Ruff, G.; Hauswirth, R.; Bellone, R.R.; Burger, D.; Rieder, S.; Wade, C.M.; Leeb, T. Accumulating mutations in series of haplotypes at the KIT and MITF loci are major determinants of white markings in Franches-Montagnes horses. *PLoS ONE* **2013**, *8*, e75071. [CrossRef]

19. Rieder, S.; Hagger, C.; Obexer-Ruff, G.; Leeb, T.; Poncet, P.-A. Genetic Analysis of White Facial and Leg Markings in the Swiss Franches-Montagnes Horse Breed. *J. Hered.* **2008**, *99*, 130–136. [CrossRef]

20. Rieder, S.; Taourit, S.; Mariat, D.; Langlois, B.; Guérin, G. Mutations in the agouti (ASIP), the extension (MC1R), and the brown (TYRP1) loci and their association to coat color phenotypes in horses (*Equus caballus*). *Mamm. Genome* **2001**, *12*, 450–455. [CrossRef]

21. Kalbfleisch, T.S.; Rice, E.S.; DePriest, M.S.; Walenz, B.P.; Hestand, M.S.; Vermeesch, J.R.; O'Connell, B.L.; Fiddes, I.T.; Vershinina, A.O.; Petersen, J.L.; et al. EquCab3, an Updated Reference Genome for the Domestic Horse. *bioRxiv* **2018**, 306928. [CrossRef]

22. Beeson, S.K.; Mickelson, J.R.; McCue, M.E. Exploration of fine-scale recombination rate variation in the domestic horse. *Genome Res.* **2019**, *29*, 1744–1752. [CrossRef] [PubMed]

23. Beeson, S.K.; Mickelson, J.R.; McCue, M.E. Equine recombination map updated to EquCab3.0. *Anim. Genet.* **2020**, *51*, 341–342. [CrossRef] [PubMed]

24. Brooks, S.A.; Palermo, K.M.; Kahn, A.; Hein, J. Impact of white-spotting alleles, including W20, on phenotype in the American Paint Horse. *Anim. Genet.* **2020**, *51*, 707–715. [CrossRef]

25. Choi, Y.; Sims, G.E.; Murphy, S.; Miller, J.R.; Chan, A.P. Predicting the functional effect of amino acid substitutions and indels. *PLoS ONE* **2012**, *7*, e46688. [CrossRef] [PubMed]

26. Ng, P.C.; Henikoff, S. SIFT: Predicting amino acid changes that affect protein function. *Nucleic Acids Res.* **2003**, *31*, 3812–3814. [CrossRef]

27. Diagnostics, E. Etalon DNA Test Panel Comparison Chart. Available online: https://www.etalondx.com/compare-tests (accessed on 31 January 2022).

28. Howe, K.L.; Achuthan, P.; Allen, J.; Allen, J.; Alvarez-Jarreta, J.; Amode, M.R.; Armean, I.M.; Azov, A.G.; Bennett, R.; Bhai, J.; et al. Ensembl 2021. *Nucleic Acids Res.* **2021**, *49*, D884–D891. [CrossRef]

29. Choi, Y.; Chan, A.P. PROVEAN web server: A tool to predict the functional effect of amino acid substitutions and indels. *Bioinformatics* **2015**, *31*, 2745–2747. [CrossRef] [PubMed]

30. Pires, D.E.V.; Ascher, D.B.; Blundell, T.L. DUET: A server for predicting effects of mutations on protein stability using an integrated computational approach. *Nucleic Acids Res.* **2014**, *42*, W314–W319. [CrossRef]

31. Woolf, C.M. Common white facial markings in bay and chestnut Arabian horses and their hybrids. *J. Hered.* **1991**, *82*, 167–169. [CrossRef]

2. Woolf, C.M. Common white facial markings in Arabian horses that are homozygous and heterozygous for alleles at the A and E loci. *J. Hered.* **1992**, *83*, 73–77. [CrossRef] [PubMed]
3. Hirobe, T.; Abe, H.; Wakamatsu, K.; Ito, S.; Kawa, Y.; Soma, Y.; Mizoguchi, M. Excess tyrosine rescues the reduced activity of proliferation and differentiation of cultured recessive yellow melanocytes derived from neonatal mouse epidermis. *Eur. J. Cell Biol.* **2007**, *86*, 315–330. [CrossRef] [PubMed]
4. Ito, S.; Wakamatsu, K. Human hair melanins: What we have learned and have not learned from mouse coat color pigmentation. *Pigment Cell Melanoma Res.* **2011**, *24*, 63–74. [CrossRef] [PubMed]
5. D'Mello, S.A.N.; Finlay, G.J.; Baguley, B.C.; Askarian-Amiri, M.E. Signaling Pathways in Melanogenesis. *Int. J. Mol. Sci.* **2016**, *17*, 1144. [CrossRef]

Article

Mitochondrial DNA Variation Contributes to the Aptitude for Dressage and Show Jumping Ability in the Holstein Horse Breed

Laura Engel [1,*], Doreen Becker [2], Thomas Nissen [3], Ingolf Russ [4], Georg Thaller [1] and Nina Krattenmacher [1]

[1] Institute of Animal Breeding and Husbandry, Christian-Albrechts-University, 24098 Kiel, Germany; gthaller@tierzucht.uni-kiel.de (G.T.); nkrattenmacher@tierzucht.uni-kiel.de (N.K.)
[2] Institute of Genome Biology, Research Institute for Farm Animal Biology (FBN), 18196 Dummerstorf, Germany; becker.doreen@fbn-dummerstorf.de
[3] Verband der Züchter des Holsteiner Pferdes e.V., 24106 Kiel, Germany; nissen@holsteiner-verband.de
[4] Tierzuchtforschung e.V. München, 85586 Grub, Germany; ingolf.russ@tzfgen-bayern.de
* Correspondence: lengel@tierzucht.uni-kiel.de

Simple Summary: In the Holstein horse breed, maternal lineages are considered to be of major importance for the breeding success and can be examined through analysis of the maternally inherited mitochondrial DNA (mtDNA). Since mitochondrial genes are involved in energy metabolism, variation might contribute to differences in performance characteristics, as has already been pointed out in humans and racehorses with respect to endurance. No corresponding studies have yet been conducted for the athletic performance of warmblood breeds and, thus, the aim of this study was to investigate the influence of mitochondrial variation on the performance of Holstein mares. The data set used for this study was composed of both sequenced and non-sequenced mares of 75 maternal lineages as a previous study revealed that Holstein mares within a maternal lineage had identical mtDNA haplotypes regarding their non-synonymous variants. Association analyses were performed using estimated breeding values (EBVs) based on information from sport and breeding events. We observed mitochondrial single nucleotide polymorphisms (SNPs) significantly associated with one or more of the examined EBVs and identified mitochondrial haplogroups with a particular aptitude for dressage or show jumping.

Citation: Engel, L.; Becker, D.; Nissen, T.; Russ, I.; Thaller, G.; Krattenmacher, N. Mitochondrial DNA Variation Contributes to the Aptitude for Dressage and Show Jumping Ability in the Holstein Horse Breed. *Animals* **2022**, *12*, 704. https://doi.org/10.3390/ani12060704

Academic Editors: Isabel Cervantes and María Dolores Gómez Ortiz

Received: 10 January 2022
Accepted: 9 March 2022
Published: 11 March 2022

Publisher's Note: MDPI stays neutral with regard to jurisdictional claims in published maps and institutional affiliations.

Abstract: Maternal lineages are considered an important factor in breeding. Mitochondrial DNA (mtDNA) is maternally inherited and plays an important role in energy metabolism. It has already been associated with energy consumption and performances, e.g., stamina in humans and racehorses. For now, corresponding studies are lacking for sport performance of warmblood breeds. MtDNA sequences were available for 271 Holstein mares from 75 maternal lineages. As all mares within a lineage showed identical haplotypes regarding the non-synonymous variants, we expanded our data set by also including non-sequenced mares and assigning them to the lineage-specific haplotype. This sample consisting of 6334 to 16,447 mares was used to perform mitochondrial association analyses using breeding values (EBVs) estimated on behalf of the Fédération Équestre Nationale (FN) and on behalf of the Holstein Breeding Association (HOL). The association analyses revealed 20 mitochondrial SNPs (mtSNPs) significantly associated with FN-EBVs and partly overlapping 20 mtSNPs associated with HOL-EBVs. The results indicated that mtDNA contributes to performance differences between maternal lineages. Certain mitochondrial haplogroups were associated with special talents for dressage or show jumping. The findings encourage to set up innovative genetic evaluation models that also consider information on maternal lineages.

Keywords: mitochondrial DNA; mitochondrial association analysis; Holstein horse; maternal lineages; show jumping; dressage

---

1. Introduction

Since the middle of the 20th century, the Holstein horse has been intensively bred for its athletic performance and aptitude for show jumping, eventing, or dressage, with an explicit focus on show jumping ability. The World Breeding Federation for Sport Horses (WBFSH) publishes annually the most successful studbooks in the above-mentioned disciplines based on competition results. In both show jumping and eventing, the Holstein horse has been ranked among the top 10 breeds over the past 10 years, and even in dressage, a ranking in the top third has consistently been achieved [1]. Holstein horse breeders rely on maternal lineages, which are considered to contribute substantially to the breeding success, and their documentation dates back to the beginning of the 19th century. Mares with unknown parents were defined as founder mares resulting in more than 8900 different maternal lineages up to today. Due to World War II and the growing mechanization of agriculture, the number of Holstein mares decreased markedly, accompanied by a loss of maternal lineages. Furthermore, the focus of selection has shifted from the use for agriculture, carriage, and cavalry toward athletic performance. To achieve this goal, English Thoroughbred and Anglo-Norman stallions were increasingly used for breeding, while the studbook remained closed. Today, 437 maternal lineages exist, comprising 5412 active brood mares [2].

The impact of maternal lineages on breeding success can be examined through analysis of the mitochondrial DNA (mtDNA) since it is inherited maternally without recombination. The equine circular mitochondrial genome is 16,660 bp in length and comprises 37 genes, 13 of which encode for respiratory chain proteins involved in energy metabolism in addition to two ribosomal and 22 transfer RNAs essential for protein synthesis. The mitochondrial genome harbors candidate genes, variants of which may possibly influence adenosine triphosphate (ATP) synthesis during exercise [3].

The performances of the Holstein horse, e.g., the upward movement during jumping, are highly energy demanding, relying particularly on aerobic capacity [4]. Since aerobic energy production takes place in mitochondria, it is reasonable to assume that variation in mitochondrial genes could possibly have an effect on athletic performance. Therefore, we recently analyzed mtDNA sequences from 271 Holstein mares belonging to 75 maternal lineages [5]. We were able to show that considerable molecular variation among mitochondrial genomes of different maternal lineages exists in Holstein horses and identified a total of 78 haplotypes that could be assigned to eight distinct haplogroups. The mtDNA sequences of mares within a haplogroup showed high levels of similarity. Within a lineage, identical mtDNA haplotypes were found in all mares with respect to the non-synonymous substitutions [5], which is in accordance with results from studies in other breeds [6,7]. Based upon these findings, we enlarged our sample size by including both sequenced and non-sequenced mares belonging to one of the maternal lineages analyzed in our previous study [5], assuming that the non-sequenced mares have the same mitochondrial haplotype of non-synonymous variants as the sequenced mares of the same lineage. Using the resulting sample, we investigated the involvement of mitochondrial variants with regard to the performance of Holstein mares. Estimated breeding values (EBVs) were used as targets to perform mitochondrial association analyses. For the Holstein horse, two separate genetic evaluation systems were run. On the one hand, EBVs were estimated population-specific on behalf of the Holstein Breeding Association (HOL) for traits recorded at studbook inspection and mare performance test. On the other hand, EBVs were estimated across all German warmblood breeds on behalf of the Fédération Équestre Nationale (FN) based on the results from sport and breeding events. A detailed description of the breeding value estimation can be found in the work of [5,8,9] Supplementary File S1.

2. Materials and Methods

For this study, a representative sample of Holstein mares belonging to 75 lineages was used to perform an association study between mitochondrial variants and EBVs estimated on behalf of the FN (FN-EBVs) and on behalf of the Holstein Breeding Association (HOL-EBVs).

2.1. Sampling, Sequencing, and Enlargement of the Data Set

Initially, 493 mares with extensive phenotypes, i.e., mares that preferably have information from studbook inspection, mare performance test, and from sport events, were preselected, and the respective breeders were asked to collect hair samples during routine care, e.g., combing the mane or tail. Besides this, mares were selected based on the availability of genotypes that were provided by an in-house project to allow consideration of interactions between the mitochondrial and the nuclear genome in upcoming studies. For the in-house project, the maternal lineage was of no interest, and mares showed a low level of preselection and a low pedigree relationship. After two sampling periods in 2019, hair samples were available for a total of 271 mares, which were sent in by 207 breeders.

For each sample, DNA was extracted from 20 to 25 hair roots using a modified protocol according to the work of [10]. The PRIMER 3 software (Version 4.1.0, https://primer3.ut.ee/ accessed on 10 January 2022) was used to create primer pairs for amplification of the mitochondrial genome [11]. Polymerase chain reaction (PCR) amplifications were performed in a 12 µL reaction volume including 20 ng total DNA, 0.2 µM of the forward and reverse primer, 200 µM deoxyribonucleoside triphosphates (dNTPs), 1.25 U of the PrimeSTAR GXL DNA-Polymerase (Takara Bio, Shiga, Japan), and the corresponding reaction buffer. The sequencing was performed in 36 reactions using the ABI 3130xl Genetic Analyzer and the BigDye® Terminator v3.1 Cycle Sequencing Kit (Applied Biosystems, Foster City, CA, USA). A more detailed description of the methods can be found in the work of [5].

Analysis of the mtDNA sequences was performed using the software Sequencher 5.0 (Gene Codes Corporation, Ann Arbor, MI, USA). If the sequences were ambiguous, the sequencing was repeated, and, if necessary, these samples were excluded from the analysis. For all samples, the repetitive part of the non-coding region was excluded due to failure in sequencing. The sequences were compared to the GenBank reference sequence X79547.1; the sequences of the mitochondrial haplotypes are provided in Supplementary File S2.

A total of 467 polymorphic sites were found, but only the 101 non-synonymous substitutions previously reported in the work of [5] were considered for the evaluation.

The sample was extended with non-sequenced mares belonging to the 75 lineages that were previously analyzed [5]. This was done based on the finding that all mares belonging to the same lineage have identical haplotypes regarding the non-synonymous mitochondrial substitutions. For 2020, HOL-EBVs for studbook inspection (SBI) and mare performance test (MPT) were estimated for 17,745 and 8167 mares, respectively. FN-EBVs were available for 13,920 mares. The mares were born between 1944 and 2017 and assigned to maternal lineages, which made up 17.4% of all lineages of the current breeding population and to which 56.4% of all active broodmares can be assigned. The number of mares per lineage in the current data sets ranges from 16 to 998 for the FN-EBVs, 28 to 1248 for the EBVs for studbook inspection (SBI-EBVs), and from five to 520 for the EBVs for mare performance test (MPT-EBVs). In both genetic evaluation systems, the majority of lineages was represented by 50 to 200 mares. Based upon previous results from the work of [5], the lineages can be assigned to eight different haplogroups defined by the authors of [12]. Table 1 shows the number of lineages, the number of sequenced mares, and the total number of mares per haplogroup in this sample.

Table 1. Overview of the data set and distribution of lineages among haplogroups.

Haplogroup [a]	Number of Lineages	Number of Sequenced Mares	Number of Total Mares with [b]		
			FN Breeding Values [c]	SBI Breeding Values	MPT Breeding Values
B	24	72	3091	3952	1618
D	4	13	290	352	120
G	9	33	2099	2590	1057
I	5	14	817	1053	420
K	1	3	112	150	42
L	16	78	3646	4650	1716
N	13	11	2209	2770	1016
P	3	47	707	930	345
Total	75	271	12,971	16,447	6334

[a] Haplogroups are assigned according to the work of [12]. [b] Number of mares with breeding values with a reliability $\geq$ 30%. [c] Number of mares varies between traits, as indicated in Table 2. The values shown here are those of the trait "dressage and show jumping competitions of young horses" (ABP) jumping with the highest number of mares.

2.2. Phenotypic Data

HOL-EBVs and FN-EBVs were used as targets for this study. All breeding values were routinely standardized with a mean of 100 and a standard deviation of 20 points by FN and HOL. For this study, only breeding values with a reliability $\geq$30% were used, leading to a loss of 1298, 1833, and 949 mares with breeding values for SBI, MPT, and FN, respectively. Reliabilities for HOL-EBVs were provided summarized for all SBI and MPT traits, respectively. Heritability estimates for all traits were moderate, except for the traits forehand and hindquarters that show low heritability estimates.

2.3. Data Analysis

Association testing was performed in PLINK version 1.9 software [13] using the ASSOC function. We excluded 50 mtSNPs that did not reach the minor allele frequency (MAF) of more than 1%, thus, leaving 51 mtSNPs for the association analysis. The mtSNPs were named according to their position (bp) on the mitochondrial genome as they have not been previously reported in other breeds. The Bonferroni-corrected significance threshold was $p < 5 \times 10^{-4}$, but we defined a significance threshold of $p < 5 \times 10^{-8}$ to avoid too many false positive results. The pairwise linkage disequilibrium (LD) was calculated as the correlation between mtSNP pairs (r^2) using the Haploview software [14]. Values of $r^2 = 1$ and $r^2 = 0$ are illustrated in black and white, respectively, with different shades of gray for the intermediate values. For mtSNPs in perfect LD ($r^2 = 1$), only one mtSNP was stated in all further evaluations. The results were visualized with the software Synthesis-View [15]. Further statistical analyses were performed using R version 4.0 [16].

3. Results

A sample of 5302–12,971 mares with FN-EBVs, 16,447 mares with SBI-EBVs, and 6334 mares with MPT-EBVs was used, where only EBVs with a reliability above 30% were considered. Table 2 provides an overview of the complete data set used for the analyses and the descriptive statistics for EBVs and reliabilities, respectively. The average EBV of the genetic evaluation of the FN ranges between 82.26 ($\pm$12.01) for the trait "sport and breeding tests for young horses" (JPf) dressage and 109.37 ($\pm$18.95) for the trait "highest level in competition" HEK jumping. There are differences between the average EBVs of the different disciplines: while average EBVs for dressage traits, including basic gaits and rideability, range between 82.26 ($\pm$12.01) and 89.95 ($\pm$12.35), those for show jumping are always above 100, with one exception for the trait ABP jumping (99.09 $\pm$ 13.87). For SBI, average EBVs range between 90.82 ($\pm$22.45) and 95.34 ($\pm$19.88) for the traits HOL hindquarters and HOL type, respectively. The average EBV for MPT range from 85.10 ($\pm$22.14) for HOL free jumping to 98.81 ($\pm$23.54) for HOL walk.

Table 2. Descriptive statistics of the breeding values and reliabilities estimated on behalf of the FN and on behalf of the Holstein Breeding Association for studbook inspection and mare performance test

		Breeding Value [a]		Reliability	
Trait [b]	**n**	**Mean (SD [c])**	**Range**	**Mean (SD [c])**	**Range**
FN genetic evaluation					
TSP jumping	12,656	105.86 (13.95)	48−148	41.82 (5.50)	30−75
TSP dressage	11,045	86.58 (9.18)	56−136	35.96 (3.92)	30−66
ABP jumping	12,971	99.09 (13.87)	31−137	48.75 (7.64)	30−78
ABP dressage	12,231	84.83 (12.00)	34−147	43.41 (7.19)	30−68
ZP walk	12,482	86.55 (10.41)	39−136	44.51 (8.59)	30−71
ZP trot	12,482	82.48 (12.01)	31−141	48.12 (10.82)	30−76
ZP canter	12,724	88.45 (13.04)	29−143	48.04 (10.33)	30−75
ZP rideability	12,701	85.32 (12.65)	34−143	47.15 (9.83)	30−74
ZP free jumping	12,587	104.56 (16.01)	45−149	43.60 (7.01)	30−71
ZP parcours jumping	12,599	102.17 (13.61)	44−142	40.59 (4.59)	30−70
JPf jumping	12,918	101.45 (16.58)	31−148	48.09 (6.83)	30−79
JPf dressage	12,918	82.26 (13.74)	24−149	49.81 (11.06)	30−77
ZP jumping	12,823	103.46 (16.39)	43−149	44.20 (5.95)	30−72
ZP dressage	12,696	82.78 (13.59)	24−149	51.13 (12.38)	30−80
HEK jumping	10,711	109.37 (18.95)	50−178	48.15 (7.96)	30−81
HEK dressage	5302	89.95 (12.35)	60−161	35.03 (4.77)	30−63
Studbook inspection					
HOL type	16,447	95.34 (19.88)	10−169	58.27 (3.83)	30−83
HOL topline	16,447	94.58 (20.97)	5−168	58.27 (3.83)	30−83
HOL forehand	16,447	92.48 (20.85)	4−165	58.27 (3.83)	30−83
HOL hindquarters	16,447	90.82 (22.45)	4−173	58.27 (3.83)	30−83
HOL correctness of gaits	16,447	94.94 (19.68)	11−187	58.27 (3.83)	30−83
HOL impulsion	16,447	95.19 (19.69)	13−189	58.27 (3.83)	30−83
Mare performance test					
HOL walk	6334	98.81 (23.54)	3−214	52.13 (7.83)	30−69
HOL trot	6334	96.83 (22.19)	12−187	52.13 (7.83)	30−69
HOL canter	6334	94.19 (21.91)	4−189	52.13 (7.83)	30−69
HOL rideability	6334	94.59 (22.81)	11−200	52.13 (7.83)	30−69
HOL free jumping	6334	85.10 (22.14)	3−149	52.13 (7.83)	30−69

[a] Only breeding values with a reliability $\geq$ 30% are considered. [b] TSP = show jumping and dressage competitions, ABP = show jumping and dressage competitions of young horses, ZP = own performance tests of mares and stallions, JPf = sport and breeding tests of young horses, HEK = highest level in competition, HOL = Holstein Breeding Association. [c] SD = Standard deviation.

20 mtSNPs were significantly associated with FN-EBVs. Associations were found for all traits except for HEK dressage. The name and position of the significant SNPs on the mitochondrial genome, as well as the $-\log_{10}$ (p-values) and the estimated effect sizes, are shown in Figure 1, where the values for the different traits are indicated in different colors. The red line specifies the significance threshold of $p < 5 \times 10^{-8}$. Additionally, r^2 between mtSNP pairs is shown. Only significant mtSNPs with $p < 5 \times 10^{-8}$ are depicted.

A total of eight mtSNPs were associated with the SBI-EBVs. Two of the significant mtSNPs were also associated with FN-EBVs. The other six are exclusively significant for SBI-EBVs. For MPT, 14 mtSNPs are significantly associated with the EBVs walk, canter, and rideability. Thirteen mtSNPs of them are also significant for FN-EBVs.

Comparisons of the average EBVs of the two alleles for each significant mtSNP were performed for each trait. The results for the association between FN-EBVs and mtSNPs are shown in Table A1. It further indicates the percentage of maternal lineages in a haplogroup carrying the minor allele. For example, the minor allele of the mtSNP mtDNA_2216 occurs in all four maternal lineages of haplogroup D, and these mares have an average EBV for the trait TSP jumping that is 5.29 points higher compared to the remaining mares studied. All maternal lineages of haplogroup B carry the minor alleles of two mtSNPs that are associated with higher EBVs in all dressage traits. All maternal lineages of haplogroup D have higher EBVs in six out of eight dressage traits and additionally in TSP jumping and HEK jumping. In haplogroup L, two sub-haplogroups consisting of three and one maternal lineages, respectively, show significant differences compared to the remaining sample. One sub-haplogroup shows lower EBVs in all jumping traits except HEK jumping, while the second sub-haplogroup has lower EBVs in six dressage traits. All maternal lineages of haplogroup N are characterized by reduced EBVs in trot and canter. Additionally, about three-quarters of them also show lower EBVs in four dressage traits and in free jumping. There is one mtSNPs whose minor allele occurs in 13 maternal lineages from five different haplogroups, and that is associated with higher EBVs in HEK jumping.

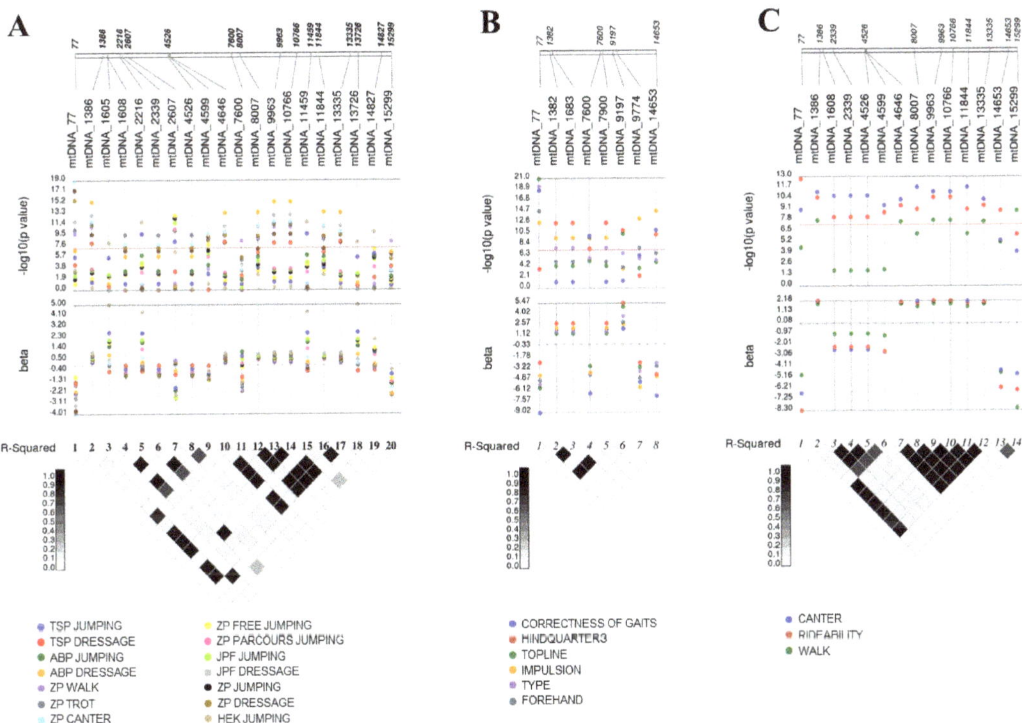

Figure 1. Mitochondrial SNPs significantly associated with breeding values. Association analyses were performed to identify mtSNPs associated with (**A**) FN breeding values, (**B**) breeding values for studbook inspection, and (**C**) breeding values for mare performance test. Only mtSNPs with $p < 5 \times 10^{-8}$ (indicated by the red line) are illustrated and are plotted according to their position on the mitochondrial genome. The values for the different traits are shown in different colors. LD between mtSNPs was measured as r^2 using Haploview software.

The results from the association analysis between SBI-EBVs and mtSNPs are shown in Table A2. All maternal lineages of haplogroup P show higher breeding values in HOL type, HOL hindquarters, and HOL impulsion. A sub-haplogroup of haplogroup G carries the minor allele of one mtSNP is associated with lower EBVs in HOL hindquarters, HOL correctness of gaits, and HOL impulsion. Besides this, only three single maternal lineages from three different haplogroups have significantly different EBVs in the six SBI traits.

Table A2 also provides the results from the association analysis between MPT-EBVs and mtSNPs. Significant associations could be found for the three traits HOL canter, HOL walk, and HOL rideability, with all maternal lineages from haplogroup B showing higher

EBVs for these traits. For HOL canter and HOL rideability, all maternal lineages from haplogroup B have higher EBVs, while all maternal lineages from haplogroup N have lower EBVs compared to the remaining sample.

4. Discussion

This study presents the first association analysis between mitochondrial variants and EBVs estimated for a show jumping breed and revealed that the mtDNA and, thus, maternal lineages were significantly associated with sport performance. A representative sample of Holstein mares was used, of which 1.51%, 1.62%, and 1.41% of the mares for SBI, MPT, and FN, respectively, were sequenced. As mitochondrial haplotypes were found to be highly consistent within maternal lineages [5], haplotypes of their respective maternal lineage were assigned to non-sequenced mares. Pedigree errors, and thus, incorrect assignment into maternal lineages, cannot be completely ruled out, but our previous study has shown that documentation of pedigrees in the Holstein breed is very accurate [5]. FN-EBVs and HOL-EBVs were used as phenotypes, although the use of EBVs in association studies remains controversial due to presumed high false discovery rates [17]. To reduce the risk of false positive results and, nevertheless, obtain reliable results, a large number of samples ranging from 6334 to 16,447, and a stringent significance threshold that differs from the Bonferroni-corrected significance threshold was used.

4.1. Mitochondrial Variants

A total of 51 non-synonymous mtSNPs were used for the association analyses. A total of 24 different mtSNPs could be associated with the examined EBVs. As can be seen in Figure 1, many of them are highly correlated, which is not surprising since the mtDNA does not recombine. The number of associated mtSNP pairs and triplets in total LD is shown in Table 3. Additionally, it provides an overview of the mtSNPs used for the association analyses and the results, including the total number of significantly associated mtSNPs and the number of associated mtSNPs for HOL- and FN-EBVs. The associated mtSNPs are located in the 12 mitochondrial genes *s-rRNA*, *l-rRNA*, *ATP8*, *COX2*, *COX3*, *CYTB*, *ND2*, *ND3*, *ND4*, *ND4L*, *ND5*, and *ND6*. The *rRNA* genes are involved in mitochondrial protein synthesis, and the protein-coding genes encode for subunits of complex I, III, IV, and V of the mitochondrial respiratory chain [18].

Table 3. Overview of mtSNPs used for mitochondrial association analyses and description of associated mtSNPs.

Genetic Data	
Total number of substitutions	467
Non-synonymous mtSNPs	101
Non-synonymous mtSNPs with MAF > 0.01	51
Results	
Total number of associated mtSNPs	24
Number of mtSNPs associated with FN-EBVs [a]	20
Number of mtSNPs associated with HOL-EBVs [b]	20
Number of mtSNP pairs in total LD ($r^2 = 1$)	4
Number of mtSNPs triplets in total LD ($r^2 = 1$)	2

[a] Breeding values estimated by the Fédération Équestre Nationale. [b] Breeding values estimated by the Holstein Breeding Association.

Considering the FN-EBVs, the same mtSNPs are significant for the traits ZP trot and ZP canter, which is in accordance with the high genetic correlation ($r_g = 0.69$) of these traits [9]. ZP rideability also shows high concordance with the two traits with respect to the significant mtSNPs, and again, this is in line with the high positive genetic correlation of $r_g = 0.67$ for both ZP trot and ZP canter [9]. The EBVs for JPf dressage and ZP dressage are derived, among others, from the above-mentioned EBVs (Supplementary File S1, Table S1) and, predictably, show the same significant mtSNPs. It has already been reported that gaits and rideability are highly correlated with dressage traits in German warmblood breeds [19]. Furthermore, it can be observed that different mtSNPs are significant in the dressage and jumping traits. Noteworthy is, e.g., the variant mtDNA_2607, which is significantly associated with the six jumping traits (TSP jumping, ABP jumping, ZP free jumping, ZP parcours jumping, JPf jumping, and ZP jumping) but has no effect on the dressage, gait, and rideability traits. The minor allele is associated with lower EBVs in all jumping traits. Similar observations were made for the EBVs of MPT. The traits HOL canter and HOL rideability show high agreement in terms of significant SNPs, while two different mtSNPs are significant for the trait HOL walk. Additionally, a comparison between the genetic evaluations of the FN and the Holstein Breeding Association revealed that almost the same mtSNPs are significant for the traits walk, canter, and rideability, which was expected since the correlations between the FN-EBVs and HOL-EBVs for the gaits and rideability are high, ranging from 0.91 to 0.92. Nevertheless, EBVs for traits from MPT estimated on behalf of the FN and on behalf of HOL were used for the evaluations since HOL-EBVs are estimated population-specific, while FN-EBVs are estimated jointly for all German warmblood breeds. For the conformation traits HOL type, HOL topline, HOL forehand, and HOL hindquarters derived from SBI, six exclusive mtSNPs are significant that did not occur in other traits. When performing analogous analyses for other breeds, it is therefore recommended to consider population-specific EBVs besides FN-EBVs.

4.2. Mitochondrial Haplogroups and Performance

In a previous investigation, we could assign the maternal lineages to eight haplogroups, whereby all mares from one lineage belong to the same haplogroup. Within a haplogroup, the mtDNA sequences of mares are very similar [5]. Taking into account the previous and current results, some further conclusions can be drawn. In all maternal lineages of haplogroup B, variations occur that were associated with higher EBVs in the dressage traits of the FN genetic evaluation, as well as higher EBVs for the basic gaits and rideability. Consequently, a pronounced dressage ability could be attributed to the 24 lineages of this haplogroup. Haplogroup D shows partly the same variations as haplogroup B resulting in higher EBVs in the dressage, gait, and rideability traits. This is in accordance with previous results, where these two haplogroups show the lowest genetic differentiation in the overall comparison and thus, share large parts of their mtDNA sequence [5]. Additionally, the four lineages representing haplogroup D show higher EBVs for two jumping traits, and thus, this haplogroup could be characterized as more versatile. There are three sub-groups auf haplogroup L, each differing from the rest of the sample: one sub-group composed of one lineage shows lower EBVs in two dressage traits, as well as basic gaits, rideability, and five traits derived from SBI. Another sub-group of three lineages shows reduced EBVs for all jumping traits estimated by the FN, except for HEK jumping. This group seems to be less predestined for elite show jumping performance. In contrast, the third sub-group consisting of six lineages seems to be more successful in show jumping. Noteworthy, all lineages of haplogroup P seem to be particularly in line with the preferred breed type, combined with strong hindquarters and impulsion. All lineages of haplogroup N have comparatively low EBVs for trot, canter, and rideability. Additionally, 10 of the 13 lineages are characterized by lower EBVs regarding dressage and free jumping.

The overall comparison shows that there is no haplogroup that stands out for special jumping ability. It rather shows, especially regarding the trait HEK jumping, that only single maternal lineages or sub-groups show higher EBVs. This indicates that the Holstein population is on an equal level regarding its jumping ability with high mean breeding values for all jumping traits (Table 2), which is in line with the breeding goal of the Holstein breed with a strong focus on show jumping ability. Mean FN-EBVs for dressage traits were always below 100, even for the favorable allele. This is because FN-EBVs were estimated, including all German warmblood breeds, thus, including a couple of horse breeds from breeding associations that put a higher emphasis on dressage performance.

The impact of mitochondrial variation on performance has already been studied inten sively in humans, where a total of 18 mitochondrial genes have been shown to be associate with fitness and performance phenotypes [20]. For example, the work of [21] found signi icant differences in the frequencies of two haplogroups in Finnish endurance and sprir athletes. Mitochondrial haplogroups are also associated with endurance performance i Spanish and Japanese humans [22,23]. In parallel, respective research in horses has mainl focused on racing performance. Thoroughbreds [24] identified haplotypes associated wit racing performance at different distances, and [3] reported a mutation in the mitochondri 16S rRNA gene associated with low racing performance. Corresponding studies are lackin for warmblood breeds. However, there are some studies that statistically examined th influence of maternal lineages on sport performance. Polish jumping breeds and thei performance during the three-day Polish Championships for Young Horses were studie by the authors of [25]. The maternal impact, defined as the proportion of the total varianc explained by the maternal additive genetic variance, was high for all traits (jumping styl score, penalty score for each day, and overall rating), ranging from 0.11 to 0.39. In Holstei horses, the work of [26] investigated the effect of the maternal lineage on traits recorded du ing studbook inspection and mare performance test. Up to 0.9% of the phenotypic variatio could be explained by the maternal lineage for traits recorded at studbook inspection. Th strongest effect was found for hindquarters [26]. For the mare performance test, materna lineages explained up to 2.1% of the variation with the highest value for the trait cante under the rider. In racehorses, there is a commercially genetic test available (Equinom Speed Gene Test; PlusVital, Dublin, Ireland) to predict the aptitude for racing performanc based on nuclear variation. Accuracy of testing could be enlarged by considering also result from mitochondrial association studies. Furthermore, the results can be considered for th implementation of genomic selection in horses and the design of the SNP chip.

4.3. Limitations

When interpreting the results, it must be considered that only the mtDNA was ex amined. Since the majority of mitochondrial proteins are nuclear-encoded, interaction between the mitochondrial and nuclear genome are to be expected. However, the results ar striking, as mitochondrial variations cause average EBVs that differ in 1.19 to 12.27 points although the mitochondrial genome represents only a very small part of the total DN/ Furthermore, we were not able to adjust for close maternal relationships on the genomi level, where mares belonging to the same lineage or haplogroup will probably also show more similarities with regard to the nuclear genome [27]. Nevertheless, the mares in th enlarged data set were only selected based on their maternal lineage, where even distan maternal relatives share mtDNA independently from their genomic relationship. It shoulc be noted, however, that sires are not evenly distributed across maternal lineages, whicl could have led to overestimation.

Since multiple mitochondria and multiple copies of the mtDNA exist in a cell, botl original and mutant mtDNA molecules can co-occur [6]. This phenomenon, known a heteroplasmy, is specific for mtDNA and caused by de novo mutations occurring eithe in the germline or in the somatic tissues. Heteroplasmy could not be considered in thi study because we used Sanger sequencing, which is not sensitive enough to detect it. Next generation sequencing technologies with sufficient coverage would enable the detection o heteroplasmy even at low levels. Total mtDNA sequences could thus be sequenced witl higher accuracy and should be applied for further investigations of mtDNA. However, th analysis of mtDNA using Sanger sequencing did not indicate the presence of heteroplasm as mainly unambiguous signals were found in the chromatograms. In case of ambiguou signals, sequencing of the respective sequence segment was repeated and confirmed th absence of heteroplasmy. However, the proportion of mutant mtDNA in a cell coulc possibly have an impact on the expression of a phenotype. Furthermore, there migh be differences in the degree of heteroplasmy between tissues, which was not considerec in this study because only hair samples were examined [28]. Analysis of mtDNA o tissues essential for energy supply could be insightful but would only be possible in horses for slaughter. However, not all horses are registered as animals for slaughter as this is accompanied by a restriction in drug administration. Thus, the sample would be less representative. Due to the above-mentioned limitations and considering that upscaling from a low number of mares with mtSNP genotypic data to the whole population was performed, the interpretation of the results should be treated with caution. Although this study demonstrates the great potential of mitochondrial association studies and the importance of mitochondrial variation for performance traits, it also shows that more research is needed to simultaneously account for nuclear and mitochondrial relatedness and thus, preventing a misinterpretation of the results.

5. Conclusions

This study is the first study performing a mitochondrial association study in warm-blood horses for sport performance traits. A representative sample of Holstein mares from 75 maternal lineages was used to which more than half of all active broodmares can be assigned. Mitochondrial variants in 12 different genes were shown to be associated with HOL-EBVs and FN-EBVs. The entire population shows a high level of jumping ability, however, with maternal lineages that stand out both positively and negatively. An enlargement of the sample with further maternal lineages is recommended since many mtSNPs have been excluded due to low MAF, and especially mtSNPs occurring in single sub-haplogroups have been found to be associated with the EBVs. Conclusively, the results provide evidence to revise the current genetic evaluation models by including information on maternal lineages. However, further work is needed to quantify the benefit of extended genetic evaluation models.

Supplementary Materials: The following supporting information can be downloaded at: https://www.mdpi.com/article/10.3390/ani12060704/s1, File S1: Breeding value estimation for the Holstein horse, File S2: Mitochondrial haplotypes.

Author Contributions: Conceptualization of the study was performed by N.K. and D.B. Acquisition of the financial support was performed by G.T. together with I.R. and T.N. provided the data. D.B. conceived the molecular experiments and L.E. performed the lab works. L.E. analyzed the data in collaboration with N.K. and D.B. L.E. and N.K. prepared the final manuscript supported by D.B. and G.T. All authors have read and agreed to the published version of the manuscript.

Funding: This research was funded by the Tierzuchtforschung e.V. München (Grub, Germany) and the H.Wilhelm Schaumann Stiftung (Hamburg, Germany). We acknowledge financial support by Land Schleswig-Holstein within the funding programme Open Access Publikationsfonds.

Institutional Review Board Statement: Ethical review and approval were waived for this study because this study is no animal experiment according to the German Animal Welfare Act as hair samples were selected from horse owners during routine care, e.g., combing the mane or tail. This was confirmed by the animal welfare officers at the University of Kiel.

Informed Consent Statement: Not applicable.

Data Availability Statement: The data set containing the mitochondrial haplotypes analyzed during the current study can be found in the additional supporting files. The phenotypic data that support the findings of this study are available from the Holstein Breeding Association and the Fédération Équestre Nationale, but restrictions apply to the availability of these data, which were used under license for the current study, and so are not publicly available. Data are, however, available from the authors upon reasonable request and with permission of Holstein Breeding Association and the Fédération Équestre Nationale.

Acknowledgments: We are grateful to the breeders who provided hair samples for this project. We thank the Association of Holstein Horse Breeders (Kiel, Germany) and the Fédération Equestre Nationale (Warendorf, Germany) for providing the phenotypic data.

Conflicts of Interest: The authors declare no conflict of interest.

Appendix A

Table A1. Results for the association between FN breeding values (EBV) and significant mitochondrial SNPs ($p < 5 \times 10^{-8}$), average breeding values for the two alleles, and distribution of the minor allele among haplogroups.

Trait	mtSNP [a]	n	Alleles [b]	MAF	Ø EBV Major Allele (SD)	Ø EBV Minor Allele (SD)	B	D	G	I	K	L	N	P	p-Value
TSP jumping	mtDNA_2216	12,656	C/T	0.023	105.74 (13.92)	111.03 (14.23)		100							1.983×10^{-10}
	mtDNA_2607	12,656	C/T	0.039	106.01 (13.92)	102.28 (14.24)						18.75			4.293×10^{-09}
TSP dressage	mtDNA_9963	11,045	G/A	0.224	86.29 (9.11)	87.48 (9.35)	100								7.325×10^{-09}
ABP jumping	mtDNA_2607	12,971	C/T	0.039	99.25 (13.80)	95.12 (14.76)						18.75			4.380×10^{-11}
ABP dressage	mtDNA_4599	12,231	C/G	0.124	85.08 (11.96)	83.03 (12.12)							76.92		2.638×10^{-10}
	mtDNA_8007	12,231	A/G	0.269	84.34 (11.98)	86.17 (11.96)	100	100							3.821×10^{-14}
	mtDNA_9963	12,231	G/A	0.246	84.34 (11.98)	86.37 (11.92)	100								5.709×10^{-16}
ZP walk	mtDNA_77	12,482	C/T	0.017	86.63 (10.40)	82.00 (9.77)						6.25			1.620×10^{-10}
	mtDNA_7600	12,482	G/C	0.018	86.62 (10.39)	82.52 (10.48)	4.17								3.067×10^{-09}
	mtDNA_8007	12,482	A/G	0.265	86.21 (10.30)	87.50 (10.64)	100	100							9.512×10^{-10}
	mtDNA_9963	12,482	G/A	0.239	86.19 (10.30)	87.67 (10.67)	100								1.260×10^{-11}
	mtDNA_15299	12,482	C/T	0.013	86.61 (10.41)	81.84 (9.25)			11.11						4.391×10^{-09}
ZP trot	mtDNA_77	12,482	C/T	0.017	82.58 (11.99)	77.16 (12.25)						6.25			4.254×10^{-11}
	mtDNA_1608	12,482	A/T	0.173	82.79 (12.04)	81.01 (11.78)							100		3.203×10^{-10}
	mtDNA_4599	12,482	C/G	0.125	82.73 (11.98)	80.76 (12.08)							76.92		1.158×10^{-09}
	mtDNA_8007	12,482	A/G	0.268	82.10 (11.88)	83.56 (12.32)	100	100							1.755×10^{-09}
	mtDNA_9963	12,482	G/A	0.242	82.07 (11.87)	83.82 (12.37)	100								2.748×10^{-12}
ZP canter	mtDNA_77	12,724	C/T	0.017	88.59 (12.99)	80.57 (13.89)						6.25			2.200×10^{-16}
	mtDNA_1608	12,724	A/T	0.169	88.75 (13.09)	87.00 (12.73)							100		1.273×10^{-08}
	mtDNA_4599	12,724	C/G	0.123	88.73 (13.05)	86.49 (12.83)							76.92		1.994×10^{-10}
	mtDNA_8007	12,724	A/G	0.263	87.98 (12.98)	89.77 (13.12)	100	100							7.456×10^{-12}
	mtDNA_9963	12,724	G/A	0.238	87.99 (12.99)	89.92 (13.11)	100								1.260×10^{-12}
ZP rideability	mtDNA_77	12,701	C/T	0.017	85.45 (12.61)	78.01 (12.60)						6.25			2.200×10^{-16}
	mtDNA_4599	12,701	C/G	0.123	85.58 (12.63)	83.47 (12.60)							76.92		5.726×10^{-10}
	mtDNA_4646	12,701	C/T	0.241	84.96 (12.56)	86.48 (12.87)	100								6.596×10^{-09}
	mtDNA_9963	12,701	G/A	0.238	84.92 (12.57)	86.59 (17.82)	100								2.243×10^{-10}
ZP free jumping	mtDNA_2607	12,587	C/T	0.039	104.78 (15.87)	99.39 (18.22)						18.75			1.588×10^{-13}
	mtDNA_4599	12,587	C/G	0.123	104.86 (15.98)	102.42 (16.03)							76.92		1.565×10^{-08}
ZP parcours jumping	mtDNA_2607	12,599	C/T	0.039	102.34 (13.53)	98.21 (14.79)						18.75			3.446×10^{-11}

Table A1. *Cont.*

Trait	mtSNP [a]	n	Alleles [b]	MAF	Ø EBV Major Allele (SD)	Ø EBV Minor Allele (SD)	Occurrence of Minor Allele in Haplogroups (%) [c]								p-Value
							B	D	G	I	K	L	N	P	
JPf jumping	mtDNA_2607	12,918	C/T	0.039	101.65 (16.48)	96.39 (18.02)						18.75			2.553×10^{-12}
JPf dressage	mtDNA_77	12,823	C/T	0.017	82.38 (13.71)	75.76 (14.08)						6.25			1.398×10^{-12}
	mtDNA_4599	12,823	C/G	0.123	82.54 (13.73)	80.27 (13.68)							76.92		7.841×10^{-10}
	mtDNA_8007	12,823	A/G	0.264	81.76 (13.65)	83.63 (13.91)	100	100							1.348×10^{-11}
	mtDNA_9963	12,823	G/A	0.238	81.76 (13.65)	83.87 (13.90)	100								1.136×10^{-13}
ZP jumping	mtDNA_2607	12,696	C/T	0.039	103.67 (16.28)	98.26 (18.27)						18.75			3.863×10^{-13}
ZP dressage	mtDNA_77	12,801	C/T	0.017	82.91 (13.55)	75.37 (13.86)						6.25			2.900×10^{-16}
	mtDNA_4599	12,801	C/G	0.123	83.05 (13.59)	80.87 (13.41)							76.92		2.268×10^{-09}
	mtDNA_8007	12,801	A/G	0.264	82.34 (13.46)	84.03 (13.87)	100	100							5.207×10^{-10}
	mtDNA_9963	12,801	G/A	0.238	82.32 (13.46)	84.27 (13.89)	100								4.479×10^{-12}
HEK jumping	mtDNA_1605	10,711	G/A	0.011	109.26 (18.93)	119.26 (18.74)				20.00					6.616×10^{-09}
	mtDNA_2216	10,711	C/T	0.022	109.28 (18.88)	117.92 (20.22)		100							1.877×10^{-12}
	mtDNA_14827	10,711	G/A	0.085	109.01 (18.92)	113.25 (18.94)		100	11.11	20.00		37.50	7.69		1.061×10^{-10}

[a] The following mtSNPs show perfect LD and only one mtSNP is declared in the subsequent evaluations: mtDNA_9963-mtDNA_1386-mtDNA_10766; mtDNA_1608-mtDNA_2339−mtDNA_4526; mtDNA_2216−mtDNA_11459; mtDNA_1605−mtDNA_13726; mtDNA_8007−mtDNA_11844; mtDNA_4646−mtDNA_13335. [b] Major/minor allele. [c] Percentage of lineages in the respective haplogroup carrying the minor allele. Haplogroups are assigned according to the work of [12].

Table A2. Results for the association between breeding values (EBV) for studbook inspection, mare performance test and significant mitochondrial SNPs ($p < 5 \times 10^{-8}$), average breeding values for the two alleles, and distribution of the minor allele among haplogroups.

Trait	mtSNP [a]	n	Alleles [b]	MAF	Ø EBV Major Allele (SD)	Ø EBV Minor Allele (SD)	Occurrence of Minor Allele in Haplogroups (%) [c]								p-Value
							B	D	G	I	K	L	N	P	
Studbook inspection															
HOL type	mtDNA_77	16,447	C/T	0.019	95.54 (19.88)	85.09 (17.44)						6.25			2.200×10^{-16}
	mtDNA_1382	16,447	C/T	0.056	95.13 (19.89)	98.90 (19.48)								100	1.977×10^{-08}
HOL topline	mtDNA_77	16,447	C/T	0.019	94.79 (20.96)	83.36 (18.23)						6.25			2.200×10^{-16}
	mtDNA_9197	16,447	T/C	0.012	94.46 (20.97)	104.37 (19.26)			11.11						4.124×10^{-11}
HOL forehand	mtDNA_77	16,447	C/T	0.019	92.66 (20.86)	83.18 (18.26)						6.25			1.688×10^{-15}
	mtDNA_7600	16,447	G/C	0.017	92.62 (20.16)	84.60 (22.16)	4.17								1.774×10^{-10}
	mtDNA_9774	16,447	G/A	0.011	92.58 (20.76)	83.67 (26.86)	4.17								2.089×10^{-08}
HOL hindquarters	mtDNA_1382	16,447	C/T	0.056	90.51 (22.42)	96.02 (22.31)								100	3.562×10^{-13}
	mtDNA_7600	16,447	G/C	0.017	90.95 (22.44)	83.46 (21.91)	4.17								3.122×10^{-08}
	mtDNA_9197	16,447	T/C	0.012	90.69 (22.45)	101.64 (20.11)			11.11						1.004×10^{-11}
	mtDNA_14653	16,447	A/G	0.021	90.99 (22.46)	83.01 (20.67)			33.33						6.133×10^{-11}

Table A2. Cont.

Trait	mtSNP [a]	n	Alleles [b]	MAF	Ø EBV Major Allele (SD)	Ø EBV Minor Allele (SD)	B	D	G	I	K	L	N	P	p-Value
HOL correctness of gaits	mtDNA_77	16,447	C/T	0.019	95.13 (19.66)	85.01 (18.09)						6.25			2.200×10^{-16}
	mtDNA_7600	16,447	G/C	0.017	95.07 (19.66)	87.40 (19.30)	4.17								9.645×10^{-11}
	mtDNA_14653	16,447	A/G	0.021	95.09 (19.69)	87.79 (17.72)			33.33						8.903×10^{-12}
HOL impulsion	mtDNA_77	16,447	C/T	0.019	95.34 (19.76)	87.12 (19.06)						6.25			2.666×10^{-13}
	mtDNA_1382	16,447	C/T	0.056	94.95 (19.74)	99.15 (18.52)								100	2.482×10^{-10}
	mtDNA_7600	16,447	G/C	0.017	95.31 (19.68)	87.83 (18.92)	4.17								2.829×10^{-10}
	mtDNA_9774	16,447	G/A	0.011	95.31 (19.62)	84.03 (23.45)	4.17								5.447×10^{-14}
	mtDNA_14653	16,447	A/G	0.021	95.37 (19.69)	86.86 (17.61)			33.33						1.897×10^{-15}
Mare performance test															
HOL canter	mtDNA_77	6334	C/T	0.016	94.41 (20.08)	81.09 (20.55)						6.25			9.199×10^{-10}
	mtDNA_1608	6334	A/T	0.160	94.99 (22.23)	89.97 (21.25)							100		2.009×10^{-11}
	mtDNA_4599	6334	C/G	0.151	91.24 (22.12)	85.94 (21.73)							76.92		2.666×10^{-10}
	mtDNA_4646	6334	C/T	0.258	93.12 (22.05)	97.24 (22.17)	100								5.784×10^{-11}
	mtDNA_8007	6334	A/G	0.277	92.99 (22.06)	97.31 (22.09)	100	100							2.000×10^{-12}
HOL walk	mtDNA_4646	6334	C/T	0.258	97.84 (22.54)	101.61 (24.12)	100								2.351×10^{-08}
	mtDNA_15299	6334	C/T	0.012	99.01 (22.99)	82.84 (17.19)			11.11						1.249×10^{-09}
HOL rideability	mtDNA_77	6334	C/T	0.016	94.85 (22.27)	78.25 (21.13)						6.25			2.088×10^{-13}
	mtDNA_1608	6334	A/T	0.160	95.31 (22.34)	90.79 (22.19)							100		6.809×10^{-09}
	mtDNA_4599	6334	C/G	0.151	95.21 (22.23)	89.88 (22.85)							76.92		4.052×10^{-12}
	mtDNA_4646	6334	C/T	0.258	93.52 (22.20)	97.64 (22.64)	100								3.055×10^{-10}
	mtDNA_8007	6334	A/G	0.277	93.50 (22.18)	97.42 (22.67)	100	100							8.262×10^{-10}
	mtDNA_14653	6334	A/G	0.021	94.83 (22.38)	82.56 (19.07)			33.33						1.183×10^{-09}

[a] The following mtSNPs show perfect LD and only one mtSNP is declared in the subsequent evaluations: mtDNA_9963-mtDNA_1386-mtDNA_10766; mtDNA_1608-mtDNA_2339−mtDNA_4526; mtDNA_4646−mtDNA_13335. [b] Major/minor allele. [c] Percentage of lineages in the respective haplogroup carrying the minor allele. Haplogroups are assigned according to the work of [12].

References

1. World Breeding Federation for Sport Horses: Breeder and Studbook Rankings. Available online: http://www.wbfsh.org/GB/Rankings/Breeder%20and%20Studbook%20rankings.aspx (accessed on 1 August 2021).
2. Fédération Équestre Nationale (FN). *Annual Report 2020*; FN-Verlag: Warendorf, Germany, 2020.
3. Lin, X.; Zheng, H.X.; Davie, A.; Zhou, S.; Wen, L.; Meng, J.; Zhang, Y.; Aladaer, Q.; Liu, B.; Liu, W.J.; et al. Association of low race performance with mtDNA haplogroup L3b of Australian thoroughbred horses. *Mitochondrial DNA Part A* **2018**, *29*, 323–33 [CrossRef] [PubMed]
4. Art, T.; Amory, H.; Desmecht, D.; Lekeux, P. Effect of show jumping on heart rate, blood lactate and other plasma biochemic values. *Equine Vet. J.* **1990**, *22*, 78–82. [CrossRef] [PubMed]
5. Engel, L.; Becker, D.; Nissen, T.; Russ, I.; Thaller, G.; Krattenmacher, N. Exploring the Origin and Relatedness of Maternal Lineage Through Analysis of Mitochondrial DNA in the Holstein Horse. *Front. Genet.* **2021**, *12*, 1242. [CrossRef] [PubMed]
6. Bowling, A.T.; Del Valle, A.; Bowling, M. A pedigree-based study of mitochondrial D-loop DNA sequence variation amon Arabian horses. *Anim. Genet.* **2000**, *31*, 1–7. [CrossRef] [PubMed]
7. Kavar, T.; Habe, F.; Brem, G.; Dovč, P. Mitochondrial D-loop sequence variation among the 16 maternal lines of the Lipizzan hors breed. *Anim. Genet.* **1999**, *30*, 423–430. [CrossRef] [PubMed]
8. Welker, V.; Stock, K.F.; Schöpke, K.; Swalve, H.H. Genetic parameters of new comprehensive performance traits for dressage an show jumping competitions performance of German riding horses. *Livest. Sci.* **2018**, *212*, 93–98. [CrossRef]
9. Jaitner, J. Genetic Evaluation for Horses. 2019. Available online: https://www.vit.de/fileadmin/DE/Zuchtwertschaetzung/FN ZWS_2019_description.pdf (accessed on 22 June 2021).
10. Miller, S.A.; Dykes, D.D.; Polesky, H.F.R.N. A simple salting out procedure for extracting DNA from human nucleated cell *Nucleic Acids Res.* **1988**, *16*, 1215. [CrossRef] [PubMed]
11. Rozen, S.; Skaletsky, H. Primer3 on the WWW for general users and for biologist programmers. In *Bioinformatics Methods an Protocols: Methods in Molecular Biology*; Krawetz, S., Misener, S., Eds.; Humana Press: Totowa, NJ, USA, 2000; Volume 13 pp. 365–386.
12. Achilli, A.; Olivieri, A.; Soares, P.; Lancioni, H.; Kashani, B.H.; Perego, U.A.; Nergadze, S.G.; Carossa, V.; Santagostino, M Capomaccio, S.; et al. Mitochondrial genomes from modern horses reveal the major haplogroups that underwent domesticatio *Proc. Natl. Acad. Sci. USA* **2012**, *109*, 2449–2454. [CrossRef]
13. Purcell, S.; Neale, B.; Todd-Brown, K.; Thomas, L.; Ferreira, M.A.; Bender, D.; Maller, J.; Sklar, P.; De Bakker, P.I.; Daly, M.J.; et a PLINK: A tool set for whole-genome association and population-based linkage analyses. *Am. J. Hum. Genet.* **2007**, *81*, 559–57 [CrossRef]
14. Barrett, J.C.; Fry, B.; Maller, J.; Daly, M.J. Haploview: Analysis and visualization of LD and haplotype maps. *Bioinformatics* **200** *21*, 263–265. [CrossRef]
15. Pendergrass, S.A.; Dudek, S.M.; Crawford, D.C.; Ritchie, M.D. Synthesis-View: Visualization and interpretation of SNP associatio results for multi-cohort, multi-phenotype data and meta-analysis. *BioData Min.* **2010**, *3*, 1–13. [CrossRef] [PubMed]
16. R Core Team. *R: A Language and Environment for Statistical Computing*; R Foundation for Statistical Computing: Vienna, Austria, 2020.
17. Ekine, C.C.; Rowe, S.J.; Bishop, S.C.; de Koning, D.J. Why breeding values estimated using familial data should not be used fo genome-wide association studies. *G3 Genes Genomes Genet.* **2014**, *4*, 341–347. [CrossRef] [PubMed]
18. Boore, J.L. Animal mitochondrial genomes. *Nucleic Acids Res.* **1999**, *27*, 1767–1780. [CrossRef] [PubMed]
19. Lührs-Behnke, H.; Röhe, R.; Kalm, E. Genetic analyses of riding test and their connections with traits of stallion performance an breeding mare tests. *Züchtungskunde* **2006**, *78*, 119–128.
20. Bray, M.S.; Hagberg, J.M.; Perusse, L.; Rankinen, T.; Roth, S.M.; Wolfarth, B.; Bouchard, C. The human gene map for performanc and health-related fitness phenotypes: The 2006-2007 update. *Med. Sci. Sports Exerc.* **2009**, *41*, 34–72. [CrossRef] [PubMed]
21. Niemi, A.K.; Majamaa, K. Mitochondrial DNA and ACTN3 genotypes in Finnish elite endurance and sprint athletes. *Eur. J. Hum Genet.* **2005**, *13*, 965–969. [CrossRef] [PubMed]
22. Mikami, E.; Fuku, N.; Takahashi, H.; Ohiwa, N.; Scott, R.A.; Pitsiladis, Y.P.; Higuchi, M.; Kawahara, T.; Tanaka, M. Mitochondria haplogroups associated with elite Japanese athlete status. *Br. J. Sports Med.* **2011**, *45*, 1179–1183. [CrossRef] [PubMed]
23. Nogales-Gadea, G.; Pinós, T.; Ruiz, J.R.; Marzo, P.F.; Fiuza-Luces, C.; López-Gallardo, E.; Ruiz-Pesini, E.; Martín, M.A.; Arenas, J Morán, M.; et al. Are mitochondrial haplogroups associated with elite athletic status? A study on a Spanish cohort. *Mitochondrio* **2011**, *11*, 905–908. [CrossRef] [PubMed]
24. Harrison, S.P.; Turrion-Gomez, J.L. Mitochondrial DNA: An important female contribution to thoroughbred racehorse perfor mance. *Mitochondrion* **2006**, *6*, 53–66. [CrossRef]
25. Próchniak, T.; Rozempolska-Rucińska, I.; Zięba, G. Maternal effect on sports performance traits in horses. *Czech J. Anim. Sci.* **2019** *64*, 361–365. [CrossRef]
26. Krattenmacher, N.; Tetens, J.; Hedt, S.; Stamer, E.; Thaller, G. The role of maternal lineages in horse breeding: Effects o conformation and performance traits. In Proceedings of the 10th World Congress on Genetics Applied to Livestock Productior Vancouver, BC, Canada, 17–22 August 2014. Available online: http://wcgalp.org/system/files/proceedings/2014/role-maternal lineages-horse-breeding-effects-conformation-and-performance-traits.pdf (accessed on 8 June 2021).

7. Børte, S.; Zwart, J.A.; Skogholt, A.H.; Gabrielsen, M.E.; Thomas, L.F.; Fritsche, L.G.; Surakka, I.; Nielsen, J.B.; Zhou, W.; Wolford, B.N.; et al. Mitochondrial genome-wide association study of migraine–the HUNT Study. *Cephalalgia* **2020**, *40*, 625–634. [CrossRef] [PubMed]

8. Li, M.; Schröder, R.; Ni, S.; Madea, B.; Stoneking, M. Extensive tissue-related and allele-related mtDNA heteroplasmy suggests positive selection for somatic mutations. *Proc. Natl. Acad. Sci. USA* **2015**, *112*, 2491–2496. [CrossRef] [PubMed]

animals

Article

Racing Performance of the Quarter Horse: Genetic Parameters, Trends and Correlation for Earnings, Best Time and Time Class

Ricardo Faria [1,2,*], António Vicente [3,4,5,6] and Josineudson Silva [7,8]

1   Faculdade de Ciências Agrárias e Veterinárias, Universidade Estadual Paulista (Unesp), Jaboticabal 14884-900, SP, Brazil
2   Hi-Tech Equine (HT Equine, Unipessoal LDA), 7330-313 Marvão, Portugal
3   Escola Superior Agrária de Santarém (ESAS), Instituto Politécnico de Santarém, 2001-904 Santarém, Portugal
4   CIISA—Centre for Interdisciplinary Research in Animal Health, Faculty of Veterinary Medicine, University of Lisbon, 1300-477 Lisboa, Portugal
5   AL4AnimalS—Associate Laboratory for Animal and Veterinary Sciences, 1300-477 Lisboa, Portugal
6   CIEQV—Life Quality Research Centre, Instituto Politécnico de Santarém, 2001-904 Santarém, Portugal
7   Faculdade de Medicina Veterinária e Zootecnia, Universidade Estadual Paulista (Unesp), Botucatu 18618-307, SP, Brazil
8   National Council for Science and Technological Development, Brasilia 71605-001, DF, Brazil
*   Correspondence: fariasky@gmail.com

Simple Summary: The main goal in selecting racehorses for breeding is usually monetary earning and animals that earn the highest amounts of money are usually selected for breeding. These animal are not always the fastest on the racecourse because there are several competitions where the animal were faster but the prize money for the winners was lower. This practical strategy meant that th fastest racehorses were not always selected for breeding. The present study evaluated several trait through the results of heritability, genetic correlations and trends. Results from this study indicat that when we evaluated the earnings trait together with other important traits, the best values wer observed. The selection process for the best sires and dams should be performed through two stage first, by evaluating the animals that obtained the best times and, in the second stage, by evaluatin the animals that had the highest monetary earnings. The design of breeding programs using th earnings trait in conjunction with another racing performance trait can change the results observe and improve genetic gains for speed racing Quarter Horses around the world.

Abstract: The aim of this study was to evaluate the sprint racing performance of Quarter Horses i Brazil. Estimating genetic parameters, trends and correlations were obtained by single- and two-trai analyses using Bayesian inference (earnings to 2 years of equestrian age, best time and time clas at distances of 301 m and 402 m). The data comprised a period of 38 equestrian years (1978 t 2015) with 23,482 sprint race records from 5861 animals. The heritability estimates were of low t moderate magnitude, ranging from 0.10 to 0.37 (single-trait) and from 0.15 to 0.41 (two-traits), and th repeatability was 0.31 to 0.46. The additive, residual and phenotypic correlations between earnings t 2 years of equestrian age and the other traits (best time and time class in distances 301 m and 402 m were high (−0.95, −0.96, 0.69 and 0.92), low (−0.29, −0.37, 0.26 and 0.27) and moderate (−0.4 −0.47, 0.37 and 0.47), respectively. There is a positive genetic trend for all traits considered. Howeve evaluation of the last 10 equestrian years (2006 to 2015) showed negative trends (genetic loss) an trends close to zero (genetic stagnation). The design of breeding programs using the earnings trait i conjunction with other racing performance traits can enhance changes in the genetic gains as a whol in speed-racing Quarter Horses. These findings suggest that the traits studied should be included i breeding selection programs for racing Quarter Horses.

Keywords: equine; heritability; racecourse; racehorse; speed

Citation: Faria, R.; Vicente, A.; Silva, J. Racing Performance of the Quarter Horse: Genetic Parameters, Trends and Correlation for Earnings, Best Time and Time Class. *Animals* **2023**, *13*, 2019. https://doi.org/10.3390/ani13122019

Academic Editors: Markku Saastamoinen, Isabel Cervantes and María Dolores Gómez Ortiz

Received: 26 February 2023
Revised: 11 June 2023
Accepted: 15 June 2023
Published: 17 June 2023

1. Introduction

The genetic origin of the Quarter Horse (QH) breed dates back to the 17th century, when English mares were bred with stallions from the regions of present-day Saudi Arabia and Turkey. The QH has become one of the largest and multifunctional horse populations in the world [1]. The use of QHs in family outings, working with cattle, Western competitions and sprint races has been preserved since the formation of the breed (17th century) until the 21st century. The official start of modern sprint QH races occurred in 1943 in the United States (Tucson, Arizona). In Brazil, information about the beginning of QH races is lacking, but zootechnical records date back to July 1978 (Sorocaba, São Paulo).

In evaluations of the racing performance of QHs in several studies, different methods have been used concerning the phenotypic traits of finish time [2], such as the ANOVA method and the effects of age and sex, with a range of 0.00 to 0.38 for heritability (h^2). For the traits of time and final rank [3], the animal model by REML was used and included random animal and permanent environmental effects with racing, sex, age and origin as fixed effects; the resulting h^2 ranged from 0.13 to 0.17. For race time and speed index [4], estimates were obtained by the MTGSAM program using animal models including the random additive genetic effect, random permanent environmental effect and the fixed effects of sex, age and race, with h^2 ranging from 0.14 to 0.41. Earnings [5] were estimated by Bayesian inference (GIBBS2F90), and included the effects of sex, year of racing and animal in all analyses, with h^2 ranging from 0.19 to 0.21. For the time class [6], the analyses used the same method as the previous study [5], and h^2 values ranged from 0.45 to 0.56. On the other side, evaluating the racing performance of the most well-known breed of horse (Thoroughbred), many phenotypic traits for a single purpose (racing performance) have been presented over the last several decades (1950 to 2010). In the review [7], 15 phenotypic traits for racing performance were presented and distributed in three categories, time and its several variations (final, best and average time), handicap or similar performance ratings (categorical traits), earnings (annual, star, average and log earnings) and morphology (standard and measures).

Genetic evaluations for racehorses based on three different traits (money earnings, time and score) using the same database and animal model, to our knowledge, have not yet been observed in the literature. The possibility of evaluating three traits of different categories (earnings, time and time class as categorical traits) with the same database will provide more complete information, with discriminatory analysis. It also allows the genetic evolution of traits over the years and the relationship between the earnings trait and other traits to be verified. This is especially important for the earnings trait (i.e., money earned by the horse for its owner) since it is considered the most important trait.

The use of earnings as a trait in horses was first reported in 1948 [8], who called it the Average Earnings Index. This trait was altered in 1977 [9] and named the Standard Starts Index. The trait was later denominated as earnings in 1980 [10]. In the QH breed, earnings showed moderate heritability estimates (0.19 to 0.28) and positive genetic trends [5], suggesting its inclusion in selection programs. The phenotype of final time is widely used in the genetic evaluations of racehorses [7]. Within this context, the best time trait, which represents the fastest time of all races of a given animal, is the trait most used as a selection criterion [11]. The categorical trait time class unites animals in categories based on the time obtained by the winner of the race. Thus, the minimal differences between competitors are disregarded, suggesting that they are of environmental and non-genetic origin. For the time class trait in QH [6], high heritability and repeatability estimates were reported, suggesting the application of this trait in selection programs. The authors suggested that these animals be evaluated at a younger age and separately for each distance.

The aim of this study was to evaluate the sprint racing performance of Quarter Horses in Brazil by single- and two-trait analyses, estimating genetic parameters and trends for earnings at 2 years of equestrian age (E2), best time (BT) and time class (TC) at distances of 301 m (BT301 and TC301) and 402 m (BT402 and TC402), using the same database in the evaluation of the different traits.

2. Materials and Methods

2.1. Data Description

The sprint race records of QHs were provided by the Sorocaba Jockey Club an
contained information for the period from July 1978 to June 2016. The races occurred in fiv
racecourses in the cities of Sorocaba, São Paulo, Ribeirão Preto, Jaú and Avaré, all of ther
in the state of São Paulo, Brazil. There were 23,482 records from 5861 animals (42.2% male
compiled from 5138 races. The records corresponded to distances of 275, 301, 320, 365 an
402 m (1072; 6579; 2726; 5682 and 7423 records, respectively). The data analyzed accour
for a considerable proportion (single measures 57.8% of the total animals and repeate
measures 55.2% of total records) of QH racehorses in Brazil (July 1978 to June 2016). For
better interpretation of the data, graphs were created showing the distribution of record
by trait (Supplementary Figure S1a,b).

The equestrian age class of the horse is determined by the equestrian year regardles
the date of birth. In the southern hemisphere, the equestrian year starts on July 1 and las
until June 30 of the following year. In the present study, equestrian age classes of 2, 3, 4 c
more years were used. To evaluate all QH animals with performance in each of the trait
the results of the genetic values and trends were analyzed separately by sex, breedin
animals and equestrian ages. In the groups of horses evaluated (Table 1), 42.2% were male
and, of these, 154 are breeding stallions (parents of 1676 athlete horses) with an average c
10.2 progenies; 57.8% were females, of which 964 are brood mares (mothers of 2811 athlet
horses) with an average of 2.9 progenies, evaluating a total of 1118 breeders.

Table 1. Descriptive statistics of the five racing traits and their respective phenotypic means in sprir
racehorses of the Quarter Horse breed in Brazil.

Item	Trait						
	Single Measures				Repeated Measures		
	E2 ± sd (log10 $)	BT301 ± sd (s)	BT402 ± sd (s)	Total (No.)	TC301 ± sd (pt)	TC402 ± sd (pt)	Total (No.)
Mean trait	3.05 ± 0.91	17.33 ± 0.68	23.06 ± 1.13	-	3.63 ± 1.05	3.42 ± 1.18	-
No of animals [a] and records [b]	1900 [a]	3365 [a]	3318 [a]	5861	6579 [b]	7423 [b]	23,482
No of males [a] and records [b]	860 [a]	1436 [a]	1438 [a]	2474	2909 [b]	3415 [b]	10,335
No of females [a] and records [b]	1040 [a]	1929 [a]	1880 [a]	3387	3670 [b]	4008 [b]	13,127
No of stallions [c] and records [d]	51 [c]	92 [c]	132 [c]	154	296 [d]	503 [d]	1227
No of mares [c] and records [d]	314 [c]	538 [c]	691 [c]	964	1177 [d]	1772 [d]	4982

E2, earnings at 2 years of equestrian age; BT301 and BT402, best time at 301 and 402 m; TC301 and TC402, tim
class at 301 and 402 m; sd, standard deviation; log10 $, base 10 logarithm to dollars; s, seconds; pt, points; Nc
number; [a], total animals, males and females evaluated in single traits E2, BT301 and BT402 (single measures
[b], total records evaluated in repeated trait TC301 and TC402 (repeated measures) belonging to the animal
evaluated in the BT301 and BT402 traits, respectively; [c], total breeding stallions and broodmares with racin
records in Brazil and evaluated in single traits E2, BT301 and BT402 (single measures); [d], total records evaluate
in repeated trait TC301 and TC402 (repeated measures) belonging to the breeding stallions and broodmares wit
racing records in Brazil and evaluated in the BT301 and BT402 traits, respectively.

2.2. Description of the Traits and Data Files

2.2.1. Earnings

Earnings represent the sum of monetary prizes earned by a horse over its lifetim
career or at a specific age or period. In the present study, considering that 74.9% o
the horses participated in their first race at 2 years of equestrian age and following th
recommendations by Silva et al. [5], earnings at 2 years of equestrian age (E2) was defined a
the trait. The prize money won at 2 years of equestrian age was obtained at different time
(from 1978 to 2015) and in different Brazilian currencies [12]. The values obtained [5] wer
then converted to international currency (USD), and the monetary update was obtained fo
June 2019 [13]. Since the phenotypic values of E2 showed no normal distribution, log1C
transformation was applied [14]. For the analysis of E2, only the records of races tha

had at least 1 or more monetary prizes were considered [15], resulting in a data file of 1900 animals, whose total prize moneys were summed when the animals had competed at the age of 2 equestrian years (Table 1).

The justification for evaluating E2 (2 years of age) was the possibility of selecting animals for breeding at the beginning of their performance career, the large number of records at 2 years of equestrian age available and the conclusions of other studies and authors [5,16]. The monetary prizes of the 1900 animals (E2) were related to 2963 classifications obtained at 2 years of age (37.8% of the total monetary prizes distributed between the years of 1978 and 2015 in Brazil). The monetary prize distributions of the 2963 classifications were obtained from the 1st to the 8th place, distributed as follows: 1st place (1054 monetary prizes), 2nd place (956 monetary prizes) and 3rd place (822 monetary prizes), and the 4th, 5th, 6th, 7th and 8th places have a collective total of 131 monetary prizes.

2.2.2. Best Time

The best time (BT) of each animal was defined as the fastest final time of all races at a given distance in which the animal competed. In the present study, the distances of 301 m (BT301) and 402 m (BT402) were evaluated, with 3365 and 3318 individual records (animals), respectively (Table 1). The final times were recorded by automatic electronic timing in seconds.

2.2.3. Time Class

The time class (TC) trait was divided into five different classes based on the final times (in seconds) of each race [6]. The scores for the formation of the classes were attributed within each race based on the percent difference in relation to the winner's time. Class 5 includes animals with a final time <0.10% in relation to the winner's final time, including the winner. Classes 4, 3, 2 and 1 include animals with a final time of 0.11% to 1.0%, 1.01% to 3.0%, 3.01% to 5.0% and >5.0% in relation to the winner's final time, respectively. The phenotypic values of TC are repeated measures of the animals and were evaluated at distances of 301 m (TC301), with 6579 records, and 402 m (TC402), with 7423 records (Table 1).

The distance of 301 m (BT301 and TC301) was chosen because it provided the largest number of records for the 21st century and for the first race of the animals. The distance of 402 m (BT402 and TC402), corresponding to a quarter of a mile, was chosen because of its importance for the breed and because it had the largest number of total records. The selection of the traits is based on the importance they have, as determined in other studies [5–7,16].

2.3. Analyses

Single-trait analyses were performed to obtain the variance components and to explore the results compared to two-trait analysis. The latter was carried out considering E2 as the anchor trait and changing the other traits one by one, i.e., computing E2 with BT301 (3967 animals), E2 with BT402 (3946 animals), E2 with TC301 (7695 records) and E2 with TC402 (9088 records).

The model used for E2 included the fixed effects of equestrian year of birth, sex and number of starts as a covariate [10], in addition to additive genetic and residual effects. For BT301 and BT402, the model included the fixed effects of equestrian year of birth, sex, equestrian age class and racecourse, as well as additive genetic and residual effects. For TC301 and TC402, the model included the systematic effects of sex, equestrian age and racecourse, and the random effects of animal, the permanent environment effects of the animal and residual effects.

The pedigree file used in all analyses contained 11,425 animals. The quality of the genealogical information was 5.5 ± 1.2 known equivalent generations [17], with 1855 sires and 4858 dams.

All models used for the evaluation of E2, BT301 and BT402 (linear) and of TC301 and TC402 (threshold) can be written in matrix form as:

$$y = X\beta + Z_1\alpha + Z_2c + e$$

where **y** is the vector of observations; β is the vector of fixed effects; α is the vector of the direct additive genetic effects of the animal; **c** is the vector of permanent environmental effects of the animal (only for TC301 and TC402); **e** is the vector of random residual effects and **X**, Z_1 and Z_2 are incidence matrices that relate the observations to the fixed effects, random direct additive genetic effects and uncorrelated permanent environmental effects, respectively.

For the threshold model (TC301 and TC402), it was assumed that the underlying scale shows a continuous normal distribution described as:

$$U|\theta \sim N\left(W\theta, I\sigma_e^2\right)$$

where **U** is the vector of the underlying scale of order **r**; $\theta' = (\beta', \alpha', c')$ is the vector of the location parameters of order **s**, with β being defined as systematic effects, α as additive genetic effect and **c** as the permanent environmental effect of the animal; **W** is the known incidence matrix of order **r** by **s**; **I** is the identity matrix of order **r** by **r**; and σ_e^2 is the residual variance.

In the analysis of categorical variables, vectors β, α and **c** are location parameters with a conditional distribution **y** | β, α, **c**. A uniform prior distribution was assumed for β, which reflects vague prior knowledge about this vector. Inverse Wishart distributions were attributed to the remaining components. Thus, the distribution of **y**, given the scale parameters, was assumed to be:

$$(y|\beta, \alpha, c, R) \sim N[X\beta + Z_\alpha\alpha + W_cc, I_N R]$$

TC301 and TC402 are categorical traits that are determined by unobservable continuous variables on an underlying scale, in which the initial threshold values are fixed: **t1 < t2** ... **< tj − 1**, with **t0** = − ∞ and **tj** = ∞, where j is the number of categories. The categories or scores of **yi** (TC trait) for each animal **i** are defined by **Ui** on the underlying scale:

$$Yi = (1)\ t0 < Ui \le t1;\ (2)\ t1 < Ui \le t2;\ (3)\ t2 < Ui \le t3;\ (4)\ t3 < Ui \le t4;\ (5)\ t4 < Ui \le t5,\ for\ i = 1, ..., n,$$

where **n** is the number of observations. After specification of the thresholds **t0** to **t5**, 1 of the thresholds (**t0** to **t5**) needed to be adjusted to an arbitrary constant. In this study, **t1 = 0** was considered, with the vector of estimable thresholds being defined as **t = t2, t3** and **t4**.

According to the Bayesian approach, in the two-trait analyses involving continuous (E2) and categorical (TC) variables, the initial distributions of the random genetic, uncorrelated and residual effects were assumed to follow a multivariate normal distribution as follows:

$$p\left(\begin{bmatrix} a_1 \\ a_2 \end{bmatrix}\middle| G\right) \sim N\left(\begin{bmatrix} 0 \\ 0 \end{bmatrix}, G = G_0 \otimes A\right), p\left(\begin{bmatrix} c_1 \\ c_2 \end{bmatrix}\middle| C\right) \sim N\left(\begin{bmatrix} 0 \\ 0 \end{bmatrix}, C = C_0 \otimes I\right),$$

$$p\left(\begin{bmatrix} e_1 \\ e_2 \end{bmatrix}\middle| R\right) \sim N\left(\begin{bmatrix} 0 \\ 0 \end{bmatrix}, R = R_0 \otimes I\right)$$

where **G0** is the genetic variance and covariance matrix; **C0** is the variance matrix of uncorrelated effects; **R0** is the residual variance matrix; $\otimes$ is the direct product operator; **A** is the relationship matrix; and **I** is the identity matrix.

Initial distributions of the (co)variances were assumed as inverse Wishart distributions for random genetic, uncorrelated and residual effects of the traits studied, including the covariance between them. Uniform prior distributions were defined for the fixed effects and thresholds. The degree of freedom corresponding to the inverse Wishart distribution which indicates the level of reliability of the initial distribution (v), was flat for all initial variances, i.e., it did not reflect the degree of knowledge about the parameters (v = 0) [18].

For single-trait analysis under linear and threshold animal models and two-trait analysis under linear–linear and linear–threshold models, Gibbs chains of 1,600,000 cycles for E2 and of 1,100,000 cycles for the other traits were run, with burn-in periods of 600,000 and 100,000 cycles, respectively. These numbers were chosen after verification of the stationary stage of the chain by graphical inspection [19]. The estimates were stored every 20 samples, totaling 50,000 samples. These samples were used to compute the posterior means, standard deviations and 95% highest posterior density intervals (HPD95) of the variance components, heritability, repeatability and genetic and residual correlations. The statistical analyses were performed with the GIBBSF90 test and THRGIBBS1F90 programs [20]. The latter was used for analysis of the categorical traits TC301 and TC402.

The results of single- and two-trait analyses were evaluated regarding convergence of the chain using the POSTGIBBSF90 program [20]. The established chain lengths (1,100,000 and 1,600,000 cycles) and burn-in periods (100,000 and 600,000 cycles) were sufficient to obtain convergence in the analyses [20]. The minimum effective sample size (ESS) of the (co)variance components indicated the number of independent samples with information equivalent to that present in the dependent sample [21]. The ESS estimates obtained in the present study were higher than 200 cycles, indicating convergence of the Gibbs chain.

The posterior estimated breeding values (EBV) were obtained by BLUP procedures using OPTION fixed_var mean [20] from the genetic and residual (co)variances obtained in the final analysis of the present study. The description of the EBV for the 5 traits corresponded to 1900 (E2), 3365 (BT301 and TC301) and 3318 (BT402 and TC402) animals born between 1971 and 2014, a period comprising the first and last births of animals with racing records in Brazil, respectively. The EBV for breeding stallions and broodmares over the same period are also described. A t-test was used to compare means between groups.

Genetic trends were calculated by linear regression of EBVs on year of birth of the animals (from 1971 to 2014). The results are presented in the form of graphs of the annual means for the five traits and their respective linear trend lines.

3. Results

3.1. Heritability (h^2) and Repeatability

The values of the additive genetic variation components (σ_g^2) were higher in all two-trait analyses than the one-trait analyses (Table 2). The values of the residual variance (σ_e^2) followed the same trend of σ_g^2, with the exception of the two-trait analysis between E2 × TC301 and E2 × TC402; in addition, the values of σ_e^2 were higher than the values σ_g^2 (Table 2), the difference being more than double the σ_e^2 values in uni-trait analyses (E2, BT301 and TC301) and two-trait analyses (E2 × BT301, E2 × BT402 and E2 × TC301, BT301 × E2 and TC301 × E2). At longer distances (BT402 and TC402), smaller differences were observed between the values of σ_g^2 and σ_e^2 (Table 2). The estimates of the permanent environmental variance (σ_c^2) showed lower values than σ_g^2 and σ_e^2.

The heritability (h^2) estimates obtained in the single-trait analyses were low for E2, BT301 and TC301 and moderate for BT402 and TC402 (Table 2). The shortest HPD95 intervals were observed for BT301 and BT402 and the highest interval was observed for TC402 (Table 2). The repeatability estimates for the two TC traits were moderate.

The h^2 estimates obtained in the two-trait analyses were higher than those of single-trait analysis (Table 2). The h^2 estimates continued to be low for the E2 when evaluated with the distance of 301 m (E2 × BT301 and E2 × TC301) and moderate at the distance of 402 m (E2 × BT402 and E2 × TC402). Moderate h^2 estimates were obtained for the distance of 301 m (TC301 × E2 and BT301 × E2) and high estimates for the longest distance of 402 m (BT402 × E2 and TC402 × E2). The central measures showed symmetry in the single- and two-trait analyses, and the standard deviations showed small magnitudes, except for the TC402 trait (Table 2).

The HPD95 intervals for the h^2 estimates exhibited low to moderate variation in th two-trait analyses. The highest interval was observed for E2 $\times$ TC402 (Table 2). Th repeatability estimates for the two TC traits were higher than those obtained in singl trait analysis.

Table 2. Posterior estimates of heritability and repeatability (including the additive genetic, residu and permanent environmental variation) in single- and two-trait analyses of E2, BT301, BT402, TC3(and TC40 in sprint racehorses of the Quarter Horse breed in Brazil.

Trait			Variation (σ)			Heritability (h^2)				ESS	Repeatability $\pm$ sd
			σ_g^2	σ_e^2	σ_c^2	Mean $\pm$ sd	Median	Mode	HPD95		
E2		Single	0.05	0.44	-	0.10 $\pm$ 0.04	0.10	0.10	0.02 to 0.19	516	-
	Two	BT301	0.08	0.44	-	0.16 $\pm$ 0.04	0.15	0.17	0.09 to 0.24	339	-
		BT402	0.06	0.22	-	0.22 $\pm$ 0.03	0.22	0.22	0.18 to 0.31	938	-
		TC301	0.07	0.41	-	0.15 $\pm$ 0.05	0.15	0.20	0.07 to 0.25	520	-
		TC402	0.18	0.36	-	0.33 $\pm$ 0.14	0.33	0.33	0.08 to 0.55	235	-
BT301	Single		0.03	0.10	-	0.23 $\pm$ 0.04	0.23	0.23	0.15 to 0.32	2500	-
	Two	E2	0.04	0.13	-	0.24 $\pm$ 0.04	0.24	0.23	0.16 to 0.32	549	-
BT402	Single		0.16	0.25	-	0.39 $\pm$ 0.04	0.39	0.35	0.30 to 0.43	6250	-
	Two	E2	0.19	0.27	-	0.41 $\pm$ 0.04	0.41	0.40	0.33 to 0.49	4232	-
TC301	Single		3.04	12.18	2.48	0.17 $\pm$ 0.06	0.17	0.14	0.08 to 0.32	280	0.31 $\pm$ 0.09
	Two	E2	3.58	13.24	2.52	0.19 $\pm$ 0.06	0.19	0.20	0.10 to 0.34	581	0.32 $\pm$ 0.08
TC402	Single		1.92	3.79	0.68	0.30 $\pm$ 0.13	0.30	0.33	0.25 to 0.43	360	0.43 $\pm$ 0.19
	Two	E2	3.55	5.99	1.01	0.34 $\pm$ 0.13	0.34	0.40	0.20 to 0.55	248	0.46 $\pm$ 0.17

σ_g^2, additive genetic variation; σ_e^2, residual variance; σ_c^2, permanent environmental variance; HPD95, 95% highe posterior density; ESS, effective sample size; E2, earnings at 2 years; BT301, best time at 301 m; BT402, best time a 402 m; TC301, time class at 301 m; TC402, time class at 402 m; sd, standard deviation.

3.2. Genetic Correlations

The lowest and highest values of r_g and r_p were observed for the evaluation o E2 $\times$ TC301 and E2 $\times$ BT402, respectively. The lowest estimates of r_r were observed fo E2 $\times$ BT301, the highest values of r_r were similar or equal for the other traits (Table 3). Th negative correlations between E2 and BT indicate that when E2 increases (greater earnings BT decreases (racing faster or completing races in a shorter time).

Table 3. Genetic, residual and phenotypic correlations between E2 and BT301, BT402, TC301 an TC402 in sprint racehorses of the Quarter Horse breed in Brazil.

Trait	E2		
	$r_g \pm$ sd	$r_r \pm$ sd	$r_p \pm$ sd
BT301	-0.95 ± 0.07	-0.29 ± 0.04	-0.41 ± 0.03
BT402	-0.96 ± 0.03	-0.37 ± 0.04	-0.47 ± 0.03
TC301	0.69 ± 0.13	0.26 ± 0.05	0.37 ± 0.04
TC402	0.92 ± 0.08	0.27 ± 0.09	0.47 ± 0.07

E2, earnings at 2 years; BT301, best time at 301 m; BT402, best time at 402 m; TC301, time class at 301 m; TC402, tim class at 402 m; r_g, genetic correlation; r_r, residual correlation; r_p, phenotypic correlation; sd, standard deviation.

3.3. Estimated Breeding Values and Genetic Trends for Uni-Trait Analyses

The average EBVs above zero for traits E2, TC301 and TC402 and below zero (negative for BT301 and BT402 are considered, in both scenarios, positive EBVs, given that the lowe the value the better for the BT trait. Comparisons of EBV should be made within each trai and not as a comparison between traits.

The mean EBV between males and females (sex) indicated significant differences for the evaluated traits, except for TC301, and males had higher EBVs than females in all traits (Table 4). Between non-breeding and breeding animals, the EBVs of non-breeding animals were significantly higher (p-value < 0.05) for traits BT301 and BT402, higher in TC301 (not significant, p-value > 0.05), equal in the E2 trait and significantly lower than breeding animals in TC402 (p-value < 0.05). Among sires and dams (stallions and mares), the EBVs were significantly different and higher in stallions for E2 and TC402 and in mares for BT301 and BT402, with higher EBVs also being seen in mares for TC301, but the differences were not significant. The EBVs within each trait regarding equestrian age (for E2; all other ages were obtained from pedigree relationship) decreased with advancing age for all traits, indicating the highest and lowest EBVs were observed for animals at 2 and 4 years of equestrian age, respectively. With the exception of TC301, EBV differences between all ages were significantly different (Table 4).

Table 4. Mean estimated breeding values of E2, BT301, BT402, TC301 and TC402 univariate traits for all animals and according to sex, breeding animal and equestrian age in racehorses of the Quarter Horse breed in Brazil.

Item			Trait Mean ± sd				
			E2 (log10 $)	BT301 (s)	BT402 (s)	TC301 (pt)	TC402 (pt)
Sex		Males	0.08 ± 0.20 [a]	−0.34 ± 0.46 [a]	−1.02 ± 0.90 [a]	0.66 ± 1.23 [a]	1.97 ± 1.67 [a]
		Females	0.07 ± 0.19 [b]	−0.26 ± 0.48 [b]	−0.88 ± 0.91 [b]	0.53 ± 1.23 [a]	1.76 ± 1.69 [b]
	Nonbreeding		0.08 ± 0.20 [a]	−0.32 ± 0.48 [a]	−0.96 ± 0.91 [a]	0.60 ± 1.22 [a]	1.82 ± 1.66 [b]
	Breeding		0.08 ± 0.22 [a]	−0.20 ± 0.47 [b]	−0.83 ± 0.88 [b]	0.58 ± 1.31 [a]	1.96 ± 1.81 [a]
BA		Stallions	0.08 ± 0.23 [a]	−0.16 ± 0.52 [b]	−0.76 ± 1.01 [b]	0.56 ± 1.37 [a]	1.99 ± 2.21 [a]
		Mares	0.07 ± 0.20 [b]	−0.20 ± 0.46 [a]	−0.84 ± 0.85 [a]	0.61 ± 1.29 [a]	1.95 ± 1.73 [b]
Age		2 years	0.08 * ± 0.21 [a]	−0.32 ± 0.48 [a]	−1.00 ± 0.90 [a]	0.63 ± 1.23 [a]	1.96 ± 1.68 [a]
		3 years	0.06 * ± 0.16 [b]	−0.23 ± 0.46 [b]	−0.78 ± 0.90 [b]	0.47 ± 1.20 [b]	1.55 ± 1.68 [b]
		4 years	0.04 * ± 0.15 [c]	−0.13 ± 0.41 [c]	−0.55 ± 0.85 [c]	0.32 ± 1.21 [b]	1.26 ± 1.59 [c]
	Total		0.08 ± 0.19	−0.29 ± 0.47	−0.94 ± 0.90	0.58 ± 1.23	1.85 ± 1.69

E2, earnings at 2 years; log10 $, base 10 logarithm to dollars; BT301, best time at 301 m; s, seconds; BT402, best time at 402 m; TC301, time class at 301 m; pt, points; TC402, time class at 402 m; sd, standard deviation; BA, breeding animal; different letters [a], [b] or [c] indicate significant differences p-value < 0.05; *, values obtained by the EBV of the pedigrees that started at age 3 years and 4 years or more.

The genetic trends over 44 years (1971 to 2014) obtained by univariate trait analyses and according to year of birth of the animals is presented in Figure 1. The negative genetic trends for BT301 and BT402 are favorable (Figure 1a) since lower values of BT are desirable. Oscillations were observed in animals born in the first 10 years (until 1980), followed by considerable improvements until the beginning of the 21st century (Figure 1a,c,e). Analysis of the last 10 years of birth (Figure 1b,d,f) showed values opposite to all equestrian years evaluated, indicating an annual decrease in the EBVs of the racing population of the QH breed in Brazil.

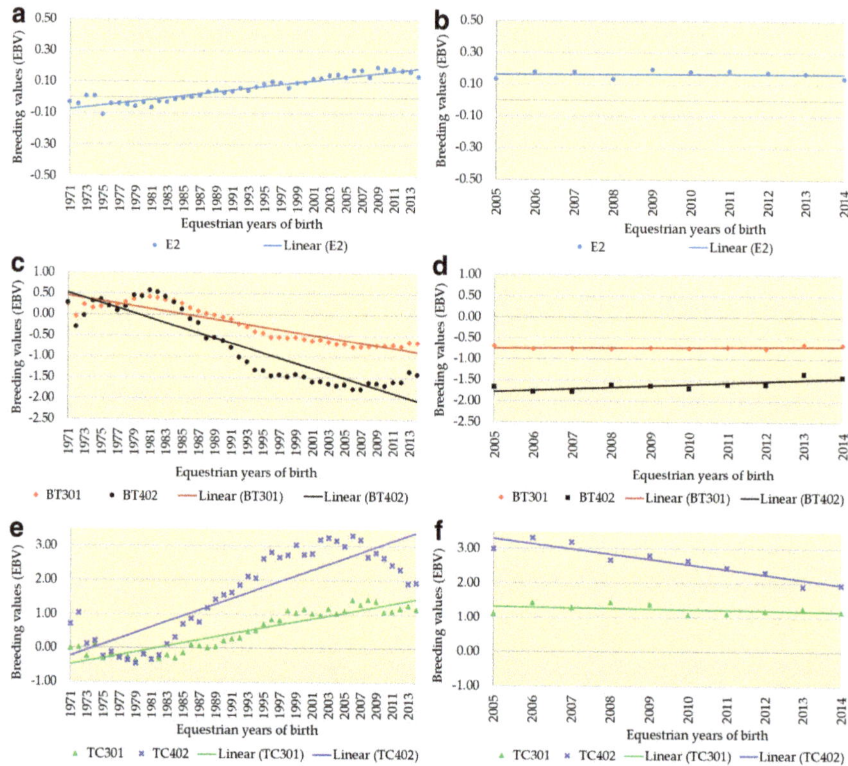

Figure 1. Linear genetic trends over 44 equestrian years of birth (1971 to 2014) and the last 10 year (2005 to 2014) for univariate traits Earnings at 2 years (E2) in (**a,b**); Best time at 301 m (BT301) an 402 m (BT402) in (**c,d**); and Time class at 301 m (TC301) and 402 m (TC402) in (**e,f**) obtained fo racehorses of the Quarter Horse breed in Brazil.

4. Discussion

The h^2 estimates of the present study, obtained through single-trait analysis (Table 2 are close to those reported in studies evaluating the performance of QHs. The earnings trai obtained an h^2 of 0.16 to 0.28 [5]. For final time, h^2 estimates of 0.20 to 0.38 [2,22], of 0.26 t 0.41 [4], and of 0.17 [3] were reported. Using multi-trait analysis [6], an h^2 of 0.45 to 0.5 was estimated for TC. Earnings, a trait used to evaluate the performance of Thoroughbred racehorses in Italy [23], France [24], Germany [25] and the Czech Republic [26], had h^2 estimates of 0.12 to 0.13, 0.02 to 0.06 and 0.10 to 0.19, respectively, which are equally low a those obtained in the present study (Table 2).

The higher h^2 estimates obtained in the two-trait analyses of BT402 × E2 and TC402 × E (Table 2) suggest that a combined evaluation provides more genetic information. The bivariate models are theoretically more accurate given that they consider the informatior on both traits to simultaneously estimate the random effects [27].

Evaluating BT in Thoroughbred racehorses obtained high (0.58 to 0.77), moderat (0.32 to 0.40) and low (0.10 to 0.23) h^2 estimates for 2 years, 3 years and considering al ages, respectively [28]. In Brazil, a low h^2 was reported (0.12) for BT in Thoroughbred animals [29]. Using earnings and BT data of Arabian racehorses also yielded low (0.14 t 0.17) and moderate (0.23 to 0.36) h^2 estimates, respectively [30].

The BT and E2 traits use racing performance, but with different definitions, which we can consider to be complementary to each other; i.e., it does not matter if a horse is the fastest without making money (qualifying) or if it earns a lot of money without being the fastest. By evaluating in model bivariate, they complement each other and allow for the obtaining of major h^2 values (Table 2). in addition to making the genetic values available to breeders for both traits and considered the value of genetic correlation (Table 3).

The h^2 estimates reported in the literature and those of the present study (Table 2) were similar, suggesting that the differences were possibly due to the different analysis methods, models and populations used in each study. The higher standard deviations in the analysis of the TC402 trait are possibly due to the greater variability in the competitors in the 402 m race, which has the highest participation; that is, due to the visibility of this event, owners strive to ensure that their animals win these races through the best training, nutrition and veterinary care. At the same time, however, there are breeders who do not carry out such preparation for the competition resulting in horses who finish with greater time differences from the winners. Thus, placing animals without the capacity to win in the same races as highly competitive horses results in greater variability when we evaluate data from the longest distance races (402 m). Among the characteristics evaluated, those with higher h^2 values (Table 2) indicate a greater possibility of genetic gains over the generations, suggesting their use as a selection criterion in the breeding programs of speed racing horses.

The repeatability estimates reported for QH in Brazil evaluating the distances of 301 and 402 m were 0.28 and 0.42, respectively, for the speed index, and 0.36 and 0.68, respectively, for final time [4]. In the United States, Mexico and Canada, the repeatability estimates at the distance of 402 m were 0.32 for final time; the authors did not provide results for the distance of 301 m but reported a repeatability of 0.36 evaluating animals at 2 years of age [2]. The moderate repeatability estimates obtained in the present study (Table 2) and in the cited literature indicate that the capacity of the animal to repeat the phenotypic value at each distance was moderate. For the selection and culling of animals, the largest number of possible measures should be taken into consideration [4]. On the other hand, multi-trait analysis of TC at different distances and the evaluation of a large number of repeated measures per QH animal [6] obtained high repeatability estimates (0.78 to 0.97), ensuring greater accuracy at the time of selection and culling of racehorses when compared to the results of single- and two-trait analyses of TC in the present study. These results indicate a more robust analysis can be obtained when including repeated records of horses in different distances. Considering that only 101 m (average difference of 5 s) separates the shortest (301 m) and the longest (402 m) distance, it is normal for the same animal to compete in both distances. Thus, it is common for animals, progenies and ancestors of QHs to have records at different distances.

Correlation estimates allow the behavior of a trait to be evaluated when selection is performed for another trait, and thus, to obtain genetic gain (indirect selection) in one trait through another that is difficult to measure or whose heritability is low [31]. The high genetic correlations (r_g) observed in the present study (Table 3) were also observed in the literature [3–5] evaluating the racing performance of QHs. The high r_g estimates suggest that selection for E2 can have benefits for BT at both distances and for TC at the longer distance. Considering the direct selection for BT and TC traits, this is the most efficient way to obtain improvement in these traits. Furthermore, joint selection with the E2 characteristic, in addition to increasing the values of both BT and TC, allows the selection of animals early.

The magnitudes of the residual correlations (Table 3) suggest that the same environmental factors affect the evaluated traits in different ways. The phenotypic correlations (r_p) followed the trends of r_g (Table 3), but the estimates were lower. The results obtained indicate that genetic factors exert a greater influence than environmental factors in QH racehorses in Brazil.

The performance of an elite horse is the sum of its genetic capacity (transmitted by its ancestors) and environmental influences (management, nutrition, training, horseman and other random effects). The EBVs are estimates of the animal's genetic capacity and can be used for the calculation of sports performance [6], morphology [32], health [33] and for the selection and culling of breeding animals.

The EBVs obtained (Table 4) indicate variability in the QH population, confirmed by the standard deviations. Faria et al. [6] suggested that QHs with superior EBVs would not be selected for breeding in view of the lower EBVs of breeding stallions compared to the average of the QH population. In the present study, this statement can be confirmed for breeding stallions by BT301, BT402 and TC301; for the remaining traits (E2 and BT402) EBVs were higher for males. Differences presented in Table 4 indicate that breeding females, compared to stallions, have genetic equality or superiority, possibly given the higher number of offspring per stallion, diluting their genetic gain. In the case of brood mares, the evaluation of EBVs for single traits (E2, BT301 and BT402) indicated that the best females are not being selected as breeding animals. However, analysis of the repeated measures (TC301 and TC402) showed that females with higher EBVs produce offspring in the population, i.e., they are selected as breeding animals. The results obtained for TC in the present study indicate that it is a valuable trait for the assessment and selection of elite QH racehorses for breeding.

The lack of selection of horses with superior EBVs for breeding impairs the evolution of racehorses of the QH breed, as observed in the present study by the EBV values between non-breeding and breeding animals (Table 4) and by the analysis of the genetic trend observed in the last 10 years (Figure 1). Breeding animals should be selected based on the correct and precise evaluations of horses, and breeders should request the help of researchers in the evaluation and selection of breeding animals, thus ensuring phenotypic improvements in QHs whose sprint race records have stagnated since 2009 [34].

There was a decrease in EBVs for all traits with advancing equestrian age (Table 4), suggesting that the animals can be selected early for breeding, i.e., immediately after completing 2 years of equestrian age. This approach makes it possible to reduce the long generational intervals (>10 years) of this population [12]; in fact, the generational interval is one of the limiting factors in the genetic improvement of horse breeding programs.

The genetic trends (Figure 1) indicated greater genetic progress in the longer distance races (BT402 and TC402). In North America [35], the authors reported opposite results with greater progress at the shorter distance evaluated (320 m) and lower progress at 402 m. The difference between the shorter and longer distances in North America was explained by the large number of records at 320 m, suggesting that the training of the animals focuses on this distance. In the present study, the same reason possibly explains the higher genetic trends at the distance of 402 m, which possessed a larger number of records (Table 1).

The genetic trends have decreased over the last 10 years (2005 to 2014) (Figure 1), a finding that suggests possible stagnation in the genetic evolution of the racing line of the QH breed in Brazil, comparable to that reported for the Thoroughbred breed [7]. The stagnation and genetic losses observed in the last decade (Figure 1) must be recovered through targeted mating, with an increase in the number of breeding animals with origin different from the ancestors of the QH breed in Brazil, as suggested in the populational study [36], who observed a genetic bottleneck in the current population of the racing line of the QH breed in Brazil.

5. Conclusions

Overall, this study emphasizes the lack of genetic response in recent years of the selection/evaluation of QHs bred in Brazil. The decline in genetic response was confirmed by the absence of significant genetic gains for the traits analyzed. The earnings trait exhibits genetic variability and genetic correlations favorable with a racing performance trait (best time and time class), suggesting that including the earnings trait in selection programs along with other racing performance traits may resulting in positive changes for the genetic trends. Quarter Horse associations, technicians and breeders must apply different strategies in selection and improvement programs that will allow for the genetic evolution of this population in conjunction with the monetary gains of the owners.

Supplementary Materials: The following supporting information can be downloaded at: https://www.mdpi.com/article/10.3390/ani13122019/s1, Figure S1: Distribution of data from Table 1 by traits E2, earnings at 2 years of equestrian age; BT301 and BT402, best time at 301 and 402 m; TC301 and TC402, time class at 301 and 402 m.

Author Contributions: Conceptualization, R.F. and J.S.; methodology, R.F. and J.S.; software, R.F.; validation, R.F., A.V. and J.S.; formal analysis, R.F.; investigation, R.F., A.V. and J.S.; resources, R.F.; data curation, R.F.; writing—original draft preparation, R.F.; writing—review and editing, A.V. and J.S.; visualization, R.F., A.V. and J.S.; supervision, J.S.; review, R.F., A.V. and J.S. All authors have read and agreed to the published version of the manuscript.

Funding: This study was financed in part by the Coordenação de Aperfeiçoamento de Pessoal de Nível Superior (CAPES), Brazil (Finance Code 001). PhD fellowship granted (PROEX—Program of Academic Excellence) (88882.180634/2018-01) to R.A.S. Faria and Conselho Nacional de Desenvolvimento Científico e Tecnológico (CNPq, Brazil) productivity grant to J. A.II V. Silva.

Institutional Review Board Statement: Not applicable.

Informed Consent Statement: Not applicable.

Data Availability Statement: The data are not available for privacy reasons and belong to the Brazilian Association of Quarter Horse Breeders (ABQM) and Jockey Clube de Sorocaba (JCS) in Brazil.

Acknowledgments: Thanks to the Brazilian Association of Quarter Horse Breeders (ABQM) and the Jockey Clube de Sorocaba (JCS) for providing the data. We also acknowledge the valuable comments and suggestions of peer reviewers.

Conflicts of Interest: The authors declare no conflict of interest. The funders had no role in the design of the study; in the collection, analyses, or interpretation of data; in the writing of the manuscript; or in the decision to publish the results.

References

1. ABQM. Associação Brasileira de Criadores Do Cavalo Quarto de Milha. Available online: https://www.abqm.com.br/pt/conteudos/quarto-de-milha/quarto-de-milha-no-brasil (accessed on 3 March 2020).
2. Buttram, S.T.; Wilson, D.E.; Willham, R.L. Genetics of Racing Performance in the American Quarter Horse: III. Estimation of Variance Components. *J. Anim Sci.* **1988**, *66*, 2808–2816. [CrossRef]
3. Villela, L.C.V.; Mota, M.D.S.; Oliveira, H.N. Genetic Parameters of Racing Performance Traits of Quarter Horses in Brazil. *J. Anim. Breed. Genet.* **2002**, *119*, 229–234. [CrossRef]
4. Corrêa, M.J.M.; Mota, M.D.S. Genetic Evaluation of Performance Traits in Brazilian Quarter Horse. *J. Appl. Genet.* **2007**, *48*, 145–151. [CrossRef] [PubMed]
5. Silva, A.P.A.; Curi, R.A.; Langlois, B.; Silva, J.A.V. Genetic Parameters for Earnings in Quarter Horse. *Genet. Mol. Res.* **2014**, *13*, 5840–5848. [CrossRef] [PubMed]
6. Faria, R.A.S.; Maiorano, A.M.; Correia, L.E.C.d.S.; Santana, M.L., Jr.; Silva, J.A.I.V. Time Class for Racing Performance of the Quarter Horse: Genetic Parameters and Trends Using Bayesian and Multivariate Threshold Models. *Livest. Sci.* **2019**, *225*, 116–122. [CrossRef]
7. Thiruvenkadan, A.K.; Kandasamy, N.; Panneerselvam, S. Inheritance of Racing Performance of Thoroughbred Horses. *Livest. Sci.* **2009**, *121*, 308–326. [CrossRef]
8. Estes, A. Statistics on prominent sires, adjusted for changing dollar. *Blood Horse* **1948**, *52*, 470.
9. Clarke, S. Standard starts index. *Blood Horse* **1977**, *103*, 1612.
10. Langlois, B. Heritability of Racing Ability in Thoroughbreds—A Review. *Livest. Prod. Sci.* **1980**, *7*, 591–605. [CrossRef]
11. Hintz, R.L. Genetics of Performance in the Horse. *J. Anim Sci.* **1980**, *51*, 582–594. [CrossRef]
12. Faria, R.A.S. *Phenotypic and Genetic Diversity of the Quarter Horse Racing*; São Paulo State University—FCAV: São Paulo, Brazil, 2019.
13. MAPA. Ministerio Da Agricultura, Pecuária e Abastecimento. Available online: http://www.agricultura.gov.br/ (accessed on 5 February 2019).
14. Snedecor, G.W.; Cochran, W.G. *Statistical Methods*, 6th ed.; Iowa State University Press: Ames, IA, USA, 1967.
15. Tolley, E.A.; Notter, D.R.; Marlowe, T.J. A Review of the Inheritance of Racing Performance in Horses. *Anim. Breed. Abstr.* **1985**, *53*, 163–185.
16. Langlois, B.; Blouin, C. Annual, Career or Single Race Records for Breeding Value Estimation in Race Horses. *Livest. Sci.* **2007**, *107*, 132–141. [CrossRef]
17. Maignel, L.; Boichard, D.; Verrier, E. Genetic Variability of French Dairy Breeds Estimated from Pedigree Information. *Interbull Bull.* **1996**, *14*, 48–56.

18. Van Tassell, C.P.; Van Vleck, L.D.; Gregory, K.E. Bayesian Analysis of Twinning and Ovulation Rates Using a Multiple-Trait Threshold Model and Gibbs Sampling. *J. Anim. Sci.* **1998**, *76*, 2048–2061. [CrossRef]
19. Kass, R.E.; Carlin, B.P.; Gelman, A.; Neal, R.M. Markov Chain Monte Carlo in Practice: A Roundtable Discussion. *Am. Stat.* **1998**, *52*, 93–100. [CrossRef]
20. Misztal, I.; Tsuruta, S.; Strabel, T.; Auvray, B.; Druet, T.; Lee, D.H. BLUPF90 and Related Programs (BGF90). In Proceedings of the 7th World Congress on Genetics Applied to Livestock Production, Montpellier, France, 19–23 August 2002; Volume 28, pp. 21–22.
21. Sorensen, D.; Andersen, S.; Gianola, D.; Korsgaard, I. Bayesian Inference in Threshold Models Using Gibbs Sampling. *Genet. Sel. Evol.* **1995**, *27*, 229–249. [CrossRef]
22. Buttram, S.T.; Willham, R.L.; Wilson, D.E. Genetics of Racing Performance in the American Quarter Horse: II. Adjustment Factors and Contemporary Groups. *J. Anim Sci.* **1988**, *66*, 2800–2807. [CrossRef]
23. Galizzi Vecchiotti, G.; Pazzaglia, G. Ereditabilita 'Dell' attitudine Alla Corsa Del Puro Sangue Inglese Studiata Su Corse Piane Italiane. *Atti Soc. Ital. Sci. Vet.* **1976**, *30*, 490–492.
24. Langlois, B.; Poirel, D.; Tastu, D.; Rose, J. Analyse Statistique et Génétique Des Gains Des Pur Sang Anglais de Trois Ans Dans Les Courses Plates Françaises. *Genet. Sel. Evol.* **1975**, *7*, 387–408. [CrossRef] [PubMed]
25. Bugislaus, A.E.; Roehe, R.; Uphaus, H.; Kalm, E. Development of Genetic Models for Estimation of Racing Performances in German Thoroughbreds. *Arch. Anim. Breed.* **2004**, *47*, 505–516. [CrossRef]
26. Svobodova, S.; Blouin, C.; Langlois, B. Estimation of Genetic Parameters of Thoroughbred Racing Performance in the Czech Republic. *Anim. Res.* **2005**, *54*, 499–509. [CrossRef]
27. Zhang, J.; Schumacher, F.R. Evaluating the estimation of genetic correlation and heritability using summary statistics. *Mol. Genet. Genom.* **2021**, *296*, 1221–1234. [CrossRef] [PubMed]
28. Yorov, I.; Kissyov, M. Heritability of Some Body Measurements and Speed in Thoroughbred Horses. *Genet. Plant Breed.* **1976**, *9*, 480–487.
29. Mota, M.D.S.; Oliveira, H.N.; Silva, R.G. Genetic and Environmental Factors That Affect the Best Time of Thoroughbred Horses in Brazil. *J. Anim. Breed. Genet.* **1998**, *115*, 123–129. [CrossRef]
30. Ekiz, B.; Kocak, O. Phenotypic and Genetic Parameter Estimates for Racing Traits of Arabian Horses in Turkey. *J. Anim. Breed. Genet.* **2005**, *122*, 349–356. [CrossRef] [PubMed]
31. Frankham, R. Introduction to Quantitative Genetics (4th Edn). *Trends Genet.* **1996**, *12*, 280. [CrossRef]
32. Vicente, A.A.; Carolino, N.; Ralão-Duarte, J.; Gama, L.T. Selection for Morphology, Gaits and Functional Traits in Lusitano Horses. I. Genetic Parameter Estimates. *Livest. Sci.* **2014**, *164*, 1–12. [CrossRef]
33. Büttgen, L.; Geibel, J.; Simianer, H.; Pook, T. Simulation Study on the Integration of Health Traits in Horse Breeding Programs. *Animals* **2020**, *10*, 1153. [CrossRef]
34. AQHA. World Records for a Quarter Horse. Available online: https://www.aqha.com/media/22294/world-records.pdf (accessed on 7 January 2022).
35. Wilson, D.E.; Willham, R.L.; Buttram, S.T.; Hoekstra, J.A.; Luecke, G.R. Genetics of Racing Performance in the American Quarter Horse: IV. Evaluation Using a Reduced Animal Model with Repeated Records. *J. Anim Sci.* **1988**, *66*, 2817–2825. [CrossRef]
36. Faria, R.A.S.; Vicente, A.P.A.; Ospina, A.M.T.; Silva, J.A.I.V. Pedigree Analysis of the Racing Line Quarter Horse: Genetic Diversity and Most Influential Ancestors. *Livest. Sci.* **2021**, *247*, 104484. [CrossRef]

~ticle

~Morpho-Functional Traits in Pura Raza Menorquina Horses: ~Genetic Parameters and Relationship with Coat Color Variables

~avinia I. Perdomo-González [1], Rocío de las Aguas García de Paredes [1], Mercedes Valera [1], Ester Bartolomé [1] ~nd María Dolores Gómez [1,2,*]

[1] Escuela Técnica Superior de Ingeniería Agronómica (ETSIA), Universidad de Sevilla, Carretera de Utrera km1, 41013 Sevilla, Spain

[2] Asociación de Criadores y Propietarios de Caballos de Raza Menorquina, Edificio Sa Roqueta, C/ Bijuters, 36, Bajos, 07760 Ciutadella de Menorca, Spain

* Correspondence: pottokamdg@gmail.com

~itation: Perdomo-González, D.I.; ~arcía de Paredes, R.d.l.A.; Valera, ~.; Bartolomé, E.; Gómez, M.D. ~orpho-Functional Traits in Pura ~aza Menorquina Horses: Genetic ~arameters and Relationship with ~oat Color Variables. *Animals* **2022**, ~, 2319. https://doi.org/10.3390/ ~ni12182319

~cademic Editor: Guosheng Su

~eceived: 25 May 2022
~ccepted: 2 September 2022
~ublished: 7 September 2022

~ublisher's Note: MDPI stays neutral ~ith regard to jurisdictional claims in ~ublished maps and institutional affil-~tions.

Simple Summary: In this study, we estimated genetic parameters of 46 linear morpho-functional traits, and analyzed the relationship between two coat color traits (quality of black coat color [QB] and the quantity of white marks [WM]) and other linear morpho-functional traits within the breeding program of Pura Raza Menorquina horses, whose studbook only permits the use of black-coated animals with a small quantity of white marks as breeding stock. A total of 772 records from 333 animals were analyzed to estimate genetic parameters for 46 linear traits scored by four appraisers using seven classes. Heritability values for morpho-functional traits were low to medium and matched the range in the bibliography. Medium heritability values were obtained for both coat color traits (0.36 for QB and 0.23 for WM). Genetic correlations between coat and morpho-functional traits ranged between 0.015 and 0.816 in absolute value for QB and between 0.014 and 0.638 in absolute value for WM. The highest correlation values were obtained between QB and upper neck line (0.816) and between WM and form of the hoof (0.638). It was observed that the animal group with low and the group with high breeding values for QB and WM had a clear differentiation of the other mor-pho-functional traits.

Abstract: The studbook of Pura Raza Menorquina horses only permits the use of black-coated animals with a small quantity of white marks as breeding stock. Its breeding program uses linear morpho-functional traits as selection criteria. Our aim was to estimate the genetic parameters of linear morpho-functional traits, and reveal relationship of quality of black coat color (QB) and percentage of white marks (WM) with the other morphological and functional linear traits in this breed. A total of 46 linear traits were scored by four appraisers using seven classes, with a total of 772 records from 333 animals ($\geq$4 years old). Univariate animal models using a Bayesian approach were used, with a pedigree of 757 animals. Sex (two) and appraiser-season (13) were included as fixed effects, age as a linear covariate, and permanent environmental and additive genetic as random effect. The heritabilities of the morpho-functional traits were low to medium (0.09–0.58) and matched the range in the bibliography. Heritabilities for coat color traits were 0.36 for QB and 0.23 for WM. The highest genetic correlations were obtained between QB and upper neck line (0.816) and between WM and form of the hoof (0.638). The negative signs of most of the genetic correlations between WM and the functional traits is also remarkable, contributing to the selection of functional traits against the presence of white marks in this population. A clear genetic differentiation was observed between animals with better breeding values for QB and WM, corroborated by a study on founders. In conclusion, QB and WM could show different genetic backgrounds.

Keywords: conformation; equine; genetic correlations; heritability; linear type traits

1. Introduction

The current conformation of horses is the result of both natural and breeders' selectio and the traits evaluated for each breed depend on their breeding purposes [1]. Conformatic has long been a driving force in horse selection and breed identification, particularly as predictor of performance and susceptibility to injury [2]. However, it is also important fe aesthetics, wellness, and the durability and functionality of horses [3] because conformatic defines the limits of the range of movement and function of the horse and its ability t perform [4]. Estimated relative economic values of selection criteria for riding horses, whic are based on sale prices, indicate that conformation and movement are the most importar traits in horse breeding [5,6] and achieve the highest prices. Conformation traits can b conditioned by several factors (e.g., breed, nutrition, age, or sex) and also coat color [7]. I addition, coat color is one of the most noticeable animal features and has interested an intrigued horse breeders for centuries [8].

The genetic evaluation of conformation traits is of major importance in horse selectio because it is a tool for indirect performance selection. This is because the heritabilit estimates for conformation traits are often higher than those estimated for performanc traits. Some authors [9] have affirmed that the efficiency of indirect selection for performan traits depends on genetic variability of the conformation traits and is also related to geneti associations between conformation and functional performance traits [10,11].

The current trend for breeding sport horses combines selection for conformatio and performance. Unfortunately, most of the information on sport performance onl becomes available later in the horse's life. For this reason, the selection of riding horse is based primarily on other related traits, which are available at a younger age and hav higher heritability values and adequate correlations with the selection traits [9] (suc as conformation traits). In this context, to economize money and time, Gómez et al. [2 reco-mmended the use of a two-stage selection program, with a first selection based o the morphological features of the animal and a second selection based on the animal' performance in equestrian events for Spanish horses.

The Pura Raza Menorquina horse (PRMe) is an endangered native breed establishe in 1988. However, they were traditionally bred and located on the island of Menorc (Ba-learic Islands, Spain). Originally, they were mainly used as draught animals i agriculture and forestry. However, it is a highly versatile breed with excellent qualitie (beauty, boldness, intelligence, nobility, agility, and resistance) that make it a highly-prize animal whether for dressage or as a saddle horse, a work horse, a light cart-horse or even sport horse [12]. Nowadays, they are mainly used in Classic and Menorcan Dressage wit very good results.

The last official update of the Spanish Ministry of Agriculture database [13] reported population size of 3798 active individuals, including 624 stallions and 764 mares, of whic only 2906 were located in Spain. The remaining 23.5% of the PRMe population is located i foreign countries, contributing to the ex-situ conservation of its genetic diversity, mostly i other European countries such as France, Germany, and Italy. Although some PRMe horse are also located in the USA and Russia.

The official breeding program of PRMe horses (with conservation and selection aims and their current uses have probably influenced their body conformation and type, as we as their movements and performance. Thus, Menorcan breeders tend to select animal based on conformation to achieve good performance for Menorcan and Classic dressage [12 From a morphological point of view, they are a medium-sized breed (1.60 m average heigh at the withers) with a sleek silhouette and a sub-convex to straight profile. According t breeding regulations, only black animals with a small quantity of white marks can be use as breeding stock [14].

The main objective of this study was to estimate genetic parameters of morphologica and functional traits, and analyze relationship of quality of their black coat color (QB and the percentage of white marks (WM) with the other morphological and functiona linear traits.

2. Materials and Methods

2.1. Materials

For this analysis, we used the linear morpho-functional data of PRMe horses provided by the Asociación de Criadores y Propietarios de Caballos de Raza Menorquina (ACPCRMe). To avoid subjectivity to an ideal type in the comparison, the overall score was replaced by several simple traits, which are linearly evaluated on a scale from one biological extreme to the other; in this way, linear evaluation does not grade an animal, but rather describes it [15].

Linear type traits scored by four skilled horse appraisers with the same evaluation criteria over nine years of evaluation (2013–2021) were considered in the current analysis. A total of 772 records from 333 animals (188 males and 145 females from 118 different studs) were recorded for 46 linear morpho-functional traits in the studs, in sport competitions, and in morphological events held for PRMe horses. Thus, the average number of records per animal was 2.32 and the average number of records per stud was 5.51. The animals were at least four years old, and the average age of the recorded animals was 8.44 years. The animals analyzed represented 12.9% of the population registered in the official Menorca Horse Studbook, which were active and located in Spain in 2021, thereby constituting the nucleus for selection and the majority of the population of PRMe horses all around the world. In this way, the results obtained in this study can be considered descriptive of the current status of this genetic resource and can be used to control its development in future years.

To collect the information, a structured score sheet with a scale of seven categories (from one to seven, without half points) was used by appraisers, in which the extremes represented the biological extremes of the population for the linear traits (see Table 1). Both extremes and the central class were defined on the sheet. The traits can be grouped into two coat color traits (QB and WM), 35 morphological traits (nine for the head and neck region, 14 for the body region, and 12 for the limbs), and nine functional traits evaluated with the horse led by hand at walk (four) and trot (five).

For data collection, the QB was defined using the same criteria as Smith [16], who described two different types of black coat: non-fading (jet black, which is charcoal black with a metallic shine) and fading (black coat color without shine, fading to a reddish-brown tinge). In PRMe horses, the appraisers used seven levels for black coat, according to their quality, which covered the whole range of both types of black coat defined by the authors cited above. The WMs were defined by seven classes using the ranking of penalties established in the official PRMe Horse studbook by their locations and extensiveness.

2.2. Methods

Preliminary analyses of variance were carried out using the univariate GLM procedure in SAS software v.9.4 [17] to assess the relative importance of the non-genetic effects that could influence the morpho-functional traits analyzed in PRMe horses. Permanent environment, sex (two levels: male and female), the combination of appraiser and the season (13 levels), quality of the black coat color (grouped into three levels: good [6,7], normal [4,5], and bad [1–3] quality), and percentage of white marks (grouped into three le-vels: without WMs [1,2], few WMs [3,4], and many WMs [5–7]) were included as fixed effects; the age was also included as a linear covariate.

The owner stud of the animals was not included in the model because no statistically significant differences were detected for most of the analyzed traits. This could be caused by (1) the similar management systems of the animals in the different studs (females were free in the countryside throughout the year with ad libitum feeding, except when they were close to giving birth, and males were in individual stables with temporal access to a paddock and controlled feeding) and (2) the geographic nearness of the different studs because all of them were located on Menorca island with approximately 700 km^2 of surface and the same weather conditions.

Table 1. Description of the 46 linear morpho-functional traits analyzed in Pura Raza Menorquina Horses.

		Trait	Class				Trait	Class	
			1	7				1	7
Color	1	Quality of black coat	Pale	Very dark		24	Chest width	Very narrow	Very wide
	2	Percentage of white marks	Absence	Excess		25	Thorax width	Very narrow	Very wide
Morphological traits	3	Head width	Very narrow	Very wide	Morphological traits	26	Forearm length	Very short	Very long
	4	Head length	Very short	Very long		27	Cannon bone length	Very short	Very long
	5	Head depth	Shallow	Very deep		28	Fore knee perimeter	Very narrow	Very wide
	6	Head profile	Concave	Convex		29	Cannon bone perimeter	Very narrow	Very wide
	7	Head expression	Not much	Very much		30	Forelimb side view	Camped under	Camped out
	8	Neck length	Very short	Very long		31	Fore hoof-pastern axis side view	Horizontal	Vertical
	9	Head-neck junction	Very covered	Very marked		32	Fore-hoof slope	Horizontal	Vertical
	10	Upper neck line	Concave	Convex		33	Form of hoof	Cylindrical	Bell-shaped
	11	Lower neck line	Concave	Convex		34	Fore-hoof front view	Toe-in	Toe-out
	12	Height of withers	Very short	Very high		35	Forelimb front view	Closed	Open
	13	Shape of withers	Not prominent	Very prominent		36	Hock side view	Closed	Open
	14	Shoulder length	Very short	Very long		37	Hock rear view	Closed	Open
	15	Shoulder angle slope	Very horizontal	Very vertical	Functional traits	38	Walking activity	Very little	Hasty
	16	Thorax depth	Shallow	Very deep		39	Walking clarity	Very little	Much
	17	Back length	Very short	Very long		40	Walking amplitude	Very little	Wide
	18	Loin length	Very short	Very long		41	Walking suppleness	Very little	Much
	19	Shape of back-loin line	Very concave	Slightly convex		42	Trotting amplitude	Very little	Wide
	20	Croup length	Very short	Very long		43	Trotting suppleness	Very little	Much
	21	Croup angle	Very horizontal	Very vertical		44	Trotting impulsion	Very little	Much
	22	Buttock length	Very short	Very long		45	Trotting equilibrium	Downhill	Uphill
	23	Withers-croup equilibrium	Downhill	Uphill		46	Trotting suspension	Very little	Much

The heritability values were calculated based on the variance components estimated using univariate animal models with a Bayesian approach via Gibbs sampling using the GIBBSF90+ module of the BLUPF90 software [18]. The following model was fitted to estimate the genetic parameters for all of the traits:

$$y = 1\mu + Xb + Zu + Wpe + e$$

where y was the vector of observations for a particular trait of the analyzed traits; μ was the overall mean; 1 is the vector of ones; b was the vector of the fixed effects (sex, appraiser-season and fixed regression on age at evaluation); u was the vector of random additive genetic effect; pe was the vector of random permanent environmental effect of the animal; and e was the vector of random residual effect. X was the incidence matrix relating observations to fixed effects, Z was the incidence matrix relating observations to additive effects, and W was the incidence matrix relating observations to random permanent environmental effects.

Also, bivariate models were implemented to estimate the genetic correlations between the two coat color traits (QB and WM) and the other morpho-functional linear traits. The bivariate models fit the same effects as the univariate models.

The Gibbs sampler was run for 1,000,000 rounds, with the first 100,000 considered as burn-in and then every 10th sample saved for later analysis. Posterior means and standard deviations were calculated to obtain estimates of (co) variance components. Convergence of the posterior parameters was assessed by visual inspection of trace plots of posterior distributions generated by the Coda R package [19].

The pedigree data for the estimation of the genetic parameters was composed of 757 animals born between 1961 and 2018 (341 males and 416 females), constituting a total of five generations of animals registered in the official Studbook of PRMe horses managed by the ACPCRMe.

Finally, four subpopulations of animals were created according to their EBVs for each coat trait analyzed. Subpopulation A included animals with an EBV belonging to the better 25% for each trait (quality of the black coat color, A_, and white marks, _A); subpopulation B included the remaining animals with an EBV belonging to the lower 75% (quality of black coat, B_, and white marks, _B). To estimate the EBVs of WMs, a change of the scale was needed because it varied between 1 (without white marks, the desirable option) and 7 (with a lot of white marks, the least desirable option). Finally, ENDOG software [20] was used to estimate the Mahalanobis distances between the four groups (AA, AB, BA, and BB) using the EBVs of all the linear morpho-functional traits included in the analysis, both morphological and functional, to evidence the relationship between the groups.

3. Results and Discussion

3.1. Descriptive Statistics of the Data

Descriptive statistics of the 46 linear traits analyzed in PRMe horses are reported in Table 2. Coat color traits showed mean values ± standard error ranging between 1.51 ± 0.036 for WM and 4.80 ± 0.046 for QB. For the morphological traits, the mean values ± standard error ranged between 3.21 ± 0.025 for the cannon bone perimeter (trait code 29) and 5.23 ± 0.042 for the head width (3), with a global average of 4.22, which was close to the central value of the scale. For the functional traits, the average values ± standard error ranged between 4.04 ± 0.031 for walking activity (38) and 4.71 ± 0.038 for walking clarity (39), with a global average of 4.46, which was also close to the central value of the scale.

In the morpho-functional evaluation of the PRMe horses, a total of seven classes were used by the appraisers to assess the traits, the same number as was used for morpho-functional evaluations in Thoroughbred horses [4] and for American Quarter Riding horses [21]. Also, the whole classes were used in 47.83% of the linear analyzed traits, being 6 of more classes used in 89.13% of the analyzed traits. Therefore, we can consider that the complete scale was used in the population analyzed, as has occurred in other horse po-pulations (Czech-Moravian Belgian horses and Silesian Noriker [22]; Old Kladru horses [23,24]; Pura Raza Español horses [25,26]; and sport horses in the Czech Republic [9 All of the variables met the assumption of a normal distribution.

Table 2. Basic statistics of the 46 linear traits analyzed in Pura Raza Menorquina horses.

	Trait	Mean ± s.e.	CV (%)	Range		Trait	Mean ± s.e.	CV (%)	Range
CT	1	4.80 ± 0.046	26.49	7		24	5.01 ± 0.046	25.63	7
	2	1.51 ± 0.036	64.95	7		25	4.83 ± 0.039	22.69	6
	3	5.23 ± 0.042	22.45	7		26	4.01 ± 0.027	18.78	6
	4	4.80 ± 0.038	21.72	6		27	3.55 ± 0.025	19.68	5
	5	4.48 ± 0.033	20.30	6		28	3.67 ± 0.029	21.80	6
	6	4.49 ± 0.032	19.53	7		29	3.21 ± 0.025	21.43	5
	7	4.82 ± 0.032	18.68	6	MT	30	3.50 ± 0.022	17.36	5
	8	4.84 ± 0.037	21.27	7		31	4.16 ± 0.029	19.10	6
	9	3.91 ± 0.040	28.11	7		32	3.98 ± 0.027	18.50	5
	10	5.16 ± 0.028	15.31	6		33	3.42 ± 0.027	21.81	6
	11	4.12 ± 0.030	20.52	6		34	4.30 ± 0.035	22.38	6
	12	4.76 ± 0.043	25.32	6		35	4.03 ± 0.024	16.79	6
MT	13	4.40 ± 0.052	32.89	7		36	4.40 ± 0.035	22.23	6
	14	4.95 ± 0.034	19.10	6		37	4.00 ± 0.038	26.51	7
	15	3.95 ± 0.042	29.18	7		38	4.04 ± 0.031	21.42	7
	16	4.66 ± 0.037	21.81	6		39	4.71 ± 0.038	22.20	6
	17	3.99 ± 0.029	19.87	5		40	4.35 ± 0.034	21.62	7
	18	3.45 ± 0.030	23.77	6		41	4.38 ± 0.037	23.19	7
	19	3.85 ± 0.030	21.36	7	FT	42	4.59 ± 0.036	21.76	7
	20	4.23 ± 0.031	20.53	7		43	4.33 ± 0.038	24.08	7
	21	4.48 ± 0.042	25.77	7		44	4.70 ± 0.041	24.02	7
	22	3.52 ± 0.037	29.52	7		45	4.50 ± 0.037	22.85	6
	23	3.89 ± 0.039	28.06	7		46	4.52 ± 0.042	25.45	7

CT are coat color traits (1 and 2), MT are morphological traits (3–11 related to head and neck, 12–25 related t body regions and 26–37 related to limbs), and FT are functional traits (38–41 related to walking and 41–46 relate to trotting). Traits names are shown in Table 1.

The coefficient of variation (CV) is considered the most important measure of variatior It generally assumes that the higher the phenotypic variation of traits, the greater th genetic variation, which guarantees a sufficient selection response in populations [23]. Th estimated CV in PRMe horses was of a medium–high level, with 26.49% for QB (1) an 64.95% for WM (2).

For morpho-functional traits, the lowest CV was observed for the upper necklin (10; 15.31%) within the morphological traits and for walking activity (38; 21.42%) withir the performance traits, showing that their phenotypic variation is limited biologically ir this population. On the other hand, the highest CV was estimated for the form of th withers (13; 32.89%) within the morphological traits and trotting suspension (46; 25.45% within the performance traits, showing that there is higher phenotypic variation in thes traits than in other traits. The CVs obtained were in the range of those reported for linea traits in Dutch Warmblood Riding horses (10.14–26.04% [10]), Heavy Draught horse (11.38–38.54% [1]), Old Kladrub horses (2.39–40.14% [22–24]), Pura Raza Español horse (6.94–51.68% [25–27]), and sport horses in the Czech Republic (2.91–20.27% [9]). In genera the analyzed population showed sufficient variability.

The influences of the different non-genetic effects are shown in Table S1. All of th analyzed non-genetic effects were statistically significant ($p < 0.05$) for some analyze traits. Permanent environment, combination of the appraiser and the season, and sex wer significant for most of the analyzed traits and were therefore included in the genetic mode used in this study. The coefficient of determination (R2) was also estimated for all of th traits, with values higher than and close to 0.55. In this way, the model accounted fo over 60% of the variance for most of the linear morpho-functional traits analyzed in the PRMe population.

3.2. Genetic Parameters

Coat color is one of the most noticeable animal features and has interested and intrigued breeders for centuries [8]. It is believed that most of the phenotypes currently observed in the modern horse are the result of domestication and selective breeding, and different breeds exist that are mainly defined by the color and patterns of their coats as a result of breeders' selection criteria, such as Paint and Appaloosa horses [28,29]. This is also the case of PRMe horses, in which black is the only coat color authorized for the breeding stock registered in the official population studbook. However, the pattern of white spots is also a major attribute, and it determines breeding practices in the PRMe production system because individuals showing large white marks on the head, limbs, or the rest of the body are barred from being used as breeding stock for the official studbook.

A horse's coat color is generally understood as a qualitative trait with Mendelian inheritance. Here, black coat color is determined by a recessive homozygote genotype at the Agouti locus and at least one dominant allele at the EXTENSION locus (aa E-) [30]. However, there are remarkable differences in color phenotypes that are not explainable by Mendelian inheritance. Different authors [16,31] indicated that there are two different types of black coat color (which they termed non-fading and fading black coats), in which the genetic determination is unknown, but age, sex, season, feeding, housing system, and body part are environmental effects that could significantly affect their expression. One plausible reason for this could be that the estimations of heritability for overall coat color reveal the dominant effect of environmental factors on total variability [31].

The heritability values of the 46 linear morpho-functional traits analyzed in the PRMe population are shown in Table 3. Although the data set may seem small, it includes the selection nucleus of the Menorca horse population and the majority of the horses all around the world.

Medium heritability values were obtained for the linear coat color traits, 0.36 for QB (1) and 0.23 for WM (2), which evidenced the level of selection that can be carried out for both linear traits related to the coat color in PRMe horses. These values were in the range of those reported for the quality of black coat color in Old Kladrub horses (0.14–0.37 [16]) and for the white marks in Hucul horses (0.68–0.69 [32]), Lipizzan horses (0.23–0.71 [33]), Swiss Franches Montagnes horses (0.52–0.69 [34]), and Arabian horses (0.77 [35]).

A previous study [16] postulated that differences in the genetic determination of fa-ding and non-fading black coat color could be influenced by modifying genes with only a minor effect, but which could be cumulative. Different hypotheses about the inheritance of white marks have also been postulated. First, Woolf [35] concluded that complex genetic systems and non-genetic factors determined the presence of common white marks in Arabian horses. Also, Stachurska and Ussing [36] postulated about the polygenic inhe-ritance of white marks that the ultimate extension of markings was influenced by genes, as well as by intrauterine factors. These authors concluded that the high heritability and QTLs involved mean that selection both towards and against white marks is effective. However, the polygenic inheritance makes it impossible to completely eradicate them because the genes affecting white marks might be recessive and marked by dominant genes and may therefore be difficult to identify. The medium heritability values obtained in PRMe horses evidenced that the influence of external factors is also very important in the phenotypic expression of these coat traits.

Besides the existence of possible pleiotropic effects associated with specific coat color and some conformation [7], performance [37], health [38], and temperamental traits [39], our results evidenced the need to analyze the relationship between QB and WM traits in PRMe horses and the linear morpho-functional traits included in this study.

Table 3. Additive, permanent environmental and residual variances, and heritability values for the 46 linear traits analyzed in Pura Raza Menorquina horses.

	Tr	σ^2_u (s.d.)	σ^2_{pe} (s.d.)	σ^2_e (s.d.)	h^2 (s.d.)		Tr	σ^2_u (s.d.)	σ^2_{pe} (s.d.)	σ^2_e (s.d.)	h^2 (s.d.)
CT	1	0.47 (0.146)	0.32 (0.115)	0.51 (0.004)	0.36 (0.098)		24	0.41 (0.137)	0.36 (0.114)	0.57 (0.039)	0.30 (0.091)
	2	0.22 (0.146)	0.64 (0.125)	0.08 (0.001)	0.23 (0.142)		25	0.30 (0.090)	0.19 (0.073)	0.43 (0.030)	0.32 (0.086)
	3	0.36 (0.067)	0.05 (0.039)	0.53 (0.035)	0.38 (0.057)		26	0.15 (0.056)	0.14 (0.047)	0.27 (0.019)	0.26 (0.088)
	4	0.45 (0.105)	0.12 (0.071)	0.46 (0.031)	0.44 (0.083)		27	0.10 (0.035)	0.10 (0.033)	0.28 (0.019)	0.21 (0.068)
	5	0.24 (0.055)	0.07 (0.042)	0.47 (0.032)	0.30 (0.062)		28	0.11 (0.042)	0.09 (0.038)	0.31 (0.021)	0.21 (0.077)
	6	0.43 (0.076)	0.07 (0.046)	0.24 (0.017)	0.58 (0.073)		29	0.09 (0.034)	0.10 (0.032)	0.25 (0.018)	0.19 (0.071)
	7	0.10 (0.061)	0.16 (0.058)	0.45 (0.031)	0.14 (0.081)	MT	30	0.03 (0.025)	0.16 (0.029)	0.18 (0.012)	0.09 (0.065)
	8	0.36 (0.105)	0.13 (0.078)	0.47 (0.032)	0.37 (0.094)		31	0.14 (0.064)	0.17 (0.055)	0.29 (0.020)	0.23 (0.098)
	9	0.31 (0.105)	0.19 (0.089)	0.65 (0.045)	0.27 (0.082)		32	0.13 (0.046)	0.10 (0.041)	0.31 (0.021)	0.24 (0.079)
	10	0.15 (0.057)	0.21 (0.049)	0.19 (0.013)	0.28 (0.093)		33	0.15 (0.066)	0.16 (0.055)	0.27 (0.019)	0.26 (0.104)
	11	0.15 (0.080)	0.14 (0.066)	0.37 (0.026)	0.23 (0.111)		34	0.26 (0.116)	0.46 (0.101)	0.25 (0.017)	0.26 (0.109)
	12	0.27 (0.132)	0.51 (0.122)	0.59 (0.040)	0.19 (0.090)		35	0.05 (0.042)	0.22 (0.042)	0.21 (0.014)	0.11 (0.083)
MT	13	0.37 (0.151)	0.58 (0.142)	0.80 (0.055)	0.21 (0.079)		36	0.09 (0.072)	0.49 (0.081)	0.41 (0.028)	0.10 (0.070)
	14	0.28 (0.077)	0.11 (0.059)	0.46 (0.031)	0.33 (0.079)		37	0.23 (0.113)	0.51 (0.106)	0.37 (0.026)	0.20 (0.094)
	15	0.13 (0.082)	0.21 (0.080)	0.73 (0.050)	0.12 (0.074)		38	0.06 (0.050)	0.26 (0.054)	0.35 (0.024)	0.09 (0.070)
	16	0.32 (0.065)	0.06 (0.043)	0.39 (0.026)	0.41 (0.069)		39	0.09 (0.065)	0.35 (0.074)	0.56 (0.039)	0.09 (0.063)
	17	0.05 (0.036)	0.13 (0.041)	0.41 (0.028)	0.09 (0.059)		40	0.31 (0.119)	0.23 (0.093)	0.39 (0.027)	0.33 (0.113)
	18	0.09 (0.034)	0.04 (0.029)	0.41 (0.027)	0.16 (0.060)		41	0.33 (0.118)	0.19 (0.093)	0.51 (0.035)	0.32 (0.102)
	19	0.07 (0.038)	0.18 (0.041)	0.30 (0.021)	0.12 (0.065)	FT	42	0.21 (0.089)	0.34 (0.080)	0.34 (0.024)	0.23 (0.092)
	20	0.17 (0.067)	0.13 (0.058)	0.46 (0.032)	0.23 (0.081)		43	0.20 (0.087)	0.27 (0.077)	0.43 (0.030)	0.22 (0.088)
	21	0.25 (0.099)	0.35 (0.092)	0.57 (0.039)	0.21 (0.078)		44	0.30 (0.129)	0.37 (0.108)	0.42 (0.029)	0.28 (0.106)
	22	0.08 (0.054)	0.19 (0.058)	0.55 (0.038)	0.10 (0.063)		45	0.25 (0.096)	0.23 (0.082)	0.48 (0.033)	0.26 (0.090)
	23	0.26 (0.119)	0.43 (0.109)	0.54 (0.038)	0.21 (0.090)		46	0.23 (0.129)	0.48 (0.116)	0.44 (0.031)	0.20 (0.103)

σ^2_u is additive genetic variance; σ^2_{pe} is permanent environmental variance; σ^2_e is residual variance; h^2 is heritability; s.d. is standard deviation; CT are coat color traits (1 and 2); MT are morphological traits (3–11 related to head and neck, 12–25 related to body regions and 26–37 related to limbs) and FT are functional traits (38–41 related to walking and 41–46 related to trotting). Trait names are shown in Table 1.

Heritability values estimated with the univariate animal models for the morphological traits obtained in the present study were of low to medium range (0.09–0.58), showing the higher values of 0.58 for head profile (6), 0.44 for head length (4), and 0.41 for thorax depth (16). In general, the heritability values we obtained were in the range of those reported for linear conformation traits in other horse breeds, which ranged between 0.03 and 0.68 (in Italian Heavy Draught Horses [1], Old Kladrub horses [24,40], sport horses in the Czech Republic [9], Belgian Warmblood horses [41], and the Pura Raza Español [7,25]).

The population of PRMe horses analyzed in this study showed similar heritability values to Belgian Warmblood and Pura Raza Español for head-neck junction (0.26 and 0.14–0.23, respectively [25,41]), to the Old Kladrub horse breed and Pura Raza Español for height of the withers (0.17 and 0.19–0.21, respectively [24,25]), to the Belgian Warmblood for shoulder length (0.31 [41]), to Pura Raza Español for loin length (0.10–0.14 [25]), form of back-loin line (0.12–0.19 [25]), and croup length (0.17–0.19 [25]), to Pura Raza Español and the Old Kladrub horse breed for chest width (0.31 [27,40]), to the Old Kladrub horse breed for forelimb side view (0.10 [40]), and to Pura Raza Español horses for hock side view (0.04–0.09 [25]). Also, the shoulder angle slope, the croup angle, and the hock rear view also showed a similar heritability value to a previous study carried out in the PRMe horses' population (0.10, 0.23, and 0.21, respectively [12]).

The heritability values estimated for the functional traits obtained in the present study were also of low to medium range, with values of 0.09 for walking activity (38) and walking clarity (39), 0.32 for walking suppleness (41), and 0.33 for walking amplitude (40). The values obtained for these kinds of linear traits were lower than those reported in the reviewed bibliography for linear functional traits in other horses, ranging between 0.18 and 0.52 in sport horses in the Czech Republic [9] and Belgian Warmblood horses [41]. Only walking amplitude showed a similar heritability value to Belgian Warmblood horses and trot amplitude was similar to sport horses in the Czech Republic and the PRMe horses analyzed (0.38 [41] and 0.20 [9], respectively).

Genetic parameters such as heritability are influenced by gene frequency, estimation method, statistical model, and trait nature. Since there may have been differences in the factors used in each study, the slight variations in values between the present study and previous studies might be attributable to those differences. The medium to high estimated heritability for the traits in the current study indicates that genetic improvement would be possible in these traits and allow us to foresee the possibility of making good genetic progress through the breeding program.

Understanding the relationships between morphological traits is extremely useful in animal breeding for determining both the breeding criteria and the possible breeding response of selection programs [25]. However, unfavorable correlated responses with other important or economic characters of the breed must also be avoided. Therefore, the correlations between the linear traits need to be estimated before including the morpho-functional traits into the breeding selection programs of these horses.

For several centuries, behavior, conformation, performance, and suitability characteristics of horses have been attributed to coat color in different populations. For example, coat color influences the conformation traits analyzed in Old Kladrub horses [23] and in Pura Raza Español horses [2,7]. In previous analyses on the same breeds, it has also been associated with certain conformation defects such as cresty neck [27] and ewe neck [42] and diseases like vitiligo or melanoma [7]. However, this is not exclusive to these populations. Its effects on horse behavior [43–45], temperament [39], and performance [37,46] have also been reported. In addition, there is still a strong belief among horse breeders that these traits account for differences between horses of different colors [47,48].

Table 4 shows the genetic correlations estimated with bivariate models for the coat color traits (QB and WM) and the other 44 morpho-functional traits analyzed in PRMe horses. Estimated genetic correlations had large standard errors as shown by large stan-dard deviation of the posterior samples.

Table 4. Genetic correlations (rg) obtained for the traits related with coat color (1: quality black coat color and 2: white marks) and the 44 morpho-functional traits analyzed in Pura Raza Menorquina horses.

	Trait	rg1 (s.d.)	rg2 (s.d.)		Trait	rg1 (s.d.)	rg2 (s.d.)
	3	−0.114 (0.209)	−0.065 (0.410)		25	0.099 (0.243)	−0.155 (0.465)
	4	0.186 (0.206)	−0.443 (0.372)		26	0.015 (0.268)	−0.177 (0.504)
	5	−0.090 (0.224)	0.295 (0.377)		27	0.047 (0.275)	0.084 (0.479)
	6	0.609 (0.184)	−0.204 (0.382)		28	−0.349 (0.430)	0.617 (0.402)
	7	0.154 (0.355)	−0.030 (0.584)		29	−0.738 (0.225)	0.448 (0.444)
	8	0.337 (0.234)	−0.404 (0.399)		30	−0.170 (0.270)	−0.175 (0.484)
	9	−0.325 (0.270)	0.074 (0.503)	MT	31	0.366 (0.255)	−0.269 (0.524)
MT	10	0.816 (0.171)	−0.434 (0.393)		32	0.226 (0.272)	0.069 (0.505)
	11	−0.441 (0.302)	0.227 (0.534)		33	0.094 (0.445)	−0.807 (0.235)
	12	−0.493 (0.323)	0.469 (0435)		34	0.311 (0.418)	−0.603 (0.414)
	13	−0.767 (0.216)	0.577 (0.372)		35	0.025 (0.325)	0.076 (0.560)
	14	0.187 (0.219)	0.224 (0.430)		36	0.310 (0.428)	0.098 (0.598)
	15	−0.419 (0.327)	0.638 (0.365)		37	0.185 (0.398)	−0.110 (0.574)
	16	0.269 (0.203)	0.329 (0.349)		38	0.123 (0.244)	0.396 (0.420)
	17	0.618 (0.298)	0.336 (0.529)		39	0.318 (0.231)	−0.574 (0.316)
	18	0.292 (0.299)	−0.045 (0.573)		40	0.325 (0.247)	−0.078 (0.492)
	19	−0.224 (0.340)	−0.379 (0.484)		41	0.243 (0.257)	−0.206 (0.475)
	20	0.307 (0.263)	−0.295 (0.479)	FT	42	0.376 (0.226)	−0.115 (0.502)
	21	−0.459 (0.296)	0.139 (0.483)		43	0.224 (0.254)	−0.199 (0.482)
	22	0.229 (0.346)	−0.576 (0.405)		44	0.241 (0.296)	−0.379 (0.498)
	23	−0.371 (0.320)	0.159 (0.518)		45	0.099 (0.243)	−0.155 (0.465)
	24	0.411 (0.227)	0.014 (0.446)		46	0.015 (0.268)	−0.177 (0.504)

MT are morphological traits (3–11 related to head and neck, 12–25 related to body regions and 26–37 related to limbs) and FT are functional traits (38–41 related to walking and 41–46 related to tro-tting). rg1 is the genetic correlation obtained for the quality of the black coat color -QB- and rg2 is the genetic correlation obtained for the white marks -WM-; s.d. is standard deviation; Traits names are shown in Table 1.

In general, both color traits showed low to high correlations with the 44 morpho-functional traits. The highest values obtained for QB were with upper neck line (10; 0.816), shape of the withers (13; −0.767), cannon bone perimeter (29; −0.738), back length (17; 0.618), and head profile (6; 0.609). The highest correlations estimated for WM were with form of the hoof (33, −0.807), shoulder angle slope (15, 0.638), fore knee perimeter (28, 0.617), fore-hoof front view (34, −0.603), shape of the withers (13, 0.577), buttock length (22, −0.576), and walking clarity (39, −0.574). The existence of an important relationship between the coat color traits and other linear traits could be caused by a pleiotropic effect which should be analyzed with the adequate tools.

It is important to note that for QB (1), 37.14% of the genetic correlations with morphological traits were negative, and none of them were correlated with functional traits. For WM (2), 48.57% of the phenotypic correlations with morphological traits were negative whereas 88.89% of the correlations with functional traits were negative. However, in the PRMe population, animals with or without a lower percentage of white marks are required as breeding stock. Therefore, these negative correlations are very interesting in the breeding program.

The genetic correlation between both coat color traits (QB and WM) was −0.272. That means that selection to improve the quality of black coat color implies a decrease in the percentage of white marks, as is desired in this population.

Finally, for the graphic representation of Mahalanobis distances, animals were grouped by their EBVs for both traits related to coat color included in the analysis taking into account all of the morpho-functional linear traits analyzed. The population's average relatedness was 2.28%, the average inbreeding was 0.6%, and 8.1% of all the pedigreed animals presented some level of inbreeding. The results are shown in Figure 1 where individuals showing the 25% best EBVs for QB color (A_) and WM (_A) were compared with the remaining 75% for the same traits (B_ and _B, respectively). A clear differentiation can be observed between animals included in the group of individuals with the top 25% of EBVs for QB and WM (AA) and the other analyzed groups because they were represented in clearly separated clusters. Also, a clear relationship can be observed between animals included in the groups of individuals in the remaining 75% for QB (B_), which are included in the same cluster, and they are also related to the animals in group AB. These results show that there is a clear differentiation of the animals transmitting a better coat color (A_) compared with the animals transmitting a worse coat color (B_), which is independent of the presence or absence of white marks. This could be explained by the existence of different founders in the different groups. Therefore, the genetic inheritance of both coat traits is independent and could be determined by different genes, although the genetic correlation obtained between both coat traits analyzed (−0.272) confirms the existence of relationship between them. Perhaps the different phenotypes for coat color can be explained by the influence of modifying genes having only a minor effect, which can be cumulative [16]. In this context, the inheritance of the quality of black coat color can have a polygenic component, while the inheritance of white marks is independent and could be determined by genetic and intrauterine effects [32,33,36,49] according to the reviewed bibliography.

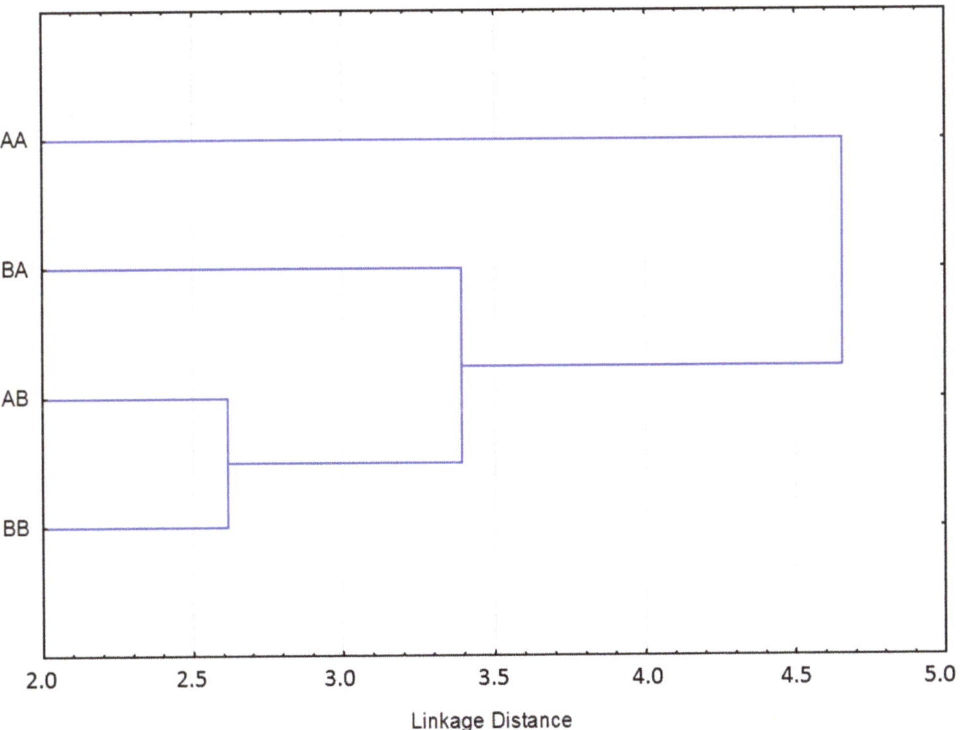

Figure 1. Mahalanobis distances between animals using their breeding values for all the morpho-functional traits analyzed in Pura Raza Menorquina Horses. Animals which have obtained the better 25% breeding value for quality of coat color (A_) and white marks (_A); animals which have obtained the remaining 75% for quality of coat color (B_) and white marks (_B).

4. Conclusions

The heritability values obtained for the morpho-functional traits related to coat color in PRMe horses indicate that selection is feasible, and that these traits could be improved in the official Breeding Program. The genetic correlations of coat color traits with linear morpho-functional traits evidence the existence of genetic and physiological mechanism controlling them, showing that selection for one of these traits could have an influence o (e.g., increase or decrease, according to the sign and value) the other traits of interest. Th existence of pleiotropic effects between the coat color traits and some morpho-function linear traits could be also suspected because of the high genetic correlations obtaine between some morpho-functional and coat color analyzed traits. The study of Mahalanob distances showed that the inheritance of the quality of coat color could be linked to th genetic group and determined by modifying genes. In this sense, a genomics analysis coul have an interesting contribution to confirming our results in the future.

Supplementary Materials: The following supporting information can be downloaded at: http //www.mdpi.com/article/10.3390/ani12182319/s1, Table S1: Analysis of the influence of th different effects on 46 morpho-functional linear traits analyzed in Pura Raza Menorquina horse using a univariate GeneralLinear Model analysis of variance with a permanent environmental effe (p values).

Author Contributions: Conceptualization, M.V. and M.D.G.; data collection and preparation, M.D.G methodology, M.V., E.B. and D.I.P.-G.; formal analysis, M.V., D.I.P.-G. and R.d.l.A.G.d.P.; geneti parameters estimation, D.I.P.-G. and R.d.l.A.G.d.P.; results validation, M.V., D.I.P.-G., R.d.l.A.G.d.I E.B. and M.D.G.; discussion, M.V., D.I.P.-G., R.d.l.A.G.d.P., E.B. and M.D.G.; writing—original dra preparation, M.D.G.; writing—review and editing, M.D.G.; supervision, M.V.; funding acquisitio M.V. and M.D.G. All authors have read and agreed to the published version of the manuscript.

Funding: This work was funded by the Asociación de Criadores y Propietarios de Caballos de Raz Menorquina (www.caballomenorquin.com, accessed on 1 January 2022).

Institutional Review Board Statement: Not applicable. No institutional animal care and us committee approval was needed due to no experimental procedures were held with animals i this study. All data used for the analyses were obtained from the pedigree information of the officia PRMe stud book. Regarding to the information from the estimation of the genetic parameters, dat were obtained from the performance control tests developed for this breed, with no additiona experimental procedures.

Informed Consent Statement: Not applicable, as this research did not involve humans.

Data Availability Statement: Restrictions apply to the availability of these data. Data were obtaine from Asociación de Criadores y Propietarios de Caballos de Raza Menorquina (ACPCRMe) and ar available from the corresponding author with the permission of ACPCRMe.

Acknowledgments: The authors wish to thank the Asociación de Criadores y Propietarios de Caballo de Raza Menorquina for making its database available.

Conflicts of Interest: The authors declare no conflict of interest.

References

1. Folla, F.; Sartori, C.; Mancin, E.; Pigozzi, G.; Mantovani, R. Genetic Parameters of Linear Type Traits Scored at 30 Months in Italia Heavy Draught Horse. *Animals* **2020**, *10*, 1099. [CrossRef] [PubMed]
2. Gómez; Molina, A.; Sánchez-Guerrero, M.; Valera, M. Prediction of adult conformation traits from shape characteristics of Pur Raza Español foals. *Livest. Sci.* **2021**, *253*, 104701. [CrossRef]
3. Forest, A. Son importance avant l'achat en fonction d'un service recherché et pour le choix des reproducteurs. In *La Conformatio d'un Cheval, Proceedings of the 7th Colloque sur le Cheval, Saint Hyacinthe, QC, Canada, 27 April 1996*; Conseil des Production Animales du Québec (CPAQ): Quebec City, QC, Canada, 1996; pp. 83–93.
4. Mawdsley, A.; Kelly, E.P.; Smith, F.H.; Brophy, P.O. Linear assessment of the Thoroughbred horse: An approach to conformatio evaluation. *Equine Vet.-J.* **1996**, *28*, 461–467. [CrossRef] [PubMed]
5. Bruns, E.; Bierbaum, M.; Frese, D.; Haring, H.J.F. Die Entwicklung von Selektionskriterienfür die Reitpferdezucht. I Schätzungrelativerökonomischer Gewichtanhand von Auktionsergebnissen. *Züchtungskunde* **1978**, *50*, 93–100.
6. Schwark, H.J.; Petzold, P.; Nörenberg, I. UntersuchungenzurAuswahl von Selektionskriterienbei der Weiterentwicklung de Pferdezucht der DDR. *Arch. Tierz.* **1988**, *31*, 279–289.
7. Sánchez-Guerrero, M.J.; Negro-Rama, S.; Demyda-Peyras, S.; Solé-Berga, M.; Azor-Ortiz, P.J.; Valera-Córdoba, M. Morphologica and genetic diversity of Pura Raza Español horse with regard to the coal colour. *Anim. Sci. J.* **2019**, *90*, 14–22. [CrossRef]

Pruvost, M.; Bellone, R.; Benecke, N.; Sandoval-Castellanos, E.; Cieslak, M.; Kuznetsova, T.; Morales-Muñiz, A.; O'Connor, T.; Reissmann, M.; Hofreiter, M.; et al. Genotypes of predomestic horses match phenotypes painted in Paleolithic works of cave art. *Proc. Natl. Acad. Sci. USA* **2011**, *108*, 18626–18630. [CrossRef] [PubMed]

Novotná, A.; Svitáková, A.; Veselá, Z.; Vostrý, L. Estimation of genetic parameters for linear type traits in the population of sport horses in the Czech Republic. *Livest. Sci.* **2017**, *202*, 1–6. [CrossRef]

Koenen, E.; van Veldhuizen, A.; Brascamp, E. Genetic parameters of linear scored conformation traits and their relation to dressage and show-jumping performance in the Dutch Warmblood Riding Horse population. *Livest. Prod. Sci.* **1995**, *43*, 85–94. [CrossRef]

Schroderus, E.; Ojala, M. Estimates of genetic parameters for conformation measures and scores in Finnhorse and Standardbred foals. *J. Anim. Breed. Genet.* **2010**, *127*, 395–403. [CrossRef]

Solé, M.; Cervantes, I.; Gutiérrez, J.P.; Gomez, M.D.; Valera, M. Estimation of genetic parameters for morphological and functional traits in a Menorca horse population. *Span. J. Agric. Res.* **2014**, *12*, 125. [CrossRef]

MAPA. Census of Pura Raza Menorquina Horses. 2022. Available online: https://www.mapa.gob.es/es/ganaderia/temas/zootecnia/razas-ganaderas/razas/catalogo-razas/equino-caballar/menorquina/default.aspx (accessed on 15 March 2022).

MAPA. Purebred Menorca Horse Breeding Program. 2022. Available online: https://www.mapa.gob.es/es/ganaderia/temas/zootecnia/en_programa_de_cria_prme_definitivo_tcm30-561612.PDF (accessed on 15 March 2022).

Giontella, A.; Sarti, F.M.; Biggio, G.P.; Giovannini, S.; Cherchi, R.; Pieramati, C.; Silvestrelli, M. Genetic Parameters and Inbreeding Effect of Morphological Traits in Sardinian Anglo Arab Horse. *Animals* **2020**, *10*, 791. [CrossRef] [PubMed]

Hofmanova, B.; Kohoutova, P.; Vostrý, L.; Vydrova, H.V.; Majzlik, I. Quantitative aspects of coat color in old kladruber black horses. *Poljoprivreda* **2015**, *21*, 224–227. [CrossRef]

SAS, Statistical Analysis Software. *Users' Guide Statistics Version 9.4*; SAS Institute Inc.: Cary, NC, USA, 2013.

Misztal, I.; Tsuruta, S.; Lourenco, D.; Aguilar, I.; Legarra, A.; Vitezica, Z. *Manual for BLUPF90 Family of Programs*; University of Georgia: Athens, GA, USA, 2018; p. 142.

Plummer, M.; Best, N.; Cowles, K.; Vines, K. CODA: Convergence diagnosis and output analysis for MCMC. *R News.* **2006**, *6*, 7–11.

Gutierrez, J.P.; Goyache, F. A note on ENDOG: A computer program for analysing pedigree information. *J. Anim. Breed. Genet.* **2005**, *122*, 172–176. [CrossRef] [PubMed]

Kuhnke, S.; Bär, K.; Bosch, P.; Rensing, M.; König, U.; Borstel, V. Evaluation of a novel system for linear conformation, gait and personality trait scoring and automatic ranking of horses at breed shows: A pilot study in America Quarter Horses. *J. Equine Vet. Sci.* **2019**, *78*, 53–59. [CrossRef]

Vostrý, L.; Capkova, Z.; Andrejsová, L.; Mach, K.; Majzlík, I. Linear type trait analysis in the Coldblood breeds: Czech-Moravian Belgian horse and Silesian Noriker. *Slovak J. Anim. Sci.* **2009**, *42*, 99–106.

Jakubec, V.; Pejfková, M.; Volenec, J.; Majzlík, I.; Vostrý, L. Analysis of linear description of type traits in the varieties and studs of the Old Kladrub horse. *Czech J. Anim. Sci.* **2007**, *52*, 299–307. [CrossRef]

Jakubec, V.; Vostrý, L.; Schlote, W.; Majzlík, I.; Mach, K. Selection in the genetic resource: Genetic variation of the linear described type traits in the Old Kladrub horse. *Arch. Anim. Breed.* **2009**, *52*, 343–355. [CrossRef]

Sánchez, M.; Gómez, M.; Molina, A.; Valera, M. Genetic analyses for linear conformation traits in Pura Raza Español horses. *Livest. Sci.* **2013**, *157*, 57–64. [CrossRef]

Sánchez, M.J.; Gómez, M.D.; Molina, A.; Valera, M. Assessment scores in morphological competitions of Pura Raza Español Horse. *Int. J. Agric. Biol.* **2014**, *16*, 557–563.

Sánchez, M.J.; Azor, P.J.; Molina, A.; Parkin, T.; Rivero, J.L.L.; Valera, M. Prevalence, risk factors and genetic parameters of cresty neck in Pura Raza Español horses. *Equine Vet. J.* **2017**, *49*, 196–200. [CrossRef] [PubMed]

Bowling, A.T. Dominant Inheritance of Overo Spotting in Paint Horses. *J. Hered.* **1994**, *85*, 222–224. [CrossRef] [PubMed]

Corbi-Botto, C.M.; Sadaba, S.A.; Francisco, E.I.; Kalemkerian, P.B.; Lirón, J.P.; Villegas-Castagnasso, E.E.; Giovambattista, G.; Peral-García, P.; Díaz, S. Genetic variability of Appaloosa horses: A study of a closed breeding population from Argentina. *Front. Agric. Sci. Eng.* **2014**, *1*, 175–178. [CrossRef]

Sponenberg, D.P. *Equine Color Genetics*; A Blackwell Publishing Company: Hoboken, NJ, USA; Iowa State Press: Ames, IA, USA, 1996.

Hofmanová, B.; Vostrý, L.; Vostrá-Vydrová, H.; Dokoupilová, A.; Majzlík, I. Estimation of genetic and non-genetic effects in-fluencing coat color in black horses. *Czech J. Anim. Sci.* **2019**, *64*, 41–48. [CrossRef]

Pasternak, M.; Krupiński, J.; Gurgul, A.; Bugno-Poniewierska, M. Genetic, historical and breeding aspects of the occurrence of the tobiano pattern and white markings in the Polish population of Hucul horses—A review. *J. Appl. Anim. Res.* **2020**, *48*, 21–27. [CrossRef]

Grilz-Seger, G.; Dobretsberger, M.; Brem, G.; Druml, T. Study on distribution of coat color related alleles and white markings in a closed population of Lipizzan horses. *Zuchtungskunde* **2020**, *92*, 76–86.

Rieder, S.; Hagger, C.; Obexer-Ruff, G.; Leeb, T.; Poncet, P.-A. Genetic Analysis of White Facial and Leg Markings in the Swiss Franches-Montagnes Horse Breed. *J. Hered.* **2008**, *99*, 130–136. [CrossRef]

Woolf, C.M. Multifactorial Inheritance of Common White Markings in the Arabian Horse. *J. Hered.* **1990**, *81*, 250–256. [CrossRef]

Stachurska, A.; Ussing, A.P. White markings in horses. *Med. Wet.* **2012**, *68*, 74–78.

37. Fegraeus, K.J.; Velie, B.D.; Axelsson, J.; Ang, R.; Hamilton, N.A.; Andersson, L.; Meadows, J.; Lindgren, G. A potential regulatory region near the EDN3 gene may control both harness racing performance and coat color variation in horses. *Physiol. Rep.* **201**, *6*, e13700. [CrossRef]

38. Bellone, R.R. Pleiotropic effects of pigmentation genes in horses. *Anim. Genet.* **2010**, *41*, 100–110. [CrossRef] [PubMed]

39. Abdel-Azeem, N.M.; Emeash, H.H. Relationship of horse temperament with breed, age, sex and body characteristic. A questionnaire-based study. *J. Basic Appl.* **2021**, *10*, 45. [CrossRef]

40. Vostrý, L.; Mach, K.; Přibyl, J. Selection of a suitable data set and model for the genetic evaluation of the linear description of conformation and type description in Old Kladruber horses. *Arch. Anim. Breed.* **2012**, *55*, 105–112. [CrossRef]

41. Rustin, M.; Janssens, S.; Buys, N.; Gengler, N. Multi-trait animal model estimation of genetic parameters for linear type and gait traits in the Belgian warmblood horse. *J. Anim. Breed. Genet.* **2009**, *126*, 378–386. [CrossRef]

42. Ripolles, M.; Sánchez-Guerrero, M.; Perdomo-González, D.; Azor, P.; Valera, M. Survey of Risk Factors and Genetic Characterization of Ewe Neck in a World Population of Pura Raza Español Horses. *Animals* **2020**, *10*, 1789. [CrossRef]

43. Ducro, B.J.; Bovenhuis, H.; Back, W. Heritability of foot conformation and its relationship to sports performance in a Dutch Warmblood horse population. *Equine Vet. J.* **2009**, *41*, 139–143. [CrossRef]

44. Finn, J.L.; Haase, B.; Willet, C.E.; van Rooy, D.; Chew, T.; Wade, C.M.; Hamilton, N.A.; Velie, B.D. The relationship between coat colour phenotype ad equine behavior: A pilot study. *Appl. Anim. Behav. Sci.* **2016**, *174*, 66–69. [CrossRef]

45. Jacobs, L.N.; Staiger, E.A.; Albright, J.D.; Brooks, S. Data from: The MC1R and ASIP coat color loci may impact behavior in the horse. *J. Hered.* **2016**, *107*, 214–219. [CrossRef]

46. Stachurska, A.; Pięta, M.; Łojek, J.; Szulowska, J. Performance in racehorses of various colours. *Livest. Sci.* **2007**, *106*, 282–28. [CrossRef]

47. Bartolomé, E.; Goyache, F.; Molina, A.; Cervantes, I.; Valera, M.; Gutiérrez, J.P. Pedigree estimation of the subpopulation contribution to the total gene diversity: The horse coat colour case. *Animal* **2010**, *4*, 867–875. [CrossRef] [PubMed]

48. Druml, T.; Baumung, R.; Sölkner, J. Mophological analysis and effect of selection for conformation in the Noriker draught horse population. *Livest. Sci.* **2008**, *115*, 118–128. [CrossRef]

49. Negro, S.; Imsland, F.; Valera, M.; Molina, A.; Solé, M.; Andersson, L. Association analysis of KIT, MUTF and PAX3 variants with White markings in Spanish horses. *Anim. Genet.* **2017**, *48*, 349–352. [CrossRef] [PubMed]

Article

Estimates of Genetic Parameters for Shape Space Data in Franches-Montagnes Horses

Annik Imogen Gmel [1,2,*], Alexander Burren [3] and Markus Neuditschko [1]

1 Animal GenoPhenomics, Agroscope, Rte de la Tioleyre 4, 1725 Posieux, Switzerland
2 Equine Department, Vetsuisse Faculty, University of Zurich, Winterthurerstrasse 260, 8057 Zurich, Switzerland
3 School of Agricultural, Forest and Food Sciences HAFL, Bern University of Applied Sciences, Länggasse 85, 3052 Zollikofen, Switzerland
* Correspondence: annik.gmel@agroscope.admin.ch or agmel@uzh.ch

Simple Summary: Equine breeding is often based on conformation traits, describing the proportions, shape and joint angles of a horse. These conformations traits are, however, mostly subjectively judged and not measured objectively, affecting the response of selection through lower heritabilities and precision. In this study, we measured joint angles, quantified the variation in shape of 608 Swiss Franches-Montagnes (FM) horses and estimated the heritability of these traits. We found that the poll angle had the highest heritability of all joint angles ($h^2 = 0.37$), and variation in shape describing the type (heavy–light) was also fairly heritable ($h^2 = 0.36$–0.37). Furthermore, the shape of the FM stallions has clearly evolved towards a lighter type from 1940 to 2018 without stabilisation in recent years, risking the loss of the light draught horse type. Phenotyping based on photographs allowed us to improve the accuracy of certain joint angle traits, and to monitor the conformational development of the FM breed.

Abstract: Conformation traits such as joint angles are important selection criteria in equine breeding, but mainly consist of subjective evaluation scores given by breeding judges, showing limited variation. The horse shape space model extracts shape data from 246 landmarks (LM) and objective joint angle measurements from triplets of LM on standardized horse photographs. The heritability was estimated for 10 joint angles (seven were measured twice with different LM placements), and relative warp components of the whole shape, in 608 Franches-Montagnes (FM) horses (480 stallions, 68 mares and 60 geldings born 1940–2018, 3–25 years old). The pedigree data comprised 6986 horses. Genetic variances and covariances were estimated by restricted maximum likelihood model (REML), including the fixed effects birth year, age (linear and quadratic), height at withers (linear and quadratic), as well as postural effects (head, neck, limb position and body alignment), together with a random additive genetic animal component and the residual effect. Estimated heritability varied from 0.08 (stifle joint) to 0.37 (poll). For the shape, the type was most heritable (0.36 to 0.37) and evolved from heavy to light over time. Image-based phenotyping can improve the selection of horses for conformation traits with moderate heritability (e.g., poll, shoulder and fetlock).

Keywords: horse; breeding; heritability; conformation; joint angles; geometric morphometrics

Citation: Gmel, A.I.; Burren, A.; Neuditschko, M. Estimates of Genetic Parameters for Shape Space Data in Franches-Montagnes Horses. *Animals* **2022**, *12*, 2186. https://doi.org/10.3390/ani12172186

Academic Editors: Isabel Cervantes and María Dolores Gómez

Received: 28 June 2022
Accepted: 23 August 2022
Published: 25 August 2022

Publisher's Note: MDPI stays neutral with regard to jurisdictional claims in published maps and institutional affiliations.

1. Introduction

Conformation traits are important in equine breeding as they have been associated with health, longevity and performance [1–8]. Conformation is a phenome and encompasses relatively objective traits such as the length of a body segment or an angle between joints, but also highly subjective traits such as the type (breed type, sex type) and the shape of the head, withers, back or croup. Many horse breeding associations perform conformation evaluations at breeding shows by subjective assessment by breeding experts on scoring sheets (e.g., subjective valuating scores, linear profiling) and/or through measurement made on the living animal (e.g., withers height, croup height, limb length). These data commonly provide the basis for the analysis of variance components and subsequent breeding value estimation in European sport horse breeds [9,10].

Despite their widespread use, essentially due to their simple implementation, conformation traits based on subjective assessments, including linear profiling, exhibited certain limitations. In many studies, the scale of the scoring scheme was not used in full, leading to poorly distributed data tending towards the optimum, while the lower extreme of the scale was avoided [11–13]. Furthermore, the inter-rater reliability between experts on the breed (judges) who routinely assessed conformation traits in horses was highly variable depending on the breed: the repeated assessments of linear profiling traits in the Pura Raza Español were highly correlated based on intra-class correlation coefficients (ICC: $0.96 < ICC < 0.99$) [14], indicating reliable data. However, the repeatability of conformation traits that were judged on 4306 Finnhorse and 294 Standardbred trotter foals was estimated using a correlation coefficient (r) that only ranged from 0.06 to 0.48 in Standardbred trotter and from 0.24 to 0.38 in Finnhorses [15], respectively. Furthermore, multiple experts simultaneously assessing the same Lipizzaner horses showed poor inter-rater reliabilities based on the kappa statistic ($0.06 < \kappa < 0.49$) [16,17]. The reliability of expert scores is seldom reported as many horse breed associations only assess the conformation of horses once in their lifetimes. Hence, specific reliability studies are rare, making it difficult to assess the quality of the scoring data.

The Franches-Montagnes (FM) horse is the last native Swiss breed. In its history that spans over one hundred years, the breed has evolved from a draught horse used for tilling fields to the light draught horse breed used for leisure known today [18]. To be able to perform these different functions, the conformation of the FM had to adapt over time, and conformation traits were used empirically to move towards modern breeding goals. Since 1990, three-year-old FM horses have been presented in hand and scored for 19 conformation and five locomotor traits on a linear profiling scale by experts of the breed. However, the distribution of the scores does not always follow a normal distribution, with the lowest scores being avoided, and the mean tending towards the perceived optimum (9) instead of the scale median (5) [19]. The same trend could be observed for gait quality traits [20,21]. Furthermore, FM experts scoring the same horses simultaneously, either in hand or on the treadmill over video recordings, had poor agreement ($ICC < 0.50$) for all scored gait quality traits [20,21]. The heritability for conformation traits in the FM ranged from 0.09 (length of shoulder, hind limb muscle) to 0.79 (height at withers, measured) [22].

The heritability of conformation traits naturally depends on the breed, the sample size used for the calculation, the method of data collection (subjective evaluation, linear profiling score or measurement), and the underlying genetic architecture. However, some tendencies are similar between breeds. Based on a meta-analysis of 30 studies on genetic parameter estimates of conformation traits in horses, height at croup and height at withers had the highest heritability for measured conformation traits ($h^2 = 0.61$ for height at croup and $h^2 = 0.58$ for height at withers) and were highly genotypically correlated ($r_g = 0.94$) [23]. The size of the horse is routinely measured, with data available from many horses and breeds, and the genetic architecture has also been shown to be highly responsive to selection as most of the variation in the height at withers is essentially determined by four genes [24]. Furthermore, objective measures are generally more heritable compared to corresponding scored traits [23]. This may be at least partially due to lower subjectivity in the data collection. However, one key source of error in measuring conformation traits is the landmark definition, i.e., which anatomical structures should be used as reference points, and whether they can be easily identified [25].

The horse shape space model proposed by Druml et al. [16] uses standardized photographs to extract shape data from landmarks on fixed anatomical structures, and semilandmarks equidistantly placed on curves so that they can then be analysed as landmarks. This method was first developed in the Lipizzaner [16,17] and then applied to the FM horse [26]. While only the overall shape was initially analysed, joint angle measurements were later extracted from the initial landmarks. When comparing linear profiling traits describing, e.g., the shoulder or croup incline with objectively measured joint angles from the horse shape space model, the two corresponding traits were not significantly associated, suggesting that the expert scores do not represent the variation that is objectively quantified using the horse shape space model [26]. In the initial horse shape space model [16], the majority of landmarks used to extract joint angle measurements were placed on the surface of the horse (i.e., in front of the joint), as that made them easier to place. However, placing the landmarks in the centre of the joint is expected to improve the predictive relationship between conformation and joint movement, provided they can be placed as accurately as those placed in front of the joints can.

The aims of the study were to estimate variance components and heritability for conformation traits extracted from the horse shape space model in the FM horse breed. Using different landmark settings, joint angles were evaluated both for their heritability and repeatability. Finally, the evolution of the conformation of the FM breed was visualized for the period from 1940 to 2018.

2. Materials and Methods

2.1. Phenotypes

One photograph for each of the 608 FM horses was selected from the archives of the Swiss National Stud Farm or collected on farms under the permits VD3096 and VD3527b. The horses were born between 1940 and 2018 (median = 2005) and aged between 3 and 25 years on the photographs (median = 3). Age was unknown for 133 horses. In total, 480 stallions, 68 mares and 60 geldings were included in the analysis. The horses were positioned in an open posture for the photograph (Figure 1). To account for individual variation, the posture of the horses on the photographs was classified based on previously described criteria [26]: head height, head position in relation to the camera, front limb position, hind limb position and body alignment to the camera.

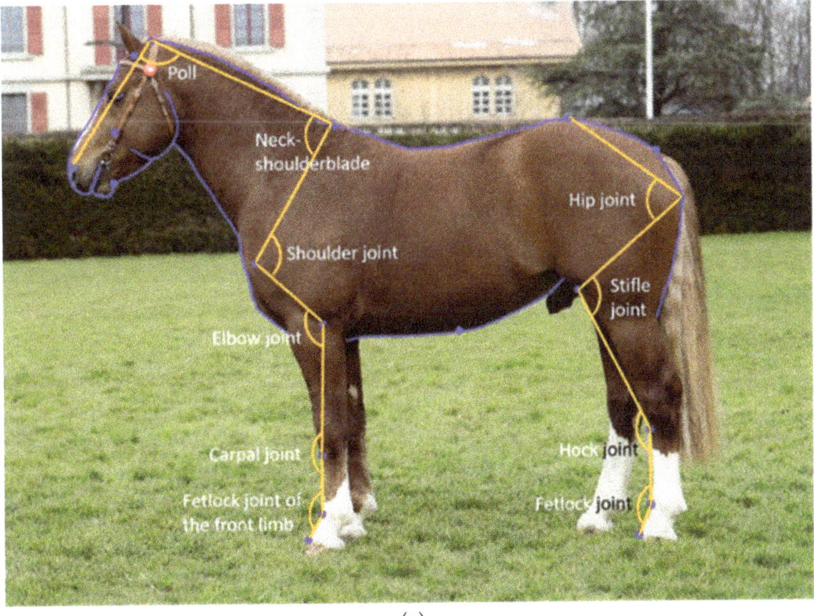

(a)

Figure 1. *Cont.*

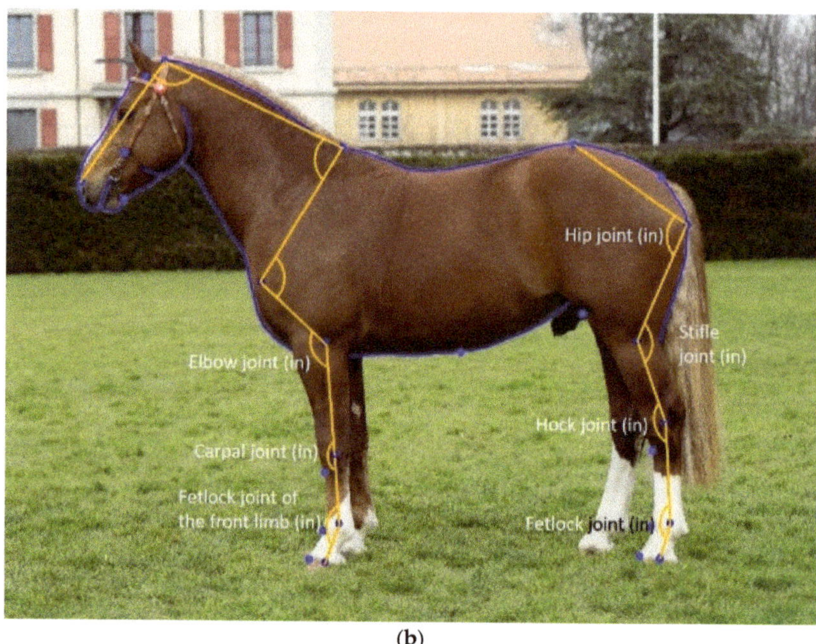

(**b**)

Figure 1. Example of the initial horse shape space model with both curves and angles derived from landmarks (**a**), and the newly proposed landmarks and additional joint angles (**b**).

Conformational data were extracted from the photographs based on the horse shape space model published by Druml et al. [16], consisting of both curves and landmarks using the digitising tool tpsDig2 version 1.78 [27]. The semi-landmarks on the curves were placed at equal distances within each curve, with the final horse model consisting of 246 landmarks (Figure S1). All photographs were digitised by the same person. The raw landmark coordinates were first normalised using a Generalised Procrustes Analysis (GPA). We then used a principal component analysis (PCA) of the normalised landmarks to convert the shape data into relative warp scores (PCs; the principal components of the partial warp matrix), explaining the main variation in the data. We considered only the first five PCs for the analysis of variance components, which were visualised in warp grids using tpsRelw version 1.70 [28].

From the proposed 246 landmarks, 10 joint angle measurements can be extracted, namely the poll angle, neck–shoulder blade angle, shoulder joint angle, elbow joint angle, carpal angle, the fetlock joint angle of the front limb, hip joint angle, stifle joint angle, hock joint angle, and fetlock joint angle of the hind limb (Figure 1a). The landmarks for measuring the limb angles (from elbow to fetlock in the front limb and from hip to fetlock in the hind limb) were placed in front of the joint [26]. As an extension to the initial model, we proposed seven additional landmarks, to measure the same angles of the limbs, but with the landmarks located in the centre of the joints when looking from the side (Figure 1b, Table 1). All angles were calculated in R [29] using a custom-made script. Mean differences in joint angle measurements due to landmark placements were analysed using a paired *t*-test. To evaluate the repeatability of landmark placements for the old versus the new proposed joint angle measurements, the photographs of 480 horses were digitised in triplicate by the same digitiser. The repeatability of the joint angle measurements was estimated with an intra-class correlation coefficient (ICC). For these 480 horses, the mean landmarks were calculated before performing the GPA with the rest of the horses digitised only once.

Table 1. Landmark placement determining the pairs of joint angles calculated from the horse shape space data.

Trait	Landmark Placement in Front of the Limb [26]	Landmark Placement within the Limb
Elbow joint angle	Greater tubercle of the humerus (point of shoulder)—lateral epicondyle of the humerus—anterior aspect of the metacarpal tuberosity of the 3rd metacarpal bone	Greater tubercle of the humerus (point of shoulder)—lateral epicondyle of the humerus—lateral aspect of the carpal ulnar bone
Carpal joint angle	Lateral epicondyle of the humerus—anterior aspect of the proximal tuberosity of the 3rd metacarpal bone—anterior aspect of the fetlock joint	Lateral epicondyle of the humerus—lateral aspect of the carpal ulnar bone—lateral aspect of the fetlock joint
Fetlock joint angle of the front limb	Anterior aspect of the proximal tuberosity of the 3rd metacarpal bone—anterior aspect of the fetlock joint—anterior aspect of the coronet	Lateral aspect of the carpal ulnar bone—lateral aspect of the fetlock—lateral aspect of the coronet
Hip joint angle	Sacral tuber of the ilium (highest point of the croup)—tuber of the ischium (point of buttock)—apex of the patella	Sacral tuber of the ilium (highest point of the croup)—tuber of the ischium (point of buttock)—lateral condyle of the tibia
Stifle joint angle	Tuber of the ischium (point of buttock)—apex of the patella—anterior aspect of the tarsus	Tuber of the ischium (point of buttock)—lateral condyle of the tibia—4th tarsal bone
Hock joint angle	Apex of the patella—anterior aspect of the tarsus—anterior aspect of the fetlock joint	Lateral condyle of the tibia—4th tarsal bone—lateral aspect of the fetlock joint
Fetlock joint angle of the hind limb	Anterior aspect of the tarsus—anterior aspect of the fetlock joint—anterior aspect of the coronet	4th tarsal bone—lateral aspect of the fetlock joint—lateral aspect of the coronet

The effects of posture, age, sex and year of birth on the measurements were evaluated using a linear model; each joint angle and PC as outcome variables; and all posture variables (head height, head position in relation to the camera, front limb position, hind limb position and body alignment), age, sex (mare, gelding or stallion) and year of birth, as fixed effects.

2.2. Evolution of Conformation Traits over Time

Finally, the evolution over time of the different measurements in stallions (to avoid the sex effect) was evaluated. A second set of linear models was computed for the stallion subsample with the effects of posture, age and year of birth as fixed effects (excluding sex). Each of the joint angles and PCs was plotted against the year of birth over time, with a trend line computed in ggplot2 as a loess function (local polynomial regression fitting). Pairs of joint angles were plotted in the same graph for comparison.

2.3. Animal Model

The following multivariate individual animal model was applied to estimate variance components (VC) for 17 joint angles. The same model was used to estimate VC for the five principal components:

$$y_{ijklmnopqr} = YOB_i + Age_j + WH_k + Head\_camera_l + Head\_height_m + Front\_limb_n + Hind\_limb_o + Body_p + a_q + e_{ijklmnopq}$$

with:

$y_{ijklmnopq}$ consecutive observation on a trait

fixed   effects:

YOB$_i$	year of birth
Age$_j$	age of horse at collection in years (linear and quadratic)
WH$_k$	height at withers (linear and quadratic)
Head_camera$_l$	head position in relation to the camera
Head_height$_m$	head height
Front_limb$_n$	front limb position
Hind_limb$_o$	hind limb position
Body$_p$	body alignment to the camera

random effect

| a$_q$ | horse q |
| e$_{ijklmnopq}$ | random residual |

Phenotypic and genetic variance components were estimated by REML using the software ASReml 4.2 [30]. For each trait, heritability (h^2) was calculated as follows:

$$h^2 = \sigma_a^2 / \left(\sigma_a^2 + \sigma_e^2 \right)$$

where σ_a^2 is the additive genetic variance and σ_e^2 is the residual variance.

For the PCs, the ASReml US variance structure was used for the assessment of genetic and residual variance. The use of the US variance structure for the joint angle traits did not result in positive definite variance structure, which is why the model did not converge. Therefore, an ASReml XFA4 variance structure was used for the additive genetic variance.

In total, the pedigree file contained 6986 horses, with 1663 sires (219 founders) and 4991 dams (436 founders). The mean pedigree completeness, considering 1 to 5 generations, was 71.59%.

3. Results

3.1. Descriptive Statistics and Comparisons between Joint Angle Measurements

The two types of fetlock joint angles of the hind limb exhibited the broadest range (42.50–43.30°, difference between maxima and minima), while both types of carpal joint angles had the smallest range of the joint angles (14.40–15.50°, Table 2). The highest repeatability was for the poll angle (ICC = 0.99), while the lowest was for the fetlock joint of the front limb (ICC = 0.66). For the pairs of joint angles (elbow joint, carpal joint, fetlock joint of the front limb, hip joint, stifle joint, hock joint, and fetlock joint of the hind limb), the joint angles measured with the new landmarks within the joints were significantly larger (less acute, $p < 0.0001$) than the initial measurements based on a paired t-test.

Table 2. Descriptive statistics and repeatability for photographs analysed in triplicate for each trait as an intra-class correlation coefficient with their 95% confidence intervals, for the joint angles.

Trait	Mean	SD	Min	Max	Range	ICC (N = 480)	Low CI	High CI
Poll	103.35	5.27	85.90	120.98	35.08	0.99	0.98	0.99
Neck–shoulder blade	83.12	5.47	66.72	101.51	34.79	0.95	0.94	0.95
Shoulder joint	103.87	4.43	90.71	118.24	27.53	0.81	0.79	0.84
Elbow joint	137.90	4.42	120.70	151.40	30.70	0.86	0.84	0.88
Elbow joint (in)	143.00	4.53	126.20	156.40	30.20	0.86	0.83	0.88
Carpal joint	180.50	1.98	173.10	187.50	14.40	0.68	0.64	0.72
Carpal joint (in)	181.30	2.06	172.00	187.50	15.50	0.67	0.63	0.71
Fetlock joint of the front limb	148.90	4.07	136.80	161.90	25.10	0.66	0.62	0.70
Fetlock joint of the front limb (in)	151.10	4.27	138.40	165.20	26.80	0.74	0.71	0.77
Hip joint	78.72	3.10	70.58	91.04	20.46	0.90	0.88	0.91
Hip joint (in)	95.88	3.49	106.24	117.19	10.95	0.92	0.91	0.93
Stifle joint	100.65	4.08	86.56	113.24	26.68	0.89	0.87	0.90
Stifle joint (in)	135.50	3.86	123.00	148.50	25.50	0.89	0.87	0.90
Hock joint	153.30	2.32	144.10	160.50	16.40	0.83	0.80	0.85
Hock joint (in)	161.80	2.78	152.00	171.70	19.70	0.78	0.75	0.81
Fetlock joint of the hind limb	157.20	5.07	135.30	177.80	42.50	0.77	0.74	0.80
Fetlock joint of the hind limb (in)	160.60	5.00	138.90	182.20	43.30	0.82	0.79	0.84

3.2. Postural Effects on the Joint Angles

Each joint angle was significantly affected by at least one postural variable (Table 3). Pairs of joint angle measurements (differing only in the landmark placement) were not affected by the same combination of postural variables, except the two types of elbow and hip joint angles.

Table 3. Summary table of the significance of the postural variables, sex, age and year of birth on the different joint angles based on linear regression.

Trait	Head height	Head in relation to camera	Position of front limb	Position of hind limb	Body alignment	Sex (gelding)	Sex (stallion)	Age	Year of Birth
Poll		***							***
Neck–shoulder blade	***		***		***			***	***
Shoulder joint			***				***	*	***
Elbow joint	***		***		*		***	**	***
Elbow joint (in)	***		***		*		***	**	***
Carpal joint					***		***	**	***
Carpal joint (in)					***	*	***	**	
Fetlock joint of the front limb				**			**		***
Fetlock joint of the front limb (in)				**				*	
Hip joint				***	***		***		***
Hip joint (in)				***	***		***		***
Stifle joint			***	***	***	*	***	**	*
Stifle joint (in)		**	**	***	***		**	*	
Hock joint			*	***	*				
Hock joint (in)			**	***	**	**	***		*
Fetlock joint of the hind limb			**						**
Fetlock joint of the hind limb (in)	*		**		*				

* = *p*-value < 0.05, ** = *p*-value < 0.01, *** = *p*-value < 0.001.

3.3. Visualisation of the Principal Components of Shape Variation

Based on the visualisation of the minima and maxima of the PCs, we can infer that PC1 represents the shape variation due to head height, and PC2 the variation due to the angle at the poll (flexion-extension) (Figure 2). PC3, PC4 and PC5 essentially represent the musculature of the neck. In addition, PC3 also represents the shape of the withers and back, PC4 and PC5 the shape of the croup and the position of the front and hind limbs.

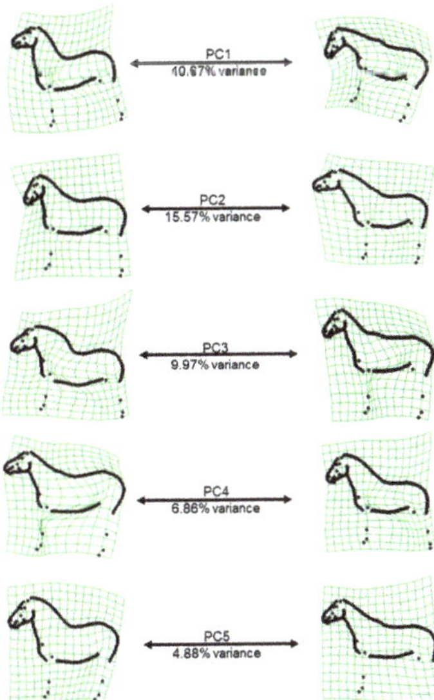

Figure 2. Representation of the extreme shapes describing the first five relative warp axes on warp grids from low (**left**) to high (**right**), with the percentage of explained variance under the arrow.

3.4. Postural Effects on the Principal Components of Shape Variation

In the shape variation (PC1 to PC5), there were no significant differences between mares and geldings, but highly significant differences between mares and stallions (Table 4). Head height significantly affected the first four PCs, while the position of the front limb, hind limb and body alignment had a significant effect on all PCs presented here. PC1, the principal component explaining most of the variance in this dataset (41%), was significantly affected by all posture variables.

Table 4. Summary table of the significance of the postural variables, sex, age and year of birth on the different joint angles based on linear regression.

Trait	Head Height	Head in Relation to Camera	Position of Front Limb	Position of Hind Limb	Body Alignment	Sex (Gelding)	Sex (Stallion)	Age	Year of Birth
PC1	***	*	*	*	*		***	**	**
PC2	**	***		*	***		***	***	
PC3	***		***	***	***			*	***
PC4	***		***	***	***			***	
PC5			*	***	***			***	

* = *p*-value < 0.05, ** = *p*-value < 0.01, *** = *p*-value < 0.001.

3.5. Evolution of the Breed

We recalculated the linear regression models for postural and other variables with only stallions (File S1). Based on the linear regression models, year of birth most significantly affected PC3, which increased linearly over time, meaning that the stallions are evolving towards a lighter type (Figure 3). Year of birth also significantly affected PC1, showing a sigmoid curve, and PC5, slightly increasing from the 2000s (Figure S2).

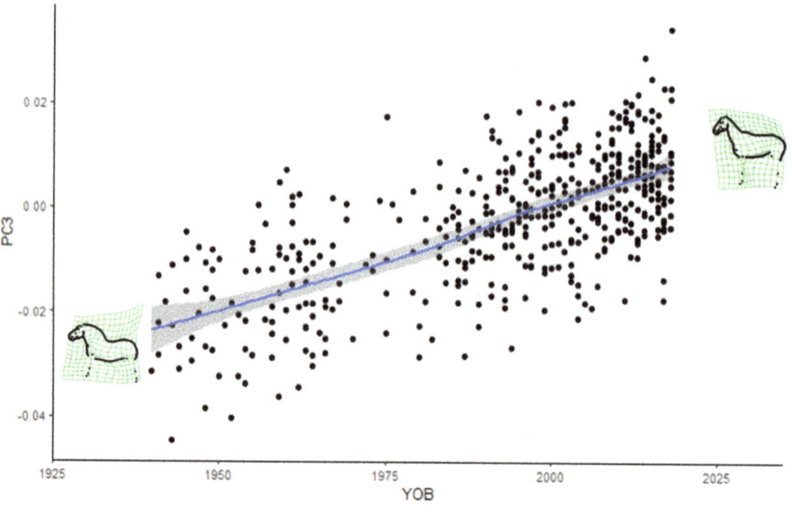

Figure 3. Evolution of the third relative warp score (PC3) in stallions born from 1940 to 2018 (YOB), with the extreme shapes on relative warp grids, with the trend line (in blue, with the confidence interval in light grey) from local polynomial regression fitting.

Year of birth had a significant effect on all joint angles except the carpal angle, the fetlock joint of the hind limb and the stifle joint (both measured with the landmarks inside the joint) as well as the hock joint (measured in front of the joint, Figures S3–S5). The poll angle decreased until the mid-1990s and then increased again slightly up to 2018. The neck–shoulder blade angle increased almost linearly over time. The shoulder joint angle increased slightly up to the mid-2000s, when it reached a plateau. Both types of elbow joint angles followed a parallel trend line, increasing from 1940 to 1960 and decreasing until 2000 to reach a plateau. The carpal joint angle (measured in front of the joint) decreased until 1980, increased until 2000, and has been decreasing again until 2018. The fetlock joint of the front limb (measured in front of the joints) and both types of hip joint measurements remained stable until the 2000s, with a decrease in recent years. The stifle joint (measured in front of the joints) increased slightly until the 1980s, decreased until the mid-2000s and is currently on a slight uphill tendency. The hock joint (measured in front) appeared to decrease around 1990 and stabilise at a lower level. The fetlock joint of the hind limb (measured in front) appeared to remain stable over time, with a slight decrease after 2000 and up to recent years. The individual plots of all measurements in stallions against the year of birth were summarised in Figures S2–S5.

3.6. Genetic Analyses of Joint Angle Measurements

For the joint angles, the highest heritability was found for the poll angle ($h^2 = 0.37 \pm 0.09$) and the lowest for the hock joint (with landmarks within the joints, $h^2 = 0.08 \pm 0.05$, Table 5). Most genetic, phenotypic and residual variances are >1, except the genetic additive variance for both types of carpal joint, the stifle joint (with landmarks in front of the joint) and both types of hock joint.

Table 5. Heritability with corresponding standard errors, additive genetic (σ_a), phenotypic (σ_p) and residual (σ_e) variances.

Trait	h^2	SE	σ_a	σ_p	σ_e
Poll	0.37	0.09	8.75	23.96	15.21
Neck–shoulder blade	0.20	0.08	4.26	21.62	17.36
Shoulder joint	0.18	0.06	2.85	15.64	12.79
Elbow joint	0.20	0.07	3.55	17.93	14.38
Elbow joint (in)	0.19	0.07	3.67	18.95	15.28
Carpal joint	0.13	0.07	0.17	1.39	1.21
Carpal joint (in)	0.27	0.09	0.52	1.88	1.36
Fetlock joint of the front limb	0.29	0.08	4.35	15.71	11.18
Fetlock joint of the front limb (in)	0.31	0.08	4.65	18.08	12.43
Hip joint	0.23	0.07	1.82	7.95	6.13
Hip joint (in)	0.17	0.07	1.50	8.66	7.16
Stifle joint	0.08	0.05	0.98	11.79	10.81
Stifle joint (in)	0.12	0.06	1.49	12.15	10.66
Hock joint	0.16	0.07	0.77	4.86	4.09
Hock joint (in)	0.06	0.04	0.41	6.58	6.17
Fetlock joint of the hind limb	0.19	0.07	4.17	21.99	17.82
Fetlock joint of the hind limb (in)	0.09	0.05	2.05	21.98	19.93

The highest genotypic and phenotypic correlations were between the two elbow angles, representing virtually the same trait ($r_p = 0.99$, $r_g = 0.99$, Table 6). Relatively high (0.70–0.90) phenotypic correlations were also present between the same joint angles measured differently (hip, stifle, hock and fetlock joint of the hind limb), except between the two carpal joint angles and the two front limb fetlock joint angles (0.45 and 0.69, respectively). The highest phenotypic correlation between two different joints was between the shoulder and elbow joints (measured from the centre of the limb, $r_p = 0.64$). In absolute values, the lowest genotypic correlation was between the hip and hock joint angles measured in front of the joint ($r_g = 0.01$). Apart from the near perfect correlation between the two elbow angles, the highest genotypic correlation was between the hip and stifle joint angles ($r_g = 0.96$).

Table 6. Genetic correlations (above the diagonal), heritabilities (in the grey-coloured diagonal) and phenotypic correlations with their standard errors in subscript below the diagonal, for joint angle data derived from the horse shape space model.

	Poll	Neck	Shoulder	Elbow	Elbow (in)	Carpal joint	Carpal Joint (in)	Fetlock Front	Fetlock Front (in)	Hip	Hip (in)	Stifle	Stifle (in)	Hock	Hock (in)	Fetlock Hind	Fetlock Hind (in)
Poll	$0.37_{0.09}$	$-0.34_{0.18}$	$0.02_{0.15}$	$0.28_{0.18}$	$0.26_{0.18}$	$-0.12_{0.19}$	$-0.08_{0.17}$	$0.10_{0.18}$	$0.21_{0.17}$	$-0.36_{0.18}$	$-0.24_{0.20}$	$-0.28_{0.22}$	$-0.04_{0.21}$	$0.05_{0.22}$	$0.13_{0.25}$	$0.14_{0.19}$	$0.14_{0.22}$
Neck	$-0.23_{0.04}$	$0.20_{0.08}$	$-0.14_{0.21}$	$-0.62_{0.21}$	$-0.58_{0.21}$	$0.31_{0.27}$	$0.06_{0.23}$	$-0.33_{0.23}$	$-0.42_{0.21}$	$0.18_{0.25}$	$-0.11_{0.27}$	$0.10_{0.34}$	$-0.43_{0.26}$	$0.25_{0.27}$	$-0.03_{0.36}$	$-0.01_{0.25}$	$-0.11_{0.3}$
Shoulder	$-0.07_{0.04}$	$0.25_{0.04}$	$0.18_{0.06}$	$0.40_{0.19}$	$0.43_{0.18}$	$-0.37_{0.18}$	$-0.37_{0.18}$	$0.20_{0.23}$	$0.37_{0.20}$	$0.30_{0.23}$	$0.32_{0.23}$	$0.40_{0.23}$	$0.37_{0.22}$	$0.25_{0.25}$	$0.28_{0.24}$	$-0.18_{0.22}$	$0.11_{0.25}$
Elbow	$-0.11_{0.04}$	$-0.05_{0.04}$	$0.63_{0.03}$	$0.20_{0.07}$	$0.99_{0.00}$	$-0.60_{0.25}$	$-0.57_{0.19}$	$0.18_{0.24}$	$0.57_{0.19}$	$0.13_{0.26}$	$0.30_{0.27}$	$0.36_{0.36}$	$0.60_{0.27}$	$0.25_{0.25}$	$0.28_{0.38}$	$-0.44_{0.24}$	$-0.04_{0.3}$
Elbow (in)	$-0.11_{0.04}$	$-0.05_{0.04}$	$0.64_{0.03}$	$0.99_{0.00}$	$0.19_{0.07}$	$-0.63_{0.24}$	$-0.61_{0.19}$	$0.22_{0.24}$	$0.62_{0.19}$	$0.18_{0.26}$	$0.32_{0.27}$	$0.41_{0.35}$	$0.60_{0.27}$	$0.32_{0.29}$	$0.35_{0.36}$	$-0.41_{0.24}$	$0.02_{0.32}$
Carpal joint	$-0.02_{0.04}$	$0.00_{0.04}$	$-0.04_{0.04}$	$-0.14_{0.04}$	$-0.13_{0.04}$	$0.13_{0.07}$	$0.51_{0.22}$	$-0.51_{0.26}$	$-0.71_{0.22}$	$-0.34_{0.31}$	$-0.43_{0.32}$	$-0.49_{0.34}$	$-0.56_{0.31}$	$-0.34_{0.31}$	$-0.53_{0.28}$	$0.02_{0.29}$	$-0.40_{0.2}$
Carpal joint (in)	$0.02_{0.04}$	$0.02_{0.04}$	$-0.11_{0.04}$	$-0.22_{0.04}$	$-0.25_{0.04}$	$0.45_{0.03}$	$0.27_{0.08}$	$-0.11_{0.22}$	$-0.49_{0.17}$	$-0.24_{0.20}$	$-0.16_{0.25}$	$-0.45_{0.30}$	$-0.24_{0.28}$	$-0.71_{0.19}$	$-0.57_{0.25}$	$0.25_{0.23}$	$-0.17_{0.2}$
Fetlock front	$0.00_{0.04}$	$-0.06_{0.04}$	$0.02_{0.04}$	$0.03_{0.04}$	$0.03_{0.04}$	$-0.15_{0.04}$	$-0.07_{0.04}$	$0.29_{0.08}$	$0.83_{0.08}$	$0.30_{0.22}$	$0.51_{0.22}$	$0.30_{0.32}$	$0.61_{0.22}$	$-0.09_{0.27}$	$0.53_{0.33}$	$0.62_{0.17}$	$0.87_{0.15}$
Fetlock front (in)	$0.05_{0.04}$	$-0.10_{0.04}$	$-0.01_{0.04}$	$0.01_{0.04}$	$0.02_{0.04}$	$-0.02_{0.04}$	$-0.16_{0.04}$	$0.69_{0.02}$	$0.31_{0.08}$	$0.26_{0.21}$	$0.42_{0.23}$	$0.40_{0.30}$	$0.61_{0.22}$	$0.32_{0.25}$	$0.73_{0.23}$	$0.33_{0.21}$	$0.76_{0.18}$
Hip	$-0.11_{0.04}$	$0.08_{0.04}$	$0.12_{0.04}$	$0.07_{0.04}$	$0.07_{0.04}$	$-0.04_{0.04}$	$-0.02_{0.04}$	$-0.02_{0.04}$	$-0.03_{0.04}$	$0.23_{0.07}$	$0.93_{0.05}$	$0.96_{0.09}$	$0.73_{0.19}$	$0.01_{0.29}$	$0.11_{0.39}$	$-0.35_{0.24}$	$0.02_{0.32}$
Hip (in)	$-0.10_{0.04}$	$0.08_{0.04}$	$0.09_{0.04}$	$0.01_{0.04}$	$0.00_{0.04}$	$-0.06_{0.04}$	$-0.01_{0.04}$	$-0.03_{0.04}$	$-0.02_{0.04}$	$0.84_{0.01}$	$0.17_{0.07}$	$0.88_{0.18}$	$0.92_{0.11}$	$-0.21_{0.30}$	$0.08_{0.41}$	$-0.25_{0.27}$	$0.11_{0.33}$
Stifle	$0.00_{0.04}$	$0.05_{0.04}$	$0.05_{0.04}$	$0.00_{0.04}$	$-0.01_{0.04}$	$-0.01_{0.04}$	$0.06_{0.04}$	$0.06_{0.04}$	$0.05_{0.04}$	$0.52_{0.03}$	$0.24_{0.04}$	$0.08_{0.05}$	$0.75_{0.17}$	$0.24_{0.38}$	$0.28_{0.42}$	$-0.42_{0.33}$	$0.05_{0.40}$
Stifle (in)	$0.03_{0.04}$	$0.02_{0.04}$	$0.02_{0.04}$	$-0.05_{0.04}$	$-0.06_{0.04}$	$-0.03_{0.04}$	$0.09_{0.04}$	$0.09_{0.04}$	$0.09_{0.04}$	$0.39_{0.04}$	$0.41_{0.03}$	$0.80_{0.02}$	$0.12_{0.06}$	$-0.22_{0.34}$	$0.15_{0.40}$	$-0.22_{0.29}$	$0.17_{0.35}$
Hock	$0.04_{0.04}$	$-0.02_{0.04}$	$0.05_{0.04}$	$0.01_{0.04}$	$0.00_{0.04}$	$0.00_{0.04}$	$0.07_{0.04}$	$0.05_{0.04}$	$0.13_{0.04}$	$0.15_{0.04}$	$0.11_{0.04}$	$0.36_{0.04}$	$0.37_{0.04}$	$0.16_{0.07}$	$0.68_{0.23}$	$-0.08_{0.29}$	$0.24_{0.35}$
Hock (in)	$0.08_{0.04}$	$-0.08_{0.04}$	$0.03_{0.04}$	$0.05_{0.04}$	$0.04_{0.04}$	$-0.01_{0.04}$	$0.06_{0.04}$	$0.09_{0.04}$	$0.11_{0.04}$	$0.10_{0.04}$	$0.08_{0.04}$	$0.34_{0.04}$	$0.54_{0.04}$	$0.74_{0.03}$	$0.06_{0.04}$	$0.37_{0.37}$	$0.72_{0.30}$
Fetlock hind	$0.05_{0.04}$	$0.08_{0.04}$	$-0.06_{0.04}$	$-0.06_{0.04}$	$-0.06_{0.04}$	$-0.03_{0.04}$	$0.00_{0.04}$	$0.19_{0.04}$	$0.12_{0.04}$	$0.04_{0.04}$	$0.02_{0.04}$	$0.03_{0.04}$	$0.00_{0.04}$	$-0.09_{0.04}$	$-0.07_{0.04}$	$0.19_{0.07}$	$0.83_{0.11}$
Fetlock hind (in)	$0.06_{0.04}$	$0.07_{0.04}$	$-0.02_{0.04}$	$-0.08_{0.04}$	$-0.08_{0.04}$	$-0.05_{0.04}$	$-0.01_{0.04}$	$0.15_{0.04}$	$0.20_{0.04}$	$0.12_{0.04}$	$0.09_{0.04}$	$0.11_{0.04}$	$0.06_{0.04}$	$0.05_{0.04}$	$-0.14_{0.04}$	$0.78_{0.02}$	$0.09_{0.05}$

3.7. Genetic Analyses of the Principal Components of Shape Variation

For the relative warp components, the variance components were very small (Table 7). PC5 had the highest heritability ($h^2 = 0.37 \pm 0.09$), and PC1 the lowest ($h^2 = 0.13 \pm 0.08$).

Table 7. Heritabilities with corresponding standard errors, additive genetic (σ_a), phenotypic (σ_p) and residual (σ_e) variances of the first five relative warp scores (PC1–PC5).

Trait	h^2	SE	σ_a	σ_p	σ_e
PC1	0.13	0.08	4.59×10^{-5}	3.64×10^{-4}	3.18×10^{-4}
PC2	0.14	0.08	3.80×10^{-5}	2.71×10^{-4}	2.33×10^{-4}
PC3	0.36	0.09	2.89×10^{-5}	7.97×10^{-5}	5.08×10^{-5}
PC4	0.29	0.10	3.04×10^{-5}	1.04×10^{-4}	7.31×10^{-5}
PC5	0.37	0.09	2.58×10^{-5}	6.93×10^{-5}	4.35×10^{-5}

The highest phenotypic correlation was between PC1 and PC3 ($r_p = 0.41$, Table 8). The fourth and fifth PCs showed low phenotypic correlations to the other PCs. The highest genetic correlation was between PC1 and PC2 ($r_g = -0.55$). The genotypic correlations ranged from 0.06 to 0.55 in absolute values.

Table 8. Genetic correlations (above the diagonal), heritabilities (in the grey-coloured diagonal) and phenotypic correlations with their standard errors (below the diagonal) for the relative warp scores derived from the horse shape space model.

	PC1	PC2	PC3	PC4	PC5
PC1	$0.13_{0.08}$	$-0.55_{0.43}$	$0.27_{0.28}$	$-0.36_{0.39}$	$0.34_{0.30}$
PC2	$-0.09_{0.04}$	$0.14_{0.08}$	$0.31_{0.29}$	$0.28_{0.34}$	$-0.51_{0.28}$
PC3	$0.41_{0.04}$	$0.06_{0.04}$	$0.36_{0.09}$	$-0.24_{0.34}$	$-0.14_{0.21}$
PC4	$0.12_{0.04}$	$0.09_{0.04}$	$-0.09_{0.04}$	$0.29_{0.10}$	$-0.06_{0.23}$
PC5	$0.08_{0.04}$	$-0.07_{0.04}$	$-0.06_{0.04}$	$-0.13_{0.04}$	$0.37_{0.09}$

4. Discussion

The joint angles with the highest heritability (h^2) were the poll and the fetlock joint of the front limb (Table 5). The high h^2 of the poll angle was consistent with results from a multi-breed genome-wide association study of the same joint angles from 300 FM and

224 Lipizzaner horse photographs [31]. The best-associated quantitative trait locus (QTL) was the poll angle on equine chromosome (ECA) 28, near the gene *ALX1*, associated with cranial morphology [32]. The genome-wide h^2 for the poll angle in the two breeds was $h^2 = 0.38$, nearly equal to the pedigree-based h^2 estimated here. In contrast, the highest genome-wide h^2 for Lipizzaner and FM was for the fetlock joint of the hind limb ($h^2 = 0.58$) and a suggestive QTL on ECA 27, whereas this trait had a much lower pedigree-based $h^2 = 0.19$ in the FM.

The placement of the landmarks for the calculation of the joint angles changed the acuteness of the joint angles. For all pairs of joint angles, the landmark placement inside the joint was significantly larger (less acute) than when the landmarks were placed in front of the joint (old model). However, the phenotypic correlation is more relevant to understand whether the landmark placement affects the anatomical meaning of the measurement. For example, the elbow joint remained virtually identical whether the third landmark was placed in front or within the carpal joint, as shown by near-perfect genetic and phenotypic correlations between the two elbow joint measurements (Table 6), and a nearly identical repeatability (ICC, Table 2) and h^2 (Table 5). The two types of elbow joint angles were also significantly affected by the same combination of posture and other external variables (age, sex and year of birth, Table 3). For this angle, it makes no difference which landmark placement to use. The two types of carpal joint angles were the least phenotypically correlated of the paired joint angles (Table 6), the angle from the old model (landmark in front of the joint) was significantly different between mares and stallions, while with the new landmark placement inside the joint, the sex difference was between mares and geldings (Table 4). The heritability for the new landmark placement was higher (Table 5), with a nearly identical ICC (Table 2). In this case, it makes more sense to measure the carpal joint with the new landmark placement. For the fetlock angle of the front limb, h^2 (Table 5) and ICC (Table 2) were higher with the new landmark placement, affected by the age of the horse and not by the birth year (Table 4), as was the case for fetlock angle measured with the old landmark placement. For the joint angles of the front limb (elbow, carpus and fetlock), the new landmark placement increases h^2 (Table 5) and ICC (or at least does not negatively affect the latter; Table 2).

The trends were less clear in the hind limbs. For the stifle and hock, ICC (Table 2) and h^2 (Table 5) concurred, so that the angle that was measured the most accurately was also the one with the highest h^2 (stifle with landmarks within the joints, hock with landmarks outside the joints). The results were more difficult to interpret for the hip and fetlock joint of the hind limb. For the hip joint, the ICC was only slightly higher, but the h^2 lower, when using the centre of the stifle joint as the third landmark. However, the hip joint angle measured with the landmarks in front of the patella was additionally affected by the head height and the sex (mares were significantly different from geldings; Table 4) which suggests that the angle is more affected by the posture (i.e., environmental effects). While the ICC was lower for the fetlock joint of the hind limb (Table 2), the h^2 was higher when using the landmarks in front of the joints (Table 5). Deciding which measurement is better for a particular joint is complicated in the case when h^2 and ICC are not in lockstep. Another way to choose the most informative landmark placement might be to compare the joint angles with kinematic parameters, to assess whether one set of landmarks is a better predictor for movements associated with gait quality traits such as hind limb protraction [33], i.e., whether a certain joint angle measurement is more functionally relevant. Considering the low amount of additional effort involved in placing the supplementary landmarks, we currently recommend assessing the ICC and h^2 of both types of joint angle measurements in other breeds using the two landmark settings, in order to optimise the results.

The first and second relative warp scores (PC1 and PC2) represented the shape variation induced by the head-neck position (PC1: head height, PC2: flexion-extension at the poll, Figure 2), which is why the posture variable head height was so strongly associated with PC1. For PC2 (as well as the poll angle), when the horse has its head turned towards the camera, the angle decreases, which explains the significant association between the variable "head in relation to camera" with PC2 and poll angle. In practice, the variation due to the head and neck position is hard to avoid, and this result was consistent with several previous horse shape space studies on FM and Lipizzaner horses [16,17,26]. While PC1 and PC2 explain most of the shape variation, they are less heritable than the three following PCs, reflecting the fact that the variation in shape mainly originates from the posture, which is an "environmental" effect. The moderate h^2 of PC3, PC4 and PC5 is consistent with previous findings that the body type (heavy–light), quantified by bone thickness up to now, is the second highest source of conformational variation in the horse after height [34]. However, these PCs are also affected by posture, especially by the position of the front and hind limbs. Therefore, posture variables should always be considered when working with conformational data, although considering all the posture variables as fixed effects in the model of analysis might have caused an over-parameterization of the model, thereby affecting the accuracy of the estimates. The majority of additive genetic, phenotypic and residual variances were large (>1; Table 5), in contrast to other studies that included several thousand horses to estimate genetic parameters [12–15,23]. Furthermore, the standard errors for the genetic correlations, in particular, were large as well (Tables 6 and 8). Reducing the postural variance in the data when photographing the horses and increasing the sample size should improve the accuracy of the estimates on the long term.

The current accuracy of data extracted from the horse shape space model still shows potential for selection on conformation traits based on this method, especially if a horse breeding association routinely records these traits. Of the 19 linear conformation traits routinely assessed by breeding experts in the FM, five describe joint angles we quantified here. We can therefore compare our h^2 estimated for measured joint angles on 608 horses against the h^2 estimates for 18,297 horses tested between 1994 and 2013 [22]. The scored trait "shoulder incline" (straight–inclined) had $h^2 = 0.09$ [22], while the measured shoulder joint angle's h^2 was twice as high (Table 5). The trait "front limb" (back-at-the-knee–over-at-the-knee) had $h^2 = 0.14$ [22], compared to the measured carpal joint angle ($h^2 = 0.13$ measured in front of the joint, $h^2 = 0.27$ within the joint), suggesting that this trait has the potential for improvement when using the measurement within the joint. As the range of the measurement is limited for both carpal joint angles, any slight error in landmark placement has a disproportionate effect on the accuracy of the measurement as is shown by the lower ICC compared to, e.g., the hip joint angles (Table 2). The scored "croup incline" (horizontal–sloping) had h^2 in the same range as the hip joint angles ($h^2 = 0.20$ [22]). The measured hock joint angles exhibited a lower h^2 than the "hock angle" trait (straight–angulated, $h^2 = 0.19$, [22]). These traits were also not associated with each other in a previous direct comparison between measurements and scores [26]. In this case, the linear profiling score may be considered more useful for selection than the hock joint measurement. The final linear profiling trait with an equivalent joint angle is the "fetlock angle" (straight–weak). However, whether this score described the front limbs only, or a combination of front and hind limbs, is not specified. Both measurements of the fetlock joint of the front limb had a higher h^2 than the scored "fetlock angle" ($h^2 = 0.11$, [22]), while the fetlock joint of the hind limb was less heritable when measured within the joint, and more heritable when measured in front of the joint (Table 5). At minimum, selection on shoulder and fetlock angles might therefore be improved by applying the horse shape space model.

Furthermore, it was demonstrated that horse shape space data can be successfully applied to assess the evolution of a breed over time. The FM stallion population evolved from a heavy to a light draught horse type. The absence of a plateau in more recent years is a cause for concern, as the breed may lose its breed-specific type of a light draught horse. Some changes in the trajectories of the trend lines seemed to coincide with recent introgressions from lighter breeds. More specifically, the changes in the poll angle starting in the 1980s may be a consequence of the Swedish Warmblood introgression in the 1970s, while changes in most of the measurements from the 1990s onward in most of the measurements may be due to the two Swiss Warmblood stallions that were introgressed in

1990. The current favoured use of stallions with high admixture proportions from the last introgressions in FM breeding may accelerate the observed changes in conformation.

Apart from the small sample size, there were some additional limitations to our study. Some stallions from the 1940s and 1950s were only distantly related to the more recent population, which might affect estimates of the variance components due to gaps in the pedigree. Furthermore, all the horses born before 2004 were stallions, due to the low availability of mare photographs in the archives of the Swiss National Stud Farm. However, this allowed us to increase our sample size and to retrace the evolution of the FM conformation traits over time. Whether the FM breeding association will implement the horse shape space model in their selection programme depends on several points: social acceptance by the breeders, practical considerations (time to take the appropriate photographs) and technical feasibility (automation of landmark placement). One further limitation to this study is that we could not provide in-depth analysis of the type of conformation favourable to specific disciplines, as was investigated in other morphometric studies [35,36]. This could be the subject of a future study.

5. Conclusions

Joint angles such as the shoulder and fetlock angles had higher heritability based on the horse shape space model than when based on the scores from linear profiling. In the front limbs, landmark placement within the centre of the joints yielded more reliable and heritable results, while the results were less consistent for the hind limb joint angles. The FM horse breed has evolved from a heavy draught breed to a much lighter type. Care must be taken to not lighten the breed beyond the breeding goal of a light draught horse.

Supplementary Materials: The following supporting information can be downloaded at: https://www.mdpi.com/article/10.3390/ani12172186/s1, File S1: Summary of fixed effects estimates for joint angles, relative warp components, postural variables, age, sex and year of birth; Figure S1: Horse shape space model with 246 landmarks on the photograph (a) and without background (b); Figure S2: Evolution of joint angle measurements of the poll angle (a), neck–shoulder blade angle (b) and shoulder joint angle (c) in Franches-Montagnes stallions born between 1940 and 2018; Figure S3: Evolution of joint angle measurements of the front limbs in Franches-Montagnes stallions born between 1940 and 2018; Figure S4: Evolution of joint angle measurements of the hindquarters in Franches-Montagnes stallions born between 1940 and 2018; Figure S5: Evolution of the first five relative warp scores in Franches-Montagnes stallions born between 1940 and 2018.

Author Contributions: Conceptualization, A.I.G. and M.N.; methodology, A.I.G. and A.B.; software, A.I.G. and A.B.; formal analysis, A.I.G. and A.B.; investigation, A.I.G. and A.B.; resources, M.N.; data curation, A.I.G.; writing—original draft preparation, A.I.G. and A.B.; writing—review and editing, M.N.; visualization, A.I.G.; supervision, M.N.; project administration, A.I.G. and M.N.; funding acquisition, A.I.G. and M.N. All authors have read and agreed to the published version of the manuscript.

Funding: This research was funded by the Swiss Federal Office of Agriculture (FOAG) under the contract number 625000469.

Institutional Review Board Statement: This study was performed according to Swiss national rules and regulations. The animal study protocol was approved by the cantonal veterinary office of the canton of Vaud under the following permits: VD3096 and VD3527b.

Informed Consent Statement: Not applicable.

Data Availability Statement: As some of the photographed horses belong to private owners, data are only available on reasonable request to the authors.

Acknowledgments: We are grateful to the breeders participating in this study, to Michael A. Weishaupt for providing comments on the landmark placement, to Christian Gerber, Hervé Sapin, Jérémie Korpès and Ludovic Taillard for presenting horses for photographs, and to the Franches-Montagnes breeding association for providing the pedigree data.

Conflicts of Interest: The authors declare no conflict of interest. The funders had no role in the design of the study; in the collection, analyses, or interpretation of data; in the writing of the manuscript; or in the decision to publish the results.

References

1. Weller, R.; Pfau, T.; Verheyen, K.; May, S.; Wilson, A. The effect of conformation on orthopaedic health and performance in cohort of National Hunt racehorses. Preliminary results. *Equine Vet. J.* **2006**, *38*, 622–627. [CrossRef]
2. Wallin, L.; Strandberg, E.; Philipsson, J. Phenotypic relationship between test results of Swedish Warmblood horses as 4-year-old and longevity. *Livest. Prod. Sci.* **2001**, *68*, 97–105. [CrossRef]
3. Stock, K.; Distl, O. Multiple-trait selection for radiographic health of the limbs, conformation and performance in Warmblood riding horses. *Animal* **2008**, *2*, 1724–1732. [CrossRef]
4. Jönsson, L.; Näsholm, A.; Roepstorff, L.; Egenvall, A.; Dalin, G.; Philipsson, J. Conformation traits and their genetic and phenotypic associations with health status in young Swedish warmblood riding horses. *Livest. Sci.* **2014**, *163*, 12–25. [CrossRef]
5. Jönsson, L.; Egenvall, A.; Roepstorff, L.; Näsholm, A.; Dalin, G.; Philipsson, J. Associations of health status and conformation with longevity and lifetime competition performance in young Swedish Warmblood riding horses. 8238 cases (1983–2005). *J. Am. Vet. Med. Assoc.* **2014**, *244*, 1449–1461. [CrossRef] [PubMed]
6. Holmström, M.; Philipsson, J. Relationships between conformation, performance and health in 4-year-old Swedish Warmblood riding horses. *Livest. Prod. Sci.* **1993**, *33*, 293–312. [CrossRef]
7. Ducro, B.; Gorissen, B.; van Eldik, P.; Back, W. Influence of foot conformation on duration of competitive life in a Dutch Warmblood horse population. *Equine Vet. J.* **2009**, *41*, 144–148. [CrossRef]
8. Dolvik, N.; Klemetsdal, G. Conformational traits of Norwegian cold-blooded trotters. Heritability and the relationship with performance. *Acta Agric. Scand. Sect. A-Anim. Sci.* **1999**, *49*, 156–162. [CrossRef]
9. Duensing, J.; Stock, K.F.; Krieter, J. Implementation and prospects of linear profiling in the Warmblood horse. *J. Equine Vet. Sci.* **2014**, *34*, 360–368. [CrossRef]
10. Koenen, E.; Aldridge, L.; Philipsson, J. An overview of breeding objectives for warmblood sport horses. *Livest. Prod. Sci.* **2004**, *88*, 77–84. [CrossRef]
11. Koenen, E.; Van Veldhuizen, A.; Brascamp, E. Genetic parameters of linear scored conformation traits and their relation to dressage and show-jumping performance in the Dutch Warmblood Riding Horse population. *Livest. Prod. Sci.* **1995**, *43*, 85–94. [CrossRef]
12. Samoré, A.; Pagnacco, G.; Miglior, F. Genetic parameters and breeding values for linear type traits in the Haflinger horse. *Livest. Prod. Sci.* **1997**, *52*, 105–111. [CrossRef]
13. Viklund, Å.; Eriksson, S. Genetic analyses of linear profiling data on 3-year-old Swedish Warmblood horses. *J. Anim. Breed. Genet.* **2018**, *135*, 62–72. [CrossRef] [PubMed]
14. Sánchez, M.; Gómez, M.; Molina, A.; Valera, M. Genetic analyses for linear conformation traits in Pura Raza Español horses. *Livest. Sci.* **2013**, *157*, 57–64. [CrossRef]
15. Schroderus, E.; Ojala, M. Estimates of genetic parameters for conformation measures and scores in Finnhorse and Standardbred foals. *J. Anim. Breed. Genet.* **2010**, *127*, 395–403. [CrossRef]
16. Druml, T.; Dobretsberger, M.; Brem, G. The use of novel phenotyping methods for validation of equine conformation scoring results. *Animal* **2015**, *9*, 928–937. [CrossRef]
17. Druml, T.; Dobretsberger, M.; Brem, G. Ratings of equine conformation–new insights provided by shape analysis using the example of Lipizzan stallions. *Arch. Anim. Breed.* **2016**, *59*, 309–317. [CrossRef]
18. Poncet, P.; Pfister, W.; Muntwyler, J.; Glowatzki-Mullis, M.; Gaillard, C. Analysis of pedigree and conformation data to explain genetic variability of the horse breed Franches-Montagnes. *J. Anim. Breed. Genet.* **2006**, *123*, 114–121. [CrossRef]
19. Pfammatter, M. *Le Système de Description Linéaire au Sein de la Race des Franches-Montagnes*; Berner Fachhochschule, Hochschule für Agrar-Forst und Lebensmittelwissenschaften: Zollikofen, Switzerland, 2017.
20. Gmel, A.I.; Gmel, G.; von Niederhäusern, R.; Weishaupt, M.A.; Neuditschko, M. Should we agree to disagree? An evaluation of the inter-rater reliability of gait quality traits in Franches-Montagnes stallions. *J. Equine Vet. Sci.* **2020**, *88*, 102932. [CrossRef]
21. Gmel, A.I.; Gmel, G.; Weishaupt, M.A.; Neuditschko, M. Gait quality scoring data of Franches-Montagnes stallions at walk and trot on a treadmill by experts of the breed and their reliability. *Data Brief* **2022**, *42*, 108123. [CrossRef]
22. Burren, A.; Bangerter, E.; Hagger, C.; Rieder, S.; Flury, C. *Züchterische Auswertung Beim Freibergerpferd*; Berner Fachhochschule, Hochschule für Agrar-Forst und Lebensmittelwissenschaften: Zollikofen, Switzerland, 2015.
23. Hossein-Zadeh, N.G. A meta-analysis of genetic parameter estimates for conformation traits in horses. *Livest. Sci.* **2021**, *250*, 104601. [CrossRef]
24. Makvandi-Nejad, S.; Hoffman, G.E.; Allen, J.J.; Chu, E.; Gu, E.; Chandler, A.M.; Loredo, A.I.; Bellone, R.R.; Mezey, J.G.; Brooks, S.A. Four loci explain 83% of size variation in the horse. *PLoS ONE* **2012**, *7*, e39929. [CrossRef] [PubMed]
25. Weller, R.; Pfau, T.; Babbage, D.; Brittin, E.; May, S.; Wilson, A. Reliability of conformational measurements in the horse using a three-dimensional motion analysis system. *Equine Vet. J.* **2006**, *38*, 610–615. [CrossRef]

Gmel, A.I.; Druml, T.; Portele, K.; von Niederhäusern, R.; Neuditschko, M. Repeatability, reproducibility and consistency of horse shape data and its association with linearly described conformation traits in Franches-Montagnes stallions. *PLoS ONE* **2018**, *13*, e0202931. [CrossRef] [PubMed]

Rohlf, F. *tpsDig2*, Version 1.78; 2001. Available online: http://life.bio.sunysb.edu/morph (accessed on 23 June 2020).

Rohlf, F. *tpsRelw*, Version 1.70; 2003. Available online: http://life.bio.sunysb.edu/morph/index.html (accessed on 2 July 2019).

R Core Team. *R. A Language and Environment for Statistical Computing*; Version 4.1.3; R Core Team: Vienna, Austria, 2013.

Gilmour, A.; Gogel, B.; Cullis, B.; Welham, S.; Thompson, R.; Butler, D.; Cherry, M.; Collins, D.; Dutkowski, G.; Harding, S. *ASReml-SA User Guide Release 4.2 Functional Specification*; VSN International Ltd., Amberside House, Wood Lane, Paradise Industrial Estate: Hemel Hempstead, UK, 2021.

Gmel, A.I.; Druml, T.; von Niederhäusern, R.; Leeb, T.; Neuditschko, M. Genome-wide association studies based on equine joint angle measurements reveal new QTL affecting the conformation of horses. *Genes* **2019**, *10*, 370. [CrossRef]

Iyyanar, P.P.; Wu, Z.; Lan, Y.; Hu, Y.-C.; Jiang, R. Alx1 Deficient Mice Recapitulate Craniofacial Phenotype and Reveal Developmental Basis of ALX1-Related Frontonasal Dysplasia. *Front. Cell Dev. Biol.* **2022**, *10*, 777887. [CrossRef] [PubMed]

Gmel, A.I.; Haraldsdóttir, E.H.; Bragança, F.S.; Cruz, A.M.; Neuditschko, M.; Weishaupt, M.A. Determining objective parameters to assess gait quality in Franches-Montagnes horses for ground coverage and over-tracking-Part 1. At walk. *J. Equine Vet. Sci.* **2022**, *115*, 104024. [CrossRef]

Brooks, S.; Makvandi-Nejad, S.; Chu, E.; Allen, J.; Streeter, C.; Gu, E.; McCleery, B.; Murphy, B.; Bellone, R.; Sutter, N. Morphological variation in the horse. Defining complex traits of body size and shape. *Anim. Genet.* **2010**, *41*, 159–165. [CrossRef]

Druml, T.; Dobretsberger, M.; Brem, G. The Interplay of Performing Level and Conformation—A Characterization Study of the Lipizzan Riding Stallions From the Spanish Riding School in Vienna. *J. Equine Vet. Sci.* **2018**, *60*, 74–82.e1. [CrossRef]

Cervantes, I.; Baumung, R.; Molina, A.; Druml, T.; Gutiérrez, J.; Sölkner, J.; Valera, M. Size and shape analysis of morphofunctional traits in the Spanish Arab horse. *Livest. Sci.* **2009**, *125*, 43–49. [CrossRef]

Cortisol Variations to Estimate the Physiological Stress Response in Horses at a Traditional Equestrian Event

Sergi Olvera-Maneu [1,*], Annaïs Carbajal [1], Paula Serres-Corral [1] and Manel López-Béjar [1,2,*]

[1] Department of Animal Health and Anatomy, Universitat Autònoma de Barcelona, Bellaterra (Cerdanyola del Vallès), 08193 Barcelona, Spain
[2] College of Veterinary Medicine, Western University of Health Sciences, 309 East Second Street, Pomona, CA 91766, USA
* Correspondence: sergi.olvera@uab.cat (S.O.-M.); manel.lopez.bejar@uab.cat (M.L.-B.)

Simple Summary: The centuries-old patronal festivals in Menorca (Spain) represent an alternative to the common use of horses. During festivals, horses are exposed to potential sources of transient stress. This study aimed to evaluate the variations in salivary cortisol concentrations to estimate the physiological stress response in horses at the Menorca patronal festivals. For this purpose, the salivary cortisol variations before, during, and after the celebrations were assessed using an enzyme immunoassay. All the samples collected during festivals were significantly higher than the control group samples ($p < 0.05$). Within twenty-four hours after the end of the celebration, cortisol concentrations returned to baseline levels and did not differ significantly from the control group ($p > 0.05$). Overall, the study found that the horses' participation in the festivals resulted in transitory and measurable stress response.

Abstract: In many countries, horses remain involved in traditional equestrian events such as those celebrated in Menorca (Balearic Islands, Spain) every year since at least the 14th century. The present study aimed to evaluate the variations in salivary cortisol concentrations to estimate the physiological stress response in horses at the Menorca patronal festivals. Two different editions (years 2016 and 2018) of the festivals in honor of the Virgin of Grace in Maó (Menorca, Spain) were studied. Nineteen and seventeen Pure Breed Menorca stallions were included in the study, respectively. The stallions were aged between seven and twelve years. During celebrations, samples were collected before the start of the festivals between 8–9 a.m. and during the festivals at 8–9 p.m. On the second day of celebrations, the samples were collected at 8–9 a.m. and 3–4 p.m. Finally, on the day after the festivals, one sample was collected at 8–9 p.m. Additionally, a control group was sampled at 8–9 a.m., 3–4 p.m., and 8–9 p.m. Salivary cortisol concentrations were assessed by using a commercial enzyme immunoassay kit specially validated to quantify salivary cortisol in horses. Salivary cortisol concentrations did not show significant differences between sampling hours in the control group ($p > 0.05$). All the samples collected during festivals were significantly higher than samples of the control group ($p < 0.05$). Within the twenty-four hours after the end of the celebrations, cortisol concentrations returned to baseline levels and did not differ significantly from the control group ($p > 0.05$). Hence, the present study describes that the participation of the horses in these particular acts generate an acute and transitory stress response. Overall, the current work provides a reasonable basis for future research on the stress physiology and well-being of horses participating in traditional celebrations or similar events.

Keywords: saliva; equid; festivals; stress; hypothalamic–pituitary–adrenal axis

Citation: Olvera-Maneu, S.; Carbajal, A.; Serres-Corral, P.; López-Béjar, M. Cortisol Variations to Estimate the Physiological Stress Response in Horses at a Traditional Equestrian Event. *Animals* **2023**, *13*, 396. https://doi.org/10.3390/ani13030396

Academic Editors: Isabel Cervantes and María Dolores Gómez Ortiz

Received: 16 December 2022
Revised: 16 January 2023
Accepted: 19 January 2023
Published: 24 January 2023

1. Introduction

In many countries, particularly in the Mediterranean region, equestrian celebrations, games, and tournaments originating from religious and historical events represent an alternative to the common use of horses [1]. In Menorca (Balearic Islands, Spain), every summer cavalry events have been held in honor of the town's patron saints since at least the XIV century. These events are popularly known as patronal festivals or celebrations, and their origin remains to be clarified. However, the principal accepted hypothesis dates their origin during the Medieval period and is strongly related to the creation of the commissions in charge of the churches. Those commissions had the purpose of going around the town, riding horses, and collecting money to upkeep the town's ecclesiastic building and celebrate acts in honor of their patron saint. At present, the main characters of the celebrations are the horses, and the most commonly used breed is the Menorca Purebred horse. This native breed is excellent for riding and performing Classical and Menorca dressage, a distinctive type of dressage in which the horse practices exercises and movements similar to those performed during the patronal festivals [2]. The centuries-old celebrations have helped to maintain and improve this endangered native horse breed, and currently, there is a census of approximately 3700 individuals, mainly located in Menorca. In the last decade, the social interest in the welfare of the participant horses has grown substantially, since the animals must face potentially stressful situations during the acts. Hence, evaluating their stress response could be the primary approach to success in the horses' performance and well-being, as described in other common equestrian disciplines [3]. As well documented in athletic animals such as horses, following physical exercise, specific changes in metabolic reactions occur in athletes leading to several changes in the body, mainly in the circulatory, respiratory, endocrine, and neuromuscular systems. Changes taking place in these systems simultaneously and in an integrated manner are aimed at maintaining homeostasis in the body [4,5].

The stress response evaluation in horses can be performed by assessing behavioral and/or physiological indicators [6]. Cortisol, the final product released into the blood after activating the hypothalamic–pituitary–adrenal axis (HPA), has become one of the most used physiological indicators to measure the stress response, not only in horses but also in other species [7–9]. Different studies have reported a positive correlation between cortisol concentrations and heart rate, respiratory rate, rectal temperature [10], eye temperature [11,12], or blood lactic acid [13], all of them indicators of the intensity and effort of the exercise, and therefore indicators of the physiological stress response in horses. Cortisol secretion is characterized by a circadian rhythm that peaks in the early morning, with the nadir phase occurring in the evening [14,15]. Furthermore, cortisol has many supportive and advantageous physiological roles, including maintaining and restoring body homeostasis [8,9]. Hence, cortisol participates in the tolerance and adaptation of the horse to short-term exercise demands [3,9].

The measurement of blood cortisol in horses has long been the most common method for measuring the HPA axis activity. However, salivary cortisol analysis has recently become increasingly popular to assess the adrenocortical response [7,16,17]. Salivary cortisol represents the biologically active form of the hormone [7,18], and a strong association between levels in blood and saliva has been described in horses. After an adrenocorticotropic hormonal challenge in horses, Peeters et al. [16] described that total blood cortisol concentrations might account for 80% of salivary cortisol concentrations and vice versa. Additionally, the salivary cortisol measurement offers a non-invasive and easy tool to perform a repeated and "non-disturbing" sampling for the animal [19,20], ideal in cases where blood collection may be difficult or impossible. Furthermore, compared to blood sampling, saliva sampling does not require trained and qualified personnel to collect the samples [19,20].

The stress response in horses exposed to different exercises and sports activities such as competition [21–23], recreational activities [24–26], and training [27–29] has been well studied, including the measurement of salivary cortisol concentrations. To date, no studies have been performed to investigate the physiological stress response of the horses at the Menorca patronal festivals. Interestingly, Pazzola et al. [1] explored the stress-related physiological changes in horses of the "Sa Sartiglia" tournament (Sardinia, Italy), also celebrated since centuries ago and with some similarities with the Menorca festivals.

The present study aimed to evaluate the variations in salivary cortisol concentrations to estimate the stress response in stallions at the Menorca patronal festivals. We hypothesized that the participation of the horses in the celebrations would produce an acute and transitory stress response that would be reflected by an increase in the salivary cortisol concentration during the celebration days.

2. Materials and Methods

2.1. Celebrations

Every year on September 7 and 8 September, more than a hundred stallions and their riders participate in the patronal festivals in honor of the Virgin of Grace in Maó (Menorca, Spain) in the presence of thousands of people. The festivals are celebrated once a year and include different religious and popular events. Both days of the celebrations follow a similar structure. The horses are transported into the event site by trailer or walking depending on their proximity. The festivals begin with the horses' parade to the church (Figure 1a), where religious events are celebrated. When the masses are held, the horses rest for approximately 1.5–2.5 h (Figure 1b), depending on the day of the festival. Shaded areas and ad libitum running water are available for the horses during the resting period. Once the religious acts are ended, the horses head towards the village's main square, where the most popular event, named "jaleo", is celebrated (Figure 1c). All the riders and horses pass through the main square in groups of five and make the horses rear up several times for a few seconds (Figure 1c), performing the "bot" (walking courbette). The horses enter the square three times (per day), and this process is popularly known as "jaleo round". As established by the authorities, each horse during the jaleo round must remain less than 1.5 min in the square. Between jaleo rounds, horses rest for approximately 1–1.5 h.

(a)   (b)   (c)

Figure 1. (**a**) Horses and riders during the parade to the church. Photo credits by Trevor Fryatt. (**b**) Horses resting during the celebration of the mass. Photo credits by Trevor Fryatt. (**c**) Horse performing the "bot" in the middle of the crowded main square during a jaleo round. Photo credits by Catalina Pasqual.

2.2. Animals, Sampling, and Ethics

The design of the study was planned according to the different phases and necessities of the patronal festivals. Two different years of the festivals were studied (2016 and 2018, named A and B, respectively). Nineteen and seventeen Pure Breed Menorca stallions aged between seven and twelve years were included in the study for the A and B editions, respectively. Animals included in the study were regular participants in the festivals. All individuals were healthy and lacked any history of illness during the studied period. The body condition score (BCS) was evaluated using the 0 (emaciated) to 5 (extremely fat) BCS scale [30]. For all the stallions evaluated, the BCS was 3 (moderate to good body condition).

Saliva was collected using sterile mounted swabs (Sugi®, Eschenburg, Germany), specially designed to absorb secretions, rubbing the cheek mucosa and under the tongue for approximately twenty seconds. Saliva was recovered from the swab, placing it into a sterile syringe, and pressing the plunger until the obtention of the maximum possible volume. Samples were stored under freezing at $-20\,^{\circ}C$ and processed and analyzed within six months from collection. We followed the timeline presented in Figure 2. The first sample was taken on the morning of 7 September, between 8–9 a.m. (F1), before the start of the festival. The following sample was collected between 8–9 p.m. (F2). On 8 September, the samples were taken between 8–9 a.m. (F3) and between 3–4 p.m. (F4). Finally, the last sample was collected the day after the festival, on 9 September, between 8–9 p.m. (F5). The same protocol and timeline were followed in both studied editions. The horses stayed at their stables when the samples were collected before and after the celebrations (F1 and F5). In contrast, during the patronal festivals (F2, F3, and F4), the samples were collected in situ, trying to avoid the disturbance of the events.

Additionally, a control group of ten stallions, regular participants at the patronal festivals, was sampled to set the salivary cortisol baseline values. The sampling days in the control group were kept apart from any housing- and handling-related changes or disturbances. All the stallions included in the control group were healthy and lacked any history of illness during the studied period. The body condition score (BCS) was evaluated using the 0 (emaciated) to 5 (extremely fat) BCS scale [30]. For all the stallions evaluated, the BCS was also 3 (moderate to good body condition). From the control group, two replicates (one in August and another in September of 2019) were collected to avoid potential intra-seasonal influence on salivary cortisol concentrations [15]. Samples were collected between 8–9 a.m. (C1), 3–4 p.m. (C2), and 8–9 p.m. (C3). The followed sampling procedure was the same as the abovementioned. Sample C1 was used as a control for samples F1 and F3, C2 was used as a control for sample F4, and C3 was used as a control for samples F2 and F5 (Figure 2).

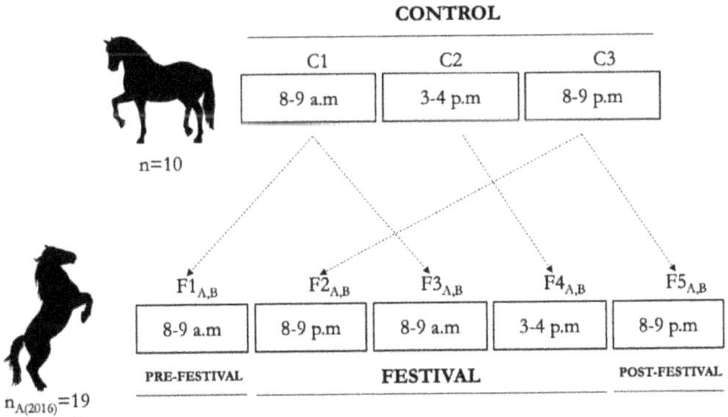

Figure 2. Schematical representation of the sampling timeline. Control samples were taken between 8 and 9 a.m. (C1), 3 and 4 p.m. (C2), and between 8 and 9 p.m. (C3). During patronal festivals samples were taken the morning of 7 September, before the start of the festivals, between 8 and 9 a.m. (F1), and during the festivals between 8 and 9 p.m. (F2). On 8 September, samples were taken between 8 and 9 a.m. (F3) and 3 and 4 p.m. (F4). Finally, the last sample was collected the day after the festivals, on 9 September, between 8 and 9 p.m. (F5). The same sampling scheme was followed in both studied editions (A and B).

During the study, all animals were managed following the principles and guidelines of the Ethics Committee on Animal and Human Experimentation from the Universitat

Autònoma de Barcelona (Barcelona, Spain) and following the Directive 2010/63/EU on the protection of animals used for scientific purposes. No other manipulation different from the salivary collection was performed. Additionally, informed consent from the owners of the horses was obtained before the start of the study.

2.3. Cortisol Quantification and Biochemical Validation of the Enzyme Immunoassay

For horses' salivary cortisol analysis, a commercial Enzyme-Linked ImmunoSorbent Assay (ELISA) kit (Neogen Corporation©, Ayr, UK) was used. According to the manufacturer, cross-reactivity of the ELISA cortisol antibody with other steroids was prednisolone 47.4%, cortisone 15.7%, 11-deoxycortisol 15.0%, prednisone 7.83%, corticosterone 4.81%, 6β-hydroxycortisol 1.37%, 17-hydroxyprogesterone 1.36%, and deoxycorticosterone 0.94%. Steroids with a cross-reactivity <0.06% are not presented. Salivary samples were analyzed at 1:1 dilution and the optical density was read using a microplate reader (Sunrise™ basic microplate reader, Tecan Austria GmbH, Grödig/Salzburg, Austria) at 450 nm.

The ELISA kit was biochemically validated for horses and saliva by following the criteria of precision, specificity, accuracy and sensitivity [31]. The validation tests were performed using a constituted pool created with 30 μL from each of the salivary samples included in the study. The precision of the test was assessed by calculating the intra-assay and inter-assay coefficients of variation from all the duplicated samples. The specificity of the test was evaluated by calculating the linearity of the dilution using 1:1, 1:2, 1:5, and 1:10 dilutions of the salivary pool diluted with the ELISA buffer provided by the kit. The accuracy was assessed through the spike-and-recovery test, calculated by adding to 100, 75, and 25 μL of the salivary pool volumes of 25, 75, and 100 μL of three different cortisol concentrations provided by the ELISA kit (0.2, 0.4, and 1 ng/mL). Finally, the sensitivity was given by the smallest cortisol concentration that the assay could detect and measure. For the present study, the intra-assay and inter-assay coefficients of variation were 6.2% and 12.2%, respectively. In the dilution test, the obtained and expected cortisol concentrations for saliva were significantly correlated (r = 0.99, $p < 0.01$). In the spike-and-recovery test, the average of recovery percentage was 107.6 ± 10.0%, and the obtained and theoretical concentrations were significantly correlated (r = 0.99, $p < 0.01$). The assay's sensitivity for salivary cortisol was 0.07 nmol/L. The biochemical validation of the assay showed reliable results and demonstrated the precision, specificity, accuracy, and sensitivity of the assay in measuring horse salivary cortisol.

2.4. Statistical Analysis

Data were analyzed using R Studio software (R version 3.4.4) and graphically represented using Graph Pad Software Inc. (GraphPad Prism, version 8.0.2; Graph Pad Software Inc., San Diego, CA, USA). The significance level in all data was set at $p < 0.05$. For the data obtained in the dilution test and the spike and recovery test of the ELISA kit biochemical validation, a Pearson's Product Moment correlation was applied between the expected and obtained values. The normality of the data was evaluated using a Shapiro–Wilk test and salivary cortisol concentrations were log transformed to achieve the normal distribution ($p > 0.05$). To analyze the effect of the month and the sampling hour in the salivary cortisol concentrations of the control group, a Linear Mixed Effect model was used considering the sampling day (D1 or D2) and the sampling hour (samples C1, C2, and C3) as fixed factors and the individuals as random factors followed by an ANOVA performed on the fitted model. To evaluate the changes in salivary cortisol concentrations during the festivals a Linear Mixed Effect model was used, setting the edition of the celebration (A and B) and the sampling hour (samples $F1_{A,B}$, $F2_{A,B}$, $F3_{A,B}$, $F4_{A,B}$, and $F5_{A,B}$) as fixed factors and the individuals as random factors. An ANOVA was performed on the fitted model. Finally, the Dunnett's post hoc test was performed between the control and the patronal festival groups. Hence, sample C1 was compared with $F1_{A,B}$ and $F3_{A,B}$, C2 was compared with $F4_{A,B}$, and finally, C3 was compared with $F2_{A,B}$ and $F5_{A,B}$.

3. Results

Salivary Cortisol Concentrations

Changes in salivary cortisol concentrations are presented in Figure 3. Salivary cortisol concentrations were not influenced by either the sampling day or the sampling hour ($p > 0.05$) in the control group. A significant effect of the sampling hour ($p < 0.05$) on salivary cortisol concentrations was detected in the patronal festival group, while no influence of the edition (A and B) was stated ($p > 0.05$). The Dunnett's post hoc test showed that samples collected in F1 (8–9 a.m.) were not statistically significant in comparison to C1 samples (8–9 a.m.) ($p > 0.05$). The first sample collected at the festivals, F2 (8–9 p.m.), was significantly different from the control group C3 (8–9 p.m.) in both editions of the study, A ($p < 0.001$) and B ($p < 0.05$). Salivary cortisol concentrations were higher in F3 (8–9 a.m.) ($p < 0.001$) compared to the control group C1 (8–9 a.m.). Sample F4 differed significantly from C2 in A ($p < 0.001$) and B ($p < 0.05$). Finally, within the twenty-four hours after the end of the celebrations, F5 (8–9 p.m.) samples were not significantly different ($p > 0.05$) compared to the samples obtained from the control group C3 (8–9 p.m.).

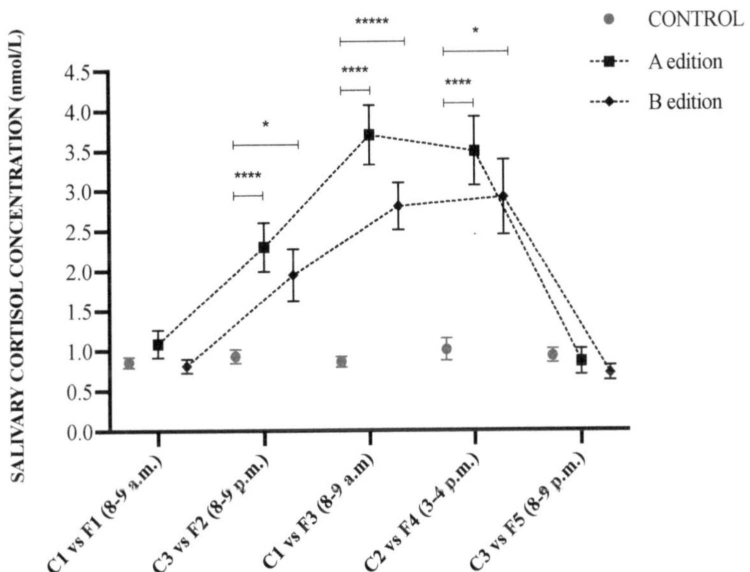

Figure 3. Mean ± SEM salivary cortisol concentrations (nmol/L) from the evaluated horses during the festival days (Fx) and the control group (Cx). Upper asterisks indicate significant differences between obtained salivary cortisol concentrations.

4. Discussion

The aim of the present study was to evaluate the physiological stress response in horses at the Menorca traditional equestrian celebrations by measuring the salivary cortisol variations. The authors hypothesized that horse's participation results in an acute and transitory stress response reflected in an increase in the salivary cortisol values during the festivals. The results corroborated the authors' hypothesis. For the first time, the generated physiological stress response in these stallions was described. Furthermore, the present results showed that cortisol levels returned to baseline within the first twenty-four hours, suggesting the participation of the horses in these festivals as an acute and transitory stressor.

To set the salivary cortisol baseline levels, a control group of ten stallions was sampled at three different hours (8–9 a.m. (C1), 3–4 p.m. (C2), and 8–9 p.m. (C3)), repeating the process in two different days (once in August and another in September of 2019). This was performed to avoid potential sources of intra-seasonal variations [32–34]. Baseline

salivary cortisol concentrations detected in the present study showed similar values t
other jumping and dressage horses [20,35]. Additionally, salivary cortisol concentrations c
the control group were not significantly influenced by the sampling day or the samplin
hour. Thus, results suggested the absence of circadian rhythm in the evaluated horse
contrary to what was reported in other investigations [14,36]. Different authors hav
suggested that changes in handling, habitat, generalized disease, and intensive exercise ca
disturb the cortisol circadian secretion pattern in horses [15,37,38]. All abovementione
disturbing factors can be discarded in this study, since during the sampling day, the horse
of the control group were kept apart from any housing- and handling-related change
or disturbances, and they were clinically healthy during the studied period. Hence, th
management conditions may not have driven the absence of circadian rhythm. Decreasin
the interval time between sample collection could help to accurately measure the presenc
of a possible circadian rhythm in these stallions [14,33]. Further research is needed to clarif
these findings.

Various studies have shown how horses anticipate a potential stressful event such a
loading transport vehicles [39], race competitions in comparison to regular training [40], o
trekking courses [41]. In the present study, the morning sample obtained before the star
of the festivals ($F1_{A,B}$) did not differ significantly from sample C1. This finding suggest
no anticipatory stress response, partially explained because the horses remained in thei
stables while the samples were taken.

In exercised horses, salivary cortisol has been demonstrated to be a valuable indicato
of stress levels [20,21,42]. In the present study, salivary cortisol concentrations obtaine
during festival days ($F2_{A,B}$, $F3_{A,B}$, and $F4_{A,B}$) were significantly higher than those obtaine
from the control group (C1, C3, and C2). The observed increase is intended to compensat
for the effects of the physical and environmental stressors to which the horses were expose
during those days (e.g., activity increase, crowded places, change in their habitual habita
presence of other stallions, transport from their origin place to the town or music, among
others). The results also showed that cortisol concentrations increased in parallel to th
duration of the festivals, suggesting that that cortisol's increase was related to the duratio
and intensity of the exercise [9,42]. Compared to the control group, salivary cortiso
increased from ~100% on the first day of the celebrations to ~300% (edition A) and ~200%
(edition B) on the second day's morning. In addition, a similar physiological stress respons
to the festivals was found in both studied editions. The observed variations during th
festivals can be compared to the results obtained by Kedzierski et al. [27], which showe
an increase of ~150% immediately after a short exercise of moderate intensity (800 m
galloped distance) in Arabian horses. Furthermore, Peeters et al. [22] showed that salivar
cortisol was ~340% higher after a cross-country road trip than just before the start. Othe
studies have reported higher increases in cortisol concentrations than those observed in th
present study, as for example the results reported by Schmidt et al. [39], who documented a
salivary cortisol increase higher than 600% during road transport. Based on these studies
it is likely that both high- and low-intensity exercise (long or short) induce a cortiso
increase. However, most studies have yet to have an exact measure of exercise intensity o
a standardized time and frequency of post-exercise cortisol measurements. Overall, th
results of the present study suggest that the stress response generated during the festival
was not exceeding the stress response generated in other activities to which domestic horse
are regularly exposed.

Throughout the year, the evaluated horses are trained to be in their best physica
condition to ensure a successful performance and human safety during the celebratio
days. Trainings include practicing the typical gaits and movements performed during
the festivals. The results of the present study are aligned with results obtained in ou
preliminary study (unpublished data), in which the stress response of horses during a
festivals' training session was evaluated. Salivary cortisol concentrations increased ~240%
after one hour of training. Therefore, the regular training of these horses would allow them
to properly prepare for the festivals' demands.

Different investigations have suggested that cortisol, the main glucocorticoid in horses, plays an essential role in metabolism and energy balance [43]. In medium–short-term action, cortisol acts as a catabolic hormone, increasing the energy production and restoring horses' homeostasis during and after exercise [9,44,45]. Hence, exercise's homeostasis disturbance would be compensated by activating the HPA axis, increasing circulating cortisol, among other hormones [9]. The present study showed that salivary cortisol concentrations were highly variable between individuals during the celebrations. This result follows other reported results of horses exposed to different types of exercise and conditions (e.g., [1,16,22,23,25]). The interindividual variability detected in salivary cortisol concentrations could be partially explained by the genetic conditioning of the horses [27], the interaction with their rider [23,46], or the experience of the animals participating in the festivals.

Sample F5 (24 h post-festivals) did not show significant differences compared to sample C3 of the control group. Thus, cortisol returned to baseline levels within the first twenty-four hours following the event. The return to salivary cortisol baseline levels has been reported to be different depending on the stimulus the horse has been exposed to. For example, after an exogenous adrenocorticotropic stimulation, salivary cortisol concentrations returned to baseline levels after 180 min [16], between 30 min and 1 h after a jumping course [20,47], or 2 h after unloading from a trailer, irrespective of the transport's duration [48]. On the other hand, we observed in our preliminary study (unpublished data) that horses returned to salivary cortisol baseline levels after a training session to participate in the festivals within two hours after finishing the exercise. In addition, in line with the present study's findings, Pazzola et al. [1] reported the return to cortisol baseline values within the day after the end of the traditional tournament of "Sa Sartiglia" (Oristano, Italy). Altogether, the return to basal levels within 24 h after the end of the event suggests the restoration of the homeostatic balance and therefore the absence of a chronic stress situation produced by the participation of these horses in the festivals.

5. Conclusions

Overall, the current study provides a basis for future research on the stress physiology of horses participating in traditional celebrations or similar events. For the first time, this study describes that the participation of the horses at the Menorca patronal festivals generates an acute stress response reflected by a transitory salivary cortisol increase. The results of the present study show how salivary cortisol concentrations increased progressively, reaching the maximal concentrations on the second day of the festivals, and returning to the baseline levels within the day after the acts. The authors encourage future studies to investigate the possible long-term consequences on the health and welfare of the horses participating in these events.

Author Contributions: Conceptualization, S.O.-M., A.C. and M.L.-B.; Data curation, S.O.-M., A.C. and P.S.-C.; Formal analysis, S.O.-M., A.C. and P.S.-C.; Funding acquisition, S.O.-M. and M.L.-B.; Investigation, S.O.-M.; Methodology, S.O.-M., A.C., and P.S.-C.; Project administration, S.O.-M. and M.L.-B.; Resources, S.O.-M. and M.L.-B.; Software, S.O.-M., A.C. and P.S.-C.; Supervision, A.C. and M.L.-B.; Validation, S.O.-M., A.C. and P.S.-C.; Visualization, S.O.-M., A.C., and P.S.-C.; Writing—original draft, S.O.-M.; Writing—review and editing, S.O.-M., A.C., P.S.-C. and M.L.-B. All authors have read and agreed to the published version of the manuscript.

Funding: This work was economically supported, partially, by the Ajuntament de Maó (Menorca, Spain).

Institutional Review Board Statement: The study was conducted following the principles and guidelines of the Ethics Committee on Animal and Human Experimentation from the Universitat Autònoma de Barcelona (Spain) and following the Directive 2010/63/EU on the protection of animals used for scientific purposes. No other manipulation different than non-invasive salivary extraction was performed. Additionally, informed consent from the owners of the horses was obtained before the initiation of the study.

Informed Consent Statement: Not Applicable.

Data Availability Statement: Data presented in this paper have not been published or stored elsewhere, but are available on request from S.O.-M.

Acknowledgments: The authors wish to thank all the owners that facilitated the participation of their horses in the study. The authors also want to deeply thank Trevor Fryatt and Catalina Pasqua for permitting the publication of their magnificent pictures to illustrate the present paper.

Conflicts of Interest: None of the authors have any competing interests that could inappropriately influence or bias the content of the paper.

References

1. Pazzola, M.; Pira, E.; Sedda, G.; Vacca, G.M.; Cocco, R.; Sechi, S.; Bonelli, P.; Nicolussi, P. Responses of hematological parameters, beta-endorphin, cortisol, reactive oxygen metabolites, and biological antioxidant potential in horses participating in a traditional tournament1. *J. Anim. Sci.* **2015**, *93*, 1573–1580. [CrossRef] [PubMed]
2. Solé, M.; Gómez, M.D.; Galisteo, A.M.; Santos, R.; Valera, M. Kinematic Characterization of the Menorca Horse at the Walk and the Trot: Influence of Hind Limb Pastern Angle. *J. Equine Vet. Sci.* **2013**, *33*, 726–732. [CrossRef]
3. Bartolomé, E.; Cockram, M.S. Potential Effects of Stress on the Performance of Sport Horses. *J. Equine Vet. Sci.* **2016**, *40*, 84–93. [CrossRef]
4. Arfuso, F.; Giudice, E.; Panzera, M.; Rizzo, M.; Fazio, F.; Piccione, G.; Giannetto, C. Interleukin-1Ra (Il-1Ra) and serum cortisol level relationship in horse as dynamic adaptive response during physical exercise. *Vet. Immunol. Immunopathol.* **2022**, *243*, 110366. [CrossRef]
5. Arfuso, F.; Giannetto, C.; Giudice, E.; Fazio, F.; Panzera, M.; Piccione, G. Peripheral Modulators of the Central Fatigue Development and Their Relationship with Athletic Performance in Jumper Horses. *Animals* **2021**, *11*, 743. [CrossRef]
6. Becker-Birck, M.; Schmidt, A.; Wulf, M.; Aurich, J.; von der Wense, A.; Möstl, E.; Berz, R.; Aurich, C. Cortisol release, heart rate and heart rate variability, and superficial body temperature, in horses lunged either with hyperflexion of the neck or with an extended head and neck position. *J. Anim. Physiol. Anim. Nutr.* **2012**, *97*, 322–330. [CrossRef]
7. Sheriff, M.J.; Dantzer, B.; Delehanty, B.; Palme, R.; Boonstra, R. Measuring stress in wildlife: Techniques for quantifying glucocorticoids. *Oecologia* **2011**, *166*, 869–887. [CrossRef]
8. Möstl, E.; Palme, R. Hormones as indicators of stress. *Domest. Anim. Endocrinol.* **2002**, *23*, 67–74. [CrossRef]
9. Ferlazzo, A.; Cravana, C.; Fazio, E.; Medica, P. The different hormonal system during exercise stress coping in horses. *Vet. World* **2020**, *13*, 847–859. [CrossRef]
10. Casella, S.; Vazzana, I.; Giudice, E.; Fazio, F.; Piccione, G. Relationship between serum cortisol levels and some physiological parameters following reining training session in horse. *Anim. Sci. J.* **2016**, *87*, 729–735. [CrossRef]
11. Yarnell, K.; Hall, C.; Billett, E. An assessment of the aversive nature of an animal management procedure (clipping) using behavioral and physiological measures. *Physiol. Behav.* **2013**, *118*, 32–39. [CrossRef]
12. Arfuso, F.; Acri, G.; Piccione, G.; Sansotta, C.; Fazio, F.; Giudice, E.; Giannetto, C. Eye surface infrared thermography usefulness as a noninvasive method of measuring stress response in sheep during shearing: Correlations with serum cortisol and rectal temperature values. *Physiol. Behav.* **2022**, *250*, 113781. [CrossRef]
13. Kędzierski, W.; Strzelec, K.; Cywińska, A.; Kowalik, S. Salivary Cortisol Concentration in Exercised Thoroughbred Horses. *J. Equine Vet. Sci.* **2013**, *33*, 1106–1109. [CrossRef]
14. Bohák, Z.; Szabó, F.; Beckers, J.F.; Melo de Sousa, N.; Kutasi, O.; Nagy, K.; Szenci, O. Monitoring the circadian rhythm of serum and salivary cortisol concentrations in the horse. *Domest. Anim. Endocrinol.* **2013**, *45*, 38–42. [CrossRef]
15. Murphy, B.A. Circadian and Circannual Regulation in the Horse: Internal Timing in an Elite Athlete. *J. Equine Vet. Sci.* **2019**, *76*, 14–24. [CrossRef]
16. Peeters, M.; Sulon, J.; Beckers, J.F.; Ledoux, D.; Vandenheede, M. Comparison between blood serum and salivary cortisol concentrations in horses using an adrenocorticotropic hormone challenge. *Equine Vet. J.* **2011**, *43*, 487–493. [CrossRef]
17. Waran, N.; Randle, H. What we can measure, we can manage: The importance of using robust welfare indicators in Equitation Science. *Appl. Anim. Behav. Sci.* **2017**, *190*, 74–81. [CrossRef]
18. Palme, R. Monitoring stress hormone metabolites as a useful, non-invasive tool for welfare assessment in farm animals. *Anim. Welf.* **2012**, *21*, 331–337. [CrossRef]
19. Contreras-Aguilar, M.D.; Hevia, M.L.; Escribano, D.; Lamy, E.; Tecles, F.; Cerón, J.J. Effect of food contamination and collection material in the measurement of biomarkers in saliva of horses. *Res. Vet. Sci.* **2020**, *129*, 90–95. [CrossRef]
20. Peeters, M.; Closson, C.; Beckers, J.F.; Vandenheede, M. Rider and Horse Salivary Cortisol Levels During Competition and Impact on Performance. *J. Equine Vet. Sci.* **2013**, *33*, 155–160. [CrossRef]
21. Christensen, J.W.; Beekmans, M.; van Dalum, M.; VanDierendonck, M. Effects of hyperflexion on acute stress responses in ridden dressage horses. *Physiol. Behav.* **2014**, *128*, 39–45. [CrossRef] [PubMed]
22. Peeters, M.; Sulon, J.; Serteyn, D.; Vandenheede, M. Assessment of stress level in horses during competition using salivary cortisol: Preliminary studies. *J. Vet. Behav.* **2010**, *5*, 216. [CrossRef]
23. Strzelec, K.; Kedzierski, W.; Bereznowski, A.; Janczarek, I.; Bocian, K.; Radosz, M. Salivary cortisol levels in horses and their riders during three-day-events. *Bull. Vet. Inst. Pulawy* **2013**, *57*, 237–241. [CrossRef]

4. Długosz, B.; Próchniak, T.; Stefaniuk-Szmukier, M.; Basiaga, M.; Łuszczyński, J.; Pieszka, M. Assessment of changes in the saliva cortisol level of horses during different ways in recreational exploitation. *Turk. J. Vet. Anim. Sci.* **2020**, *44*, 757–762. [CrossRef]
5. Strzelec, K.; Kankofer, M.; Pietrzak, S. Cortisol concentration in the saliva of horses subjected to different kinds of exercise. *Acta Vet. Brno* **2011**, *80*, 101–105. [CrossRef]
6. Kang, O.D.; Lee, W.S. Changes in salivary cortisol concentration in horses during different types of exercise. *Asian-Australas. J. Anim. Sci.* **2016**, *29*, 747–752. [CrossRef]
7. Kedzierski, W.; Cywińska, A.; Strzelec, K.; Kowalik, S. Changes in salivary and plasma cortisol levels in Purebred Arabian horses during race training session. *Anim. Sci. J.* **2014**, *85*, 313–317. [CrossRef]
8. Witkowska-Piłaszewicz, O.; Grzędzicka, J.; Seń, J.; Czopowicz, M.; Żmigrodzka, M.; Winnicka, A.; Cywińska, A.; Carter, C. Stress response after race and endurance training sessions and competitions in Arabian horses. *Prev. Vet. Med.* **2021**, *188*, 105265. [CrossRef]
9. Schmidt, A.; Aurich, J.; Möstl, E.; Müller, J.; Aurich, C. Changes in cortisol release and heart rate and heart rate variability during the initial training of 3-year-old sport horses. *Horm. Behav.* **2010**, *58*, 628–636. [CrossRef]
10. Carroll, C.L.; Huntington, P.J. Body condition scoring and weight estimation of horses. *Equine Vet. J.* **1988**, *20*, 41–45. [CrossRef]
11. Buchanan, K.L.; Goldsmith, A.R. Noninvasive endocrine data for behavioural studies: The importance of validation. *Anim. Behav.* **2004**, *67*, 183–185. [CrossRef]
12. Olvera-Maneu, S.; Carbajal, A.; Gardela, J.; Lopez-Bejar, M. Hair Cortisol, Testosterone, Dehydroepiandrosterone Sulfate and Their Ratios in Stallions as a Retrospective Measure of Hypothalamic–Pituitary–Adrenal and Hypothalamic–Pituitary–Gonadal Axes Activity: Exploring the Influence of Seasonality. *Animals* **2021**, *11*, 2202. [CrossRef]
13. Contreras-Aguilar, M.D.; Lamy, E.; Escribano, D.; Cerón, J.J.; Tecles, F.; Quiles, A.J.; Hevia, M.L. Changes in salivary analytes of horses due to circadian rhythm and season: A pilot study. *Animals* **2020**, *10*, 1486. [CrossRef]
14. Aurich, J.; Wulf, M.; Ille, N.; Erber, R.; von Lewinski, M.; Palme, R.; Aurich, C. Effects of season, age, sex, and housing on salivary cortisol concentrations in horses. *Domest. Anim. Endocrinol.* **2015**, *52*, 11–16. [CrossRef]
15. Smiet, E.; Van Dierendonck, M.C.; Sleutjens, J.; Menheere, P.P.C.A.; van Breda, E.; de Boer, D.; Back, W.; Wijnberg, I.D.; Van Der Kolk, J.H. Effect of different head and neck positions on behaviour, heart rate variability and cortisol levels in lunged Royal Dutch Sport horses. *Vet. J.* **2014**, *202*, 26–32. [CrossRef]
16. Cordero, M.; Brorsen, B.W.; McFarlane, D. Circadian and circannual rhythms of cortisol, ACTH, and α-melanocyte-stimulating hormone in healthy horses. *Domest. Anim. Endocrinol.* **2012**, *43*, 317–324. [CrossRef]
17. Irvine, C.H.G.; Alexander, S.L. Factors affecting the circadian rhythm in plasma cortisol concentrations in the horse. *Domest. Anim. Endocrinol.* **1994**, *11*, 227–238. [CrossRef]
18. Lassourd, V.; Gayrard, V.; Laroute, V.; Alvinerie, M.; Benard, P.; Courtot, D.; Toutain, P.L. Cortisol disposition and production rate in horses during rest and exercise. *Am. J. Physiol. Integr. Comp. Physiol.* **1996**, *271*, R25–R33. [CrossRef]
19. Schmidt, A.; Biau, S.; Möstl, E.; Becker-Birck, M.; Morillon, B.; Aurich, J.; Faure, J.M.; Aurich, C. Changes in cortisol release and heart rate variability in sport horses during long-distance road transport. *Domest. Anim. Endocrinol.* **2010**, *38*, 179–189. [CrossRef]
20. Bohák, Z.; Harnos, A.; Joó, K.; Szenci, O.; Kovács, L. Anticipatory response before competition in Standardbred racehorses. *PLoS ONE* **2018**, *13*, e0201691. [CrossRef]
21. Ono, A.; Matsuura, A.; Yamazaki, Y.; Sakai, W.; Watanabe, K.; Nakanowatari, T.; Kobayashi, H.; Irimajiri, M.; Hodate, K. Influence of riders' skill on plasma cortisol levels of horses walking on forest and field trekking courses. *Anim. Sci. J.* **2017**, *88*, 1629–1635. [CrossRef] [PubMed]
22. Munk, R.; Jensen, R.B.; Palme, R.; Munksgaard, L.; Christensen, J.W. An exploratory study of competition scores and salivary cortisol concentrations in Warmblood horses. *Domest. Anim. Endocrinol.* **2017**, *61*, 108–116. [CrossRef] [PubMed]
23. Viru, A.; Viru, M. Cortisol—Essential Adaptation Hormone in Exercise. *Int. J. Sport Med.* **2004**, *25*, 461–464. [CrossRef] [PubMed]
24. Peckett, A.J.; Wright, D.C.; Riddell, M.C. The effects of glucocorticoids on adipose tissue lipid metabolism. *Metabolism.* **2011**, *60*, 1500–1510. [CrossRef]
25. Rabasa, C.; Dickson, S.L. Impact of stress on metabolism and energy balance. *Curr. Opin. Behav. Sci.* **2016**, *9*, 71–77. [CrossRef]
26. von Lewinski, M.; Biau, S.; Erber, R.; Ille, N.; Aurich, J.; Faure, J.-M.; Möstl, E.; Aurich, C. Cortisol release, heart rate and heart rate variability in the horse and its rider: Different responses to training and performance. *Vet. J.* **2013**, *197*, 229–232. [CrossRef]
27. Ille, N.; Von Lewinski, M.; Erber, R.; Wulf, M.; Aurich, J.; Möstl, E.; Aurich, C. Effects of the level of experience of horses and their riders on cortisol release, heart rate and heart-rate variability during a jumping course. *Anim. Welf.* **2013**, *22*, 457–465. [CrossRef]
28. Schmidt, A.; Möstl, E.; Wehnert, C.; Aurich, J.; Müller, J.; Aurich, C. Cortisol release and heart rate variability in horses during road transport. *Horm. Behav.* **2010**, *57*, 209–215. [CrossRef]

MDPI AG
Grosspeteranlage 5
4052 Basel
Switzerland
Tel.: +41 61 683 77 34

Animals Editorial Office
E-mail: animals@mdpi.com
www.mdpi.com/journal/animals

www.ingramcontent.com/pod-product-compliance
Lightning Source LLC
LaVergne TN
LVHW070857170726
843515LV00005B/666